D1036831

SCIENTIFIC FARM ANIMAL PRODUCTION

PRODUCTION

Second Edition

SCIENTIFIC FARM ANIMAL PRODUCTION

PRODUCTION
Second Edition

Ralph Bogart
Professor Emeritus
Department of Animal Science
Oregon State University
Corvallis, Oregon

Robert E. Taylor
Department of Animal Sciences
Colorado State University
Fort Collins, Colorado

 **Burgess Publishing Company**
Minneapolis, Minnesota

Editorial: Jim Ruen, Nancy Crochiere, Gary
    Phillips, Judith Goodrich
Copy Editor: Marsha Crowder
Design and Layout: Mari Ansari, Priscilla Heimann
Production: Judy Vicars, Pat Barnes, Morris Lundin
Cover: Courtesy of Cargill, Inc.

Copyright © 1983, 1977 by Burgess Publishing Company
Printed in the United States of America

Library of Congress Cataloging in Publication Data

Bogart, Ralph, 1908–
    Scientific farm animal production.

    Bibliography: p.
    Includes index.
    1. Livestock.   I. Taylor, Robert E.   II. Title.
SF61.B63   1983   636   82-71051
ISBN 0-8087-4093-8

Burgess Publishing Company
7108 Ohms Lane
Minneapolis, MN 55435

All rights reserved. No part of this book may be
reproduced in any form whatsoever, by photograph
or mimeograph or by any other means, by broadcast
or transmission, by translation into any kind of lan-
guage, nor by recording electronically or otherwise,
without permission in writing from the publisher,
except by a reviewer, who may quote brief passages in
critical articles and reviews.

J   I   H   G   F   E   D   C   B

CREDITS

Chapter 1. *Figure 1-1:* Courtesy of American Egg Board. *Figure 1-2:* (A) Courtesy of Heifer Project International; (B) Courtesy of R. E. McDowell, Cornell University; (C) Courtesy of Winrock International; (D) Courtesy of Colorado State University. *Figure 1-3:* (A) Courtesy of California Agriculture Magazine, University of California; (B) and (C) Courtesy of American Simmental Associations; (D) Courtesy of Winrock International. *Figures 1-5 and 1-6:* Courtesy of USDA. *Figure 1-7:* (A) Courtesy of National Livestock and Meat Board; (B) Courtesy of R. E. McDowell, Cornell University; (C) USDA photo, courtesy of National Wool Growers; (D) Courtesy of Colorado

(continued on p. 401)

Dedication

The authors dedicate this book to the hundreds of students they have taught in introductory courses. The positive and rewarding interactions with these students have provided the stimulus to write this book. We hope the reading and studying of *Scientific Farm Animal Production* will be a motivational influence to many students as they seek to enhance their education.

Contents

Preface

Scientific Farm Animal Production is designed for the introductory animal science course. It is also a valuable reference book for livestock producers, vocational agriculture instructors, and other students who desire an overview of livestock production principles and management.

This book gives animal science students a preliminary overview of the breadth and depth of the livestock and poultry industry. Students need such an overview before delving into specialized principles and production courses. The book is designed to accommodate several instructional approaches to the introductory course: (1) the life-cycle principles approach, including such areas as end products, reproduction, breeding, nutrition, and marketing, (2) the species approach (teaching the course primarily by the various species), or (3) a combination of the previous two.

The text is basic and sufficiently simple for the student with an urban background or limited livestock experience, yet challenging for the student who has a livestock production background.

This book has been revised throughout to bring it up to date, and much new technical information is clearly presented. We have taken a new approach to Chapters 18, 19, 22, and 23, placing a greater emphasis on production economy of beef cattle and swine.

We have provided an excellent glossary of the terms used throughout the book so that students may become familiar with animal science terminology.

Chapter 6 presents new material of special importance. In it are photographs of carcass cross sections of beef cattle, sheep, and swine, from high-yield and low-yield (overfat) live animals. This information is vital to students in animal science because it shows the wastes present in U.S. yield grades of 4 or 5, and the high percentage of lean, edible meat in U.S. yield grade 1 animals.

Many illustrations in the form of photographs and line drawings are used throughout the book to communicate information. There are 345 individual photographs or drawings used in the text.

Study questions are presented in each chapter as a guide to important concepts and information.

Selected references are provided for each chapter to direct those students who want to learn more or to go into greater depth in certain areas.

Finally, we have included a subject index so that students may easily locate information on a particular subject.

The talents and backgrounds of the two authors have combined to give the book a strong scientific foundation tempered with practical production experience. Dr. Ralph Bogart was raised on a general livestock farm in Missouri. He received his B.S. degree in agriculture with a major in animal husbandry from the University of Missouri, his M.S. degree with a major in genetics from Kansas State, and his Ph.D. degree with a major in genetics and physiology from Cornell. He taught introductory animal husbandry classes as well as genetics and physiology at the University of Missouri for eight years, and continued his teaching at Oregon State University.

As teacher and researcher, Dr. Bogart has worked with students in the classroom as well as livestock producers throughout the world. Dr. Bogart received the Animal Breeding and Genetics Award in 1963, and the Animal Industry Service Award in 1975 from The American Society of Animal Science for his contribution to the livestock industry.

Dr. Robert Taylor was raised on an Idaho ranch. He received a B.S. degree in animal husbandry and a master's degree in animal production from Utah State University. This background, combined with his Ph.D. work in animal breeding and physiology from Oklahoma State University, has provided much depth to his knowledge of livestock production.

Dr. Taylor received teaching awards at Iowa State University (where he also managed a swine herd) and at Colorado State University. He also received the Distinguished Teaching Award in 1974 from The American Society of Animal Science in recognition of his ability to organize and present materials to students. Many of his concepts for effective teaching are used in this book.

Acknowledgments

Appreciation is extended to those persons who read material in the first edition of *Scientific Farm Animal Production* and offered suggestions for revision:

Dr. Walter Kennick, Oregon State University — The Meat We Eat.

Dr. Hokie Adams, Oregon State University and Professor Dawson Jordan, Colorado State University — Milk and Milk Products, Lactation — How Milk is Produced, Dairy Cattle Breeds and Breeding, and Feeding and Managing Dairy Cattle.

Dr. Jack Avens, Colorado State University — The Poultry Industry, and Managing Poultry.

Dr. William Hohenboken, Oregon State University — Behavior of Farm Animals.

Dr. Lauren Christian, Iowa State University — Swine Breeds and Breeding, and Feeding and Managing Swine.

Dr. Clair Terrill, USDA — How We Use Wool, Mohair, Hides, and Furs, Sheep Breeds and Breeding, Feeding and Managing Sheep, and Goats — What are Their Contributions?

Dr. Peter Cheeke, Oregon State University — How Feeds are Digested and Absorbed, The Functions of Nutrients in Simple Stomached and Ruminant Animals, and How Feeds Provide Needed Substances for Body Functions.

Dr. Melinda Burrill, California State Polytechnic University, and Dr. Elizabeth Juergensmeyer, prepared manuscript material on DNA and genetic coding.

Dr. Larry Cundiff, United States Meat Animal Research Center — Beef Cattle Breeds.

Dr. Robert Temple, formerly with International Livestock Center for Africa and previously with The Food and Agriculture Organization of the United Nations — Animal Contributions to Human Needs.

Also thanks to Dr. John Adair of Oregon State University, and Dr. Howard Enos of Colorado State University.

Special thanks are given to Dr. Frank A. Hudson, Animal Science Department, Texas
Tech University; Dr. T. M. Sutherland, Department of Animal Sciences, Colorado
State University; Dr. L. P. Wythe, Jr., Animal Science Department, Texas A & M
University; and Dr. Peter Hoffman, Department of Animal Science, Iowa State
University, who reviewed the manuscript material and made suggestions for
improvement of the final manuscript.

The authors particularly appreciate those persons or organizations that provided or
gave permission to revise illustrations or data for tables.

1

Animal Contributions to Human Needs

As human beings, our basic needs are for food, shelter, clothing, fuel, and emotional well-being. Throughout our history, we have used the resources of this earth to provide for these needs and improve our standard of living and animals have contributed significantly to our advance.

Since the domestication of such animals as dogs, horses, cattle and sheep, between 6,000 and 10,000 years ago, wide differences have developed between people in various regions of the world as to how they have used agricultural technology to advance their standard of living; but in all societies, domestic animals are a source of food, commercial products, and companionship for people.

Of particular importance among the multitude of benefits that domestic animals provide for humans are food; clothing; slaughter byproducts used for various chemical purposes and animal feeds; power; manure for fuel, buildings, and fertilizer; information on disease through studies of experimental animals; and pleasure for those who keep animals. Table 1-1 shows most useful domesticated animal species, their approximate numbers, and how they are being used by people throughout the world.

CONTRIBUTIONS TO FOOD NEEDS

When the opportunity exists, most humans consume both animal and plant products (Figure 1-1). Meat is nearly always consumed in quantity when it is available. Its availability in most countries is highly related to the economic status of the people and their agricultural technology. Vegetarianism in countries such as India may be the long-term result of intense population pressures and scarcity of feed for animals because of competition between humans and animals for food. Rising population pressures, particularly in developing regions (Southeast Asia, Africa, and Latin America) force people to consume foods primarily of plant origin (Tables 1-2 and 1-3). Some major groups in human society practice

Table 1-1. World animal species, numbers, and uses

Animal species	World numbers (millions)	Leading countries or areas with numbers (millions)	Primary uses
Ruminants			
Cattle	1,130	India (241), USSR (115), US (110)	Meat, milk, hides
Sheep	1,040	Oceania (224), Middle East (145), USSR (140), South America (115)	Wool, meat, milk, hides
Goats	392	Central Africa (89), Middle East (70), India (68), China (58)	Milk, meat, hair, hides
Buffalo	217	India (58), China (30), SE Asia (33)	Draft, milk, meat, hides, bones
Camel	⎫	North Africa, Middle East ⎫	⎫
Yak	⎬ 30	USSR, Tibet	Packing, riding, draft, meat, milk, hides, horns
Llama		South America	
Reindeer	⎭	USSR (3), Alaska ⎭	⎭
Nonruminants			
Swine	720	China (288), USSR (74), US (67)	Meat
Poultry	8,900	Meat (metric ton) — US (4.9), Japan (1.0), Brazil (0.8)	
		Eggs — US (68,280), USSR (66,000), Japan (33,000)	Meat, eggs, feathers
Horses	62	Mexico (10), US (8), Brazil (6), USSR (5.5)	⎫ Draft, packing, riding, pleasure and recreation, food, companion animals
Asses	42	Near East (8.9), Latin America (8.8)	⎬
Mules	14	Latin America (9.3), Africa (2.2)	⎭
Dogs			Protection, herding, guiding the blind, companion animals
Cats			Protection from rodents, companion animals
Rabbits		Meat (thousand metric ton) — France (270), USSR (170), Italy (100), Spain (80)	Meat, fur, angora fibers
Mink			Fur

Source: Adapted from several sources.

vegetarianism as a result of their ethics. In the Buddhist philosophy and some other religions of India, for example, all animal life is sacred and some believe in reincarnation.

Vegetarianism has never been practiced in a significant portion of society in the United States. The vegetarian movement in this country developed in the nineteenth century and was based on the view that eating only plant products was healthier than including meat in the diet. (Actually, medical surveys of vegetarians have often shown evidence of anemia and poor health). The development of vegetarianism in England and in the United States was also closely linked to temperance movements and to conservative attitudes toward sexuality. Vegetarianism was thought to cool "animal passions."

Figure 1-1. Animal food products, such as meat, milk, and eggs, are highly preferred foods in countries with high standards of living.

Table 1-2. World population and food supply

Item	Period	Developed[a] countries	Developing[b] countries	World
Population growth (annual percent increase)	1970–76	0.9	2.3	1.9
Food production (annual percent increase)	1970–76	2.3	2.7	2.4
Food production per person (annual percent increase)	1970–76	1.4	0.3	0.5
Agricultural labor force as percent of total labor force	1980	12.4	59.1	44.9
Agricultural output per agricultural worker (wheat price equivalent)	1974–76	3.30	0.25	0.55
Food supply, calories (daily per person)	1972–74	3,380	2210	2550
Food supply (percent of requirement)	1972–74	132	96	107
Food supply, all protein, grams (daily per person)	1972–74	98	57	69
Food supply, animal protein, grams (daily per person)	1972–74	54	12	24

Source: Fourth World Food Survey, FAO, 1977.

[a]Developed countries include United States, Canada, Europe, USSR, Japan, Republic of South Africa, Australia and New Zealand.

[b]Developing countries include South and Central America, Africa (except Republic of South Africa), Asia (except Japan, Communist Asia).

Table 1-3. Contributions of various food groups to the world food supply

	Developed countries		Developing countries		World	
	Calories	Protein	Calories	Protein	Calories	Protein
Cereals	30.7[a]	30.2	61.0	54.8	49.4	44.7
Roots, tubers, pulses	5.7	6.0	12.1	13.6	9.7	10.3
Nuts, oils, vegetable fats	15.4	2.5	7.8	5.5	7.6	4.2
Sugar and sugar products	22.8	6.3	11.7	5.1	8.8	2.2
Vegetables and fruits	9.8	8.7	5.2	4.7	7.8	6.9
All animal products	16.4	37.0	5.2	16.1	16.7	31.7
meat					(7.1)	(14.0)
eggs					(0.8)	(2.1)
fish					(0.8)	(4.6)
milk					(4.9)	(10.8)
animal fat					(3.1)	(0.2)

Source: Adapted from several FAO World Food Surveys.
[a]All figures expressed in percentage.

Table 1-3 shows that cereal grains are the most important source of energy in world diets. The energy derived from cereal grains, however, is twice as important in developing countries (as a group, because there are exceptions) as in developed countries. Animal products provide approximately 25% of the dietary energy supply for people in developed countries as contrasted to approximately 7% for developing countries. Animal products supply about 55% of the protein in human diets in the developed countries, but only 21% for the developing countries. The two major foods derived from animals are meat and milk.

Meat

Most of the world meat supply comes from cattle, buffaloes, swine, sheep, goats, and horses. There are, however, 20 or more additional species, unfamiliar to most Americans, that collectively contribute about 3 million metric tons of edible protein per year or approximately 10% of the estimated total protein from all meats. These include alpaca, llama, yak, deer, elk, antelope, kangaroo, rabbits, guinea pigs, capybaras, fowl other than chicken (duck, turkey, goose, guinea fowl, pigeon), and wild game exclusive of birds. For example, the USSR cans more than 50,000 metric tons of reindeer meat per year, and in West Germany the annual sales of local venison exceed $1 million. Peru derives more than 5% of its meat from guinea pigs.

Meat is important as a food for two scientifically based reasons. The first is that the assortment of amino acids in animal protein more closely matches the needs of the human body than does the assortment of amino acids in plant protein. The second is that vitamin B_{12}, which is required in human nutrition, may be obtained in adequate quantities from consumption of meat or other animal products but not from consumption of plants.

Milk

Milk is one of the largest single sources of food from animals. In the United States, 99% of the milk comes from cattle, but on a worldwide basis, milk from other species is important, too: domestic buffalo, sheep, goats, alpaca, camel, reindeer, and yak supply significant amounts of milk in certain countries (Figure 1-2). Milk and products made from milk contribute protein, energy, vitamins, and minerals for humans.

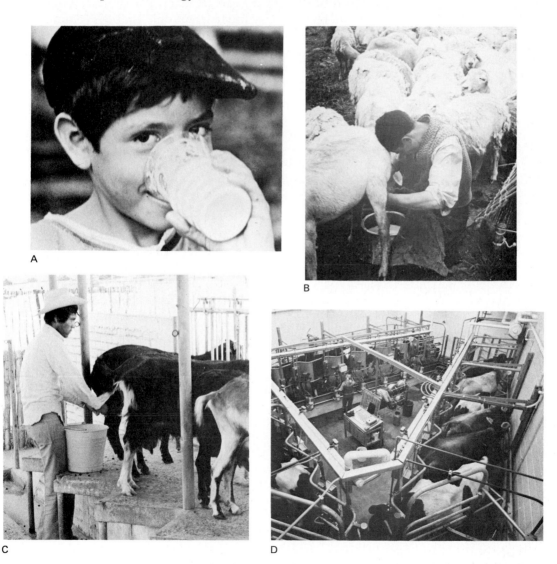

A

B

C

D

Figure 1-2. **(A)** Milk plays an important role in providing healthful nutrition for today's children the world over. **(B)** Milking native sheep in the desert of Iran. **(C)** Goats being milked in Mexico. **(D)** Milking dairy cows in the United States.

Besides the nutritional advantages, a major reason for human use of animals for food is that most countries have land areas unsuitable for growing cultivated crops. Approximately two-thirds of the world's agricultural land is permanent pasture, range, and meadow; of this, about 60% is unsuitable for producing cultivated crops that would be consumed directly by humans. This land, however, can produce roughage in the form of grass and other vegetation that is digestible by grazing **ruminant** animals, the most important of which are cattle and sheep (Figure 1-3). These animals can harvest and convert the vegetation, which is undigestible for the most part by humans, to high-quality protein food. In the United States about 385 million acres of rangeland and forest, representing 44% of the total land area, are used for grazing. Although this acreage now supports only about 40% of

Figure 1-3. Animals produce food for humans by utilizing grass, crop residues, and other forages from land which cannot produce crops to be consumed directly by humans. **(A)** Sheep grazing a steep hillside. **(B)** Cattle grazing a mountain valley in Switzerland. **(C)** Cattle produce meat from the mountains and Plains areas of the western United States. **(D)** Sheep utilizing the crop residue remaining after harvesting the corn grain.

the total cattle population, it could carry twice this amount if developed and managed intensively.

Animal agriculture therefore does not compete with human use for the production of most of the land used as permanent pasture, range, and meadow. On the contrary, the use of animals as intermediaries provides a means by which land that is otherwise unproductive for humans can be made productive.

People today are concerned about energy, protein, population pressures, and land resources as they relate to animal agriculture (Figure 1-4). The quantities of energy and protein present in foods from animals are smaller than the quantities consumed by the animals in their feed because animals are inefficient in the ratio of nutrients used to nutrients produced. More acres of cropland are required per person for a diet high in foods from animals than for a diet including only plant products. As a consequence, animal agriculture has been criticized for wasting food and land resources that could otherwise be used to feed persons with inadequate diets. Consideration must be given to economic systems and consumer preferences to understand why agriculture perpetuates what critics perceive as resource-inefficient practices. These practices relate primarily to providing food-producing animals with feed that could be eaten by humans and using land resources to produce crops specifically for animals instead of producing crops that could be consumed by humans.

Agriculture producers produce what consumers want to eat as reflected in the prices consumers can and are willing to pay. Eighty-five percent of the world's population desires food of animal origin in its diets, perhaps because foods of animal origin are considered more palatable than food from plants. In more countries, as per capita income rises, consumers tend to increase their consumption of meat and animal products, which are generally more expensive pound for pound than products derived from cereal grains. (Figure 1-5).

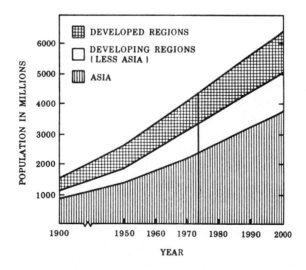

Figure 1-4. Past, present, and projected world population, 1900–2000.

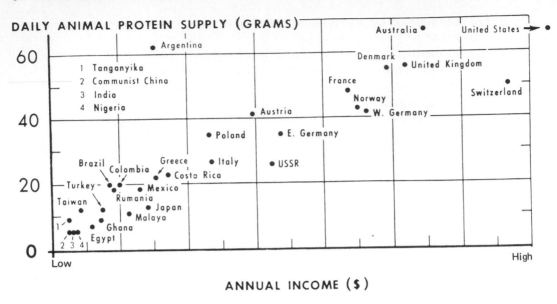

Figure 1-5. Income and supply of animal protein per person from selected countries.

If many consumers in countries in which animal products are consumed at a high rate were to decide to eat only food of plant origin, the consumption and price of foods from plants would increase, and the consumption and price of foods from animals would decrease. Agriculture would then adjust to produce greater quantities of food from plants and lesser quantities of food from animals. Ruminant animals can produce large amounts of meat without grain feeding. The amount of grain feeding in the future will be dictated by cost of grain and the price consumers are willing to pay for meat.

Some people in the United States advocate shifting from the consumption of foods from animals to foods from plants. They see this primarily as a moral issue, stating that it is unethical to let people elsewhere in the world starve when we could meet our own food needs by eating foods from plants rather than feeding the plants to animals. We could then send the balance of the plant-derived foods abroad. These people believe that grain is especially good to ship because grain can be shipped with comparative ease, because a surplus of grain exists in the United States, and because any surplus we have should be provided at no cost. Providing free food to other countries has met with limited success in the past. It puts economic pressure on the American farmer who then cannot make an adequate living and eventually raises taxes. Free food has also been distributed in inadequate quantities in developing countries because transportation and marketing systems are poor in these countries and because such distribution can upset their agricultural production. People in the United States do have a moral obligation to share their abundance with other people in the world, particularly those in developing countries. It appears that this can best be done by sharing our time and sharing our basic but not necessarily advanced technology. People need to have self-motivation to improve, need to be shown how to help themselves step by step, and need to develop fully the agricultural resources in their own countries.

The increases in the efficiency of animal production in the United States have been remarkable during the past half century (Table 1-4). This increase in efficiency has occurred primarily because people had an incentive to progress under a free enterprise system. They learned how to improve their standard of living using the available resources. This progress has taken time under the environmental conditions that motivated people to be successful.

The citizens of the United States not only should share their advanced technology with other people, but should also recognize how these achievements have been made. These achievements have been built on knowledge developed through experience and research; the extension of knowledge to producers; and the development of an industry to provide inputs, transportation, processing, and marketing. Dwindling dollars presently being spent to suppport agricultural research and extension of knowledge will not provide the technology needed for future food demands.

About 30% of the world human population and 32% of the ruminant animal population live in the developed regions of the world, but ruminants of these same regions produce two-thirds of the world's meat and 80% of the world's milk. In the developed regions, a higher percentage of the animals are used as food producers and these animals are higher in productivity on a per-animal basis than animals in the developing regions. This is the primary reason for the higher level of human nutrition in the developed countries of the world.

Table 1-4. Improvement in efficiency of producing foods of animal origin in the United States

Species and measure of productivity	Value in indicated year		
	1925	1950	1975
Beef cattle			
Liveweight marketed per breeding female (lbs)	220	310	482
Sheep			
Liveweight marketed per breeding female (lbs)	60	90	130
Dairy cattle			
Milk marketed per breeding female (lbs)	4,189	5,313	10,500
Swine			
Liveweight marketed per breeding female (lbs)	1,600	2,430	2,850
Broiler chickens			
Age to market weight (weeks)	15.0	12.0	7.5
Feed per pound of gain (lb)	4.0	3.3	2.1
Liveweight at marketing (lb)	2.8	3.1	3.8
Turkeys			
Age to market weight (weeks)	34	24	19
Feed per pound of gain (lb)	5.5	4.5	3.1
Liveweight at marketing (lb)	13.0	18.6	18.4
Laying hens			
Eggs per hen per year (number)	112	174	232
Feed per dozen eggs (lb)	8.0	5.8	4.2

Source: Food from Animals, CAST Report No. 82. March, 1980, p. 13.

There is the possibility that many so-called developing regions of the world can achieve levels of plant, animal, and eventually human food productivity similar to those of developed regions (Figure 1-6). Except perhaps in India, abundant world supplies of animal feed resources that do not compete with production of food for people are available to support expansion of animal populations and production. It has been estimated that through changes in resource allocation, an additional 8 billion acres of arable land (twice what is presently being used) and 9.2 billion acres of permanent pasture and meadow (23% more than is presently being used) could be put into production in the world. These estimates, plus the potential increase in productivity per acre and per animal in the developed countries, demonstrates the magnitude of world food production potential. This potential cannot be realized, however, without proper government planning and increased incentive to individual producers.

In the long run, each nation must assume the responsibility of more of its own food supply by efficient production, barter, or purchase, and by keeping future food production technology ahead of population increases and demand. Extensive untapped resources that can greatly enhance greater food production exist throughout the world, including an ample supply of animal products. The greatest resource is each individual human being, who can, through self-motivation, become more productive and self-reliant.

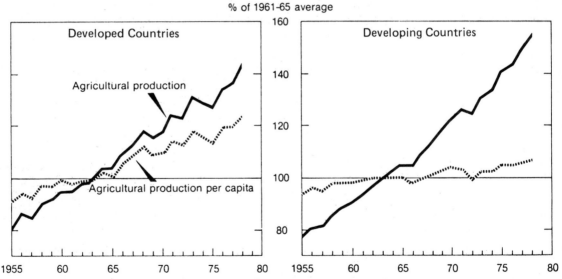

Developed countries include United States, Canada, Europe, USSR, Japan, Republic of South Africa, Australia and New Zealand.

Developing countries include South and Central America, Africa(except Republic of South Africa), Asia(except Japan, Communist Asia).

Figure 1-6. Changes in agricultural production in developed and developing countries (% of 1961–1965 average).

CONTRIBUTION TO CLOTHING AND OTHER NONFOOD PRODUCTS

Products other than food from ruminants include wool, hair, hides, and pelts. Although synthetic materials have made some inroads into markets for these products, world wool production has remained relatively stable over the past 15 years. It is important to note that in more than 100 countries, ruminant fibers are used in domestic production and cottage industries for clothing, bedding, housing and carpets.

The annual production of animal wastes from ruminants contains millions of tons of nitrogen, phosphorus, and potassium. The annual value of these wastes for fertilizer is estimated at $1 billion (Figure 1-7).

Inedible tallows and greases are animal byproducts used primarily in soaps and animal feeds and as sources of fatty acids for lubricants and industrial use. Additional tallow and grease byproducts are used in the manufacture of pharmaceuticals, candles, cosmetics, leather goods, woolen fabrics, and tin plating. The individual fatty acids can be used to produce synthetic rubber, food emulsifiers, plasticizers, floor waxes, candles, paints and varnishes, printing inks, and pharmaceuticals.

Gelatin is obtained from hides, skins, and bones and can be used in foods, films, and glues. Collagen, obtained primarily from hides, is used to make sausage casings.

CONTRIBUTION TO WORK AND POWER NEEDS

Early history of the developed world abounds with examples of the importance of animals as a source of work energy through draft work, packing, and being ridden. In the United States during the 1920s, approximately 25 million horses and mules were used primarily for draft purposes. The tractor has replaced all but a few of these draft animals. In parts of the developing world, however, animals provide as much as 99% of the power for agriculture even today.

In more than half the countries of the world, animals—mostly buffaloes and cattle, but also horses, mules, camels, and llamas—are kept primarily for work and draft purposes (Figure 1-8). About 20% of the world's human population depends largely or entirely on animals for moving goods. It should be recognized that much of the animal draft power makes a significant contribution to the production of major foods (rice and other cereal grains) in some of the heavily populated areas of the world. India, for example, has more than 200 million cattle and buffaloes, the largest number of any country. Although not slaughtered because they are considered sacred, the cattle of India contribute significantly to the food supply by providing work energy for the fields and milk for the people. It is estimated that India alone would have to spend more than $1 billion annually for gasoline to replace the animal energy it uses in agriculture.

ANIMALS FOR COMPANIONSHIP, RECREATION, AND ENTERTAINMENT

Estimates of the number of companion animals in the world are unavailable. There are an estimated 26 million family-owned dogs and 21 million family-owned cats in the United States. The pet food industry of the United States processes more than 3 million tons of cat

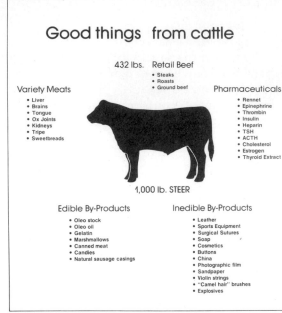

Good things from cattle

432 lbs.   Retail Beef
- Steaks
- Roasts
- Ground beef

Variety Meats
- Liver
- Brains
- Tongue
- Ox Joints
- Kidneys
- Tripe
- Sweetbreads

Pharmaceuticals
- Rennet
- Epinephrine
- Thrombin
- Insulin
- Heparin
- TSH
- ACTH
- Cholesterol
- Estrogen
- Thyroid Extract

1,000 lb. STEER

Edible By-Products
- Oleo stock
- Oleo oil
- Gelatin
- Marshmallows
- Canned meat
- Candies
- Natural sausage casings

Inedible By-Products
- Leather
- Sports Equipment
- Surgical Sutures
- Soap
- Cosmetics
- Buttons
- China
- Photographic film
- Sandpaper
- Violin strings
- "Camel hair" brushes
- Explosives

A

B

C

D

Figure 1-7. Animals produce many other products in addition to food. **(A)** Numerous pharmaceuticals and inedible byproducts are obtained from slaughter animals. **(B)** A load of cow-dung cakes enroute to market in India. About 200 million tons of cow and buffalo manure are used each year for fuel and (mostly) for cooking food. **(C)** Sheep being shorn for their yearly production of wool. **(D)** Yucatan miniature pigs being used as experimental animals to help solve the human medical problems of diabetes, obesity, and atherosclerosis. The mature weight of the miniature pig is approximately 150 pounds compared to a mature weight of several hundred pounds for normal swine.

A

B

C

Figure 1-8. Animals provide significant contributions to the draft and transportation needs of countries lacking mechanization in their agricultural technology. **(A)** Cattle used for draft in Honduras. **(B)** Water buffalos are used to cultivate many of the rice paddies in the Far East. **(C)** Camels being used to transport fertilizer to farming areas in Ethiopia.

and dog food annually valued at more than $1 billion. Many species of animals would qualify as companions where mankind derives pleasure from them. The contribution of animals as companions is significant even though it is difficult to quantify the emotional value. The animals used in rodeo, bull fighting, and other sport areas provide income for thousands of people and entertainment and recreation for millions of people (Figure 1-9).

From the livestock that provide high-quality food to the pet that gives pleasure to its owner, domestic animals serve people in diverse and vital ways. This book is intended to discuss the contributions of livestock, a group of domestic animals whose contributions to human welfare have long been and will continue to be fundamentally important to human welfare.

A

B

C

Figure 1-9. **(A)** The dog serves man in a variety of ways. This guide dog leads the way for a visually-impaired man. **(B)** Many people enjoy riding horses. **(C)** Horse racing is one of the highest attended spectator sports in the United

D

E

F

States. **(D)** Steer wrestling is exciting for the spectator, but dangerous for the participants. Fast, well-trained horses and strong participants are needed to wrestle the steers to the ground in less than five seconds. **(E)** Calf roping is a favorite rodeo event. **(F)** Bull fighting originated in Spain centuries ago. It is also popular in Mexico and several South American countries.

STUDY QUESTIONS

1. What primary contributions do animals make to people throughout the world?
2. Why is grain fed to animals in the United States rather than shipped to hungry people in other parts of the world?
3. What is the justification of having companion animals that do not contribute to the human food supply?
4. What is the relationship between the standard of living in the United States and its level of agricultural and animal productivity?
5. Discuss the contribution of power provided by animals to the food supply of many developing countries.
6. What appears to be the most feasible way to assist underdeveloped countries in improving their per capita food supply?
7. What are the contributions of animal byproducts in fulfilling human needs?
8. What advantages do ruminant animals have in contributing to the world food supply?
9. What effect does palatability and consumer economics have on the type of food that is produced in a country?
10. Why is per capita beef consumption low in India, which has the largest cattle population of any country in the world?

SELECTED REFERENCES

Baldwin, R. L. 1980. *Animals, Feed, Food and People: An Analysis of the Role of Animals in Food Production.* Boulder: Westview Press.

Cravens, W. W. 1981. Plants and animals as protein sources. *Journal of Animal Science* 53:817.

Devendra, C. 1980. Potential of sheep and goats in less developed countries. *Journal of Animal Science* 51:461.

Fitzhugh, H. A., Hodgson, H. J., Scoville, O. J., Nguyen, T. D., and Byerly, T. C. 1978. *The Role of Ruminants in Support of Man.* Winrock International Livestock Research and Training Center, Petit Jean Mountain, Morrilton, Arkansas.

Hodgson, H. J. 1979. Role of the dairy cow in world food production. *Journal of Dairy Science* 62:343.

McDowell, R. E. 1978. Contributions of animals to human welfare. *New York's Food and Life Science* 11:15.

Reid, J. T., White, O. D., Anrique, R., and Fortin, A. 1980. Nutritional energetics of livestock: Some present boundaries of knowledge and future research needs. *Journal of Animal Science* 51:1393.

_____ 1978. *Plant and Animal Products in the U.S. Food System.* (Symposium Proceedings.) National Academy of Sciences. Washington, D.C.

_____ 1980. *Food from Animals: Quantity, Quality and Safety.* Council for Agricultural Science and Technology (CAST). Report No. 82.

2

Meat

Meat from cattle, swine, sheep, goats, llamas, alpacas, and horses provides **protein**, **energy**, **vitamins**, and **minerals** for proper human nutrition. Meat is an excellent source of protein for humans because the protein of meat contains the "essential" amino acids needed by the human body. **Essential amino acids** are defined as those which the body cannot synthesize and which it must, therefore, obtain from food. Protein can be provided from plant sources but many plant proteins are low or deficient in one or more essential amino acids. It is possible to develop a meatless diet that is satisfactory for humans by carefully combining plant sources of protein and including a vitamin B_{12} supplement. It is easier, however, to provide a part of the needed protein by use of some meat or other animal product as a means of assuring that all essential amino acids, vitamin B_{12} and iron are present in the diet in the proper amount for good health. In addition to its high quality protein and vitamin B_{12}, meat also contains minerals and other nutritional vitamins.

The National Dairy Council (1979) advises against strict vegetarian diets. The suggestion is made that if persons insist on vegetarian dieting, the lacto-ovavegetarian approach should be used, particularly for young growing people. The Milk Industry Foundation (1977) states that meat provides 43% of the protein in the average American diet; dairy products, 22%; cereals, 17%; eggs, 5%; and miscellaneous foods, 13%.

Meats are named according to their source: **beef** is from cattle one year old or older; **veal** is from young calves; **pork** is from swine; **mutton** is from mature sheep; **lamb** is from young sheep.

Meat contains, on the average, approximately 17% protein, 62% water, 20% **fat**, and 1% minerals. As the percentage of fat increases, the percentage of lean decreases; therefore, fatty carcasses contain lower percentages of protein and water than lean carcasses. The lean is composed largely of protein and water. Even the connective tissues (the tissues surrounding the muscles) of meat contain considerable protein.

CONSUMPTION AND WHAT AFFECTS IT

In 1979, total meat consumption in the United States decreased to about 182 lbs per person per year for beef, veal, pork, and lamb from 190 lbs per person per year in 1974. Most of the decrease has been in beef (down from 117 lbs in 1974 to 109 lbs in 1979). Pork consumption (69 lbs per person per year) was the same in 1974 and 1979.

Argentina, Australia and New Zealand have higher per capita consumption of meat than the United States but no country produces more meat than the United States (Table 2-1). The Soviet Union is the second largest producer of meat but their per capita consumption is only 112 lbs per year. Brazil produces more than six million tons of meat annually but annual per capita consumption of only 57 lbs is low.

The kinds and amounts of meat that people eat vary with family income and geographical location. People who live in the western United States eat more beef and less pork than people in other parts of the country. People who live in the southern United States eat more pork and less beef than people in other areas, and eat but little veal and lamb. People who live in large cities of the West and East Coasts eat large quantities of lamb. Beef consumption depends on family income; the larger the income, the more beef consumed. Pork consumption, by contrast, is not highly influenced by family income. Therefore, when family income decreases, total meat consumption decreases primarily due to a reduction in the consumption of beef. In families having very low income, beef consumption may go to as low as 30 to 40 lbs per person per year, contrasted to an average of about 120 lbs. High income families eat most of the veal and lamb consumed in the United States, but total annual per capita consumption of veal and lamb is always relatively low (less than 4 lbs) in this country.

WHAT CONSUMERS DEMAND

Marked changes in demand for meat have occurred because people now want maximum lean and minimum outside fat, but much **marbling**. Today's typical consumer also favors "pan-ready" meat, that is, meat cut into serving sizes and packaged after bone and

**Table 2-1. Annual red meat production
and consumption (1979) for selected countries.**

Country	Meat production	Per capita consumption	
	tons	kgs	lbs
United States	37,000,000	83	183
Canada	3,700,000	70	164
Argentina	7,260,000	98	216
Brazil	6,820,000	26	57
France	7,800,000	73	161
West Germany	9,370,000	77	169
Poland	5,790,000	70	154
USSR	26,000,000	51	112
Australia	5,460,000	92	202
New Zealand	2,350,000	98	216

Source: Adapted from USDA figures.

excess fat have been removed. Pan-ready products require expensive labor for boning, trimming and packaging, which adds to their retail cost. Because the bone and fat have been removed, however, pan-ready products are priced thriftily in terms of cost per unit of meat. Tenderness is important to the consumer but it is not possible to visually appraise tenderness at the market.

Saturated fats have recently been named as a causative factor of several diseases, including hardening of the arteries and heart disease. The consumption of lean meats from young animals reduces the hazards that might arise from consuming fat meats. Lean meats contain fewer saturated fats than fat meats. Even more important, lean meats contain fewer calories and are an asset to diets designed to avoid **obesity** (excessive fatness). Because nervous tension, lack of exercise, and obesity are all important causes of heart disease, one is well advised to exercise regularly, restrict calorie intake, and develop a hobby that will help to relieve tensions. Consuming lean meat is one way of providing proper nutrition and keeping calorie intake at a low level.

HOW ANIMALS ARE PROCESSED AT THE SLAUGHTERHOUSE

Animals sent to slaughter are gassed with carbon dioxide or stunned and then bled. After cattle and sheep have been completely bled, their hides are removed. Hogs that have been bled are scalded to remove the hair and scurf (flakes and scales on the skin), but the skin is usually left on the carcass. In many small slaughterhouses, hogs are skinned to avoid the expense of heating water to scald them. After the hides or hair are removed, the stomach, intestines, heart, lungs, liver, and spleen are removed. The kidneys are left in place until the carcass is cut into wholesale cuts.

Dressing percentage, or yield, is the weight of the carcass in relation to the weight of the live animal. It is calculated as follows:

$$\text{Dressing percentage} = \frac{\text{carcass weight}}{\text{live weight}} \times 100$$

Dressing percentage is important in determining the value of animals received by the slaughterhouse. Several factors influence dressing percentage, including the amount of material in the digestive tract (fill), the fatness, the thickness of the hide, and, in sheep, the amount of wool. Animals carrying considerable fat yield high dressing percentages, but too much fat is wasteful because excess fat must be trimmed from the salable meat. The feet and head are always removed from carcasses of cattle and sheep before dressing percent is computed. Dressing percent in hogs may be reported with head and feet on or off.

Beef and pork carcasses are split down the center of the backbone, giving two sides that are approximately equal in weight. These sides are washed. Beef carcasses are put in the cooler to chill and age before the carcasses are cut into wholesale cuts. Pork carcasses are usually cut after chilling without being aged in the cooler. In some parts of the world where refrigeration is scarce, meat is cooked and eaten shortly after slaughter. Freshly slaughtered meat is generally tender, but after it has had time to harden in the cooler (1 to 3 days following slaughter), it is much less tender. If beef hangs in the cooler for 10 days to 2 weeks, the enzymes present in the meat tend to tenderize it. Meat keeps well for two to four weeks in a cooler at 35 °F but it is seldom kept for more than 10 days in a cooler because 10 days time is sufficient to cause tenderizing.

WHOLESALE CUTS

Carcass meat is generally cut into wholesale cuts. The wholesale cuts of beef are shown in Figure 2-1. These include loin, rib, round, flank, chuck, plate, brisket, and foreshank. The most valuable cuts of beef are the loin, rib, and round. Half of a beef carcass is cut into the hindquarter (which is composed of the round, loin, and flank) and the forequarter (which is composed of the chuck, rib, plate, brisket, and foreshank). Although the hindquarter contains only 48% of the weight of a half carcass of beef, it has about 54% to 58% of the value of the half carcass. Most beef carcasses are ribbed (cut) between the 12th and 13th ribs for grading.

The wholesale cuts of pork are shown in Figure 2-2. These include ham (leg), loin, fatback, belly (bacon), spareribs, Boston butt, picnic, clear plate, and jowl. The fat back and

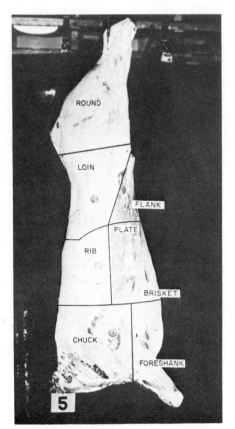

Figure 2-1. Beef carcass showing the location of the wholesale cuts of beef. This carcass came from the animal shown in Figure 2-7 and was graded at No. 1 lean cutability.

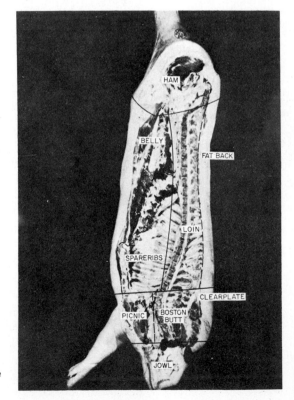

Figure 2-2. Pork carcass showing the location of the wholesale cuts of pork.

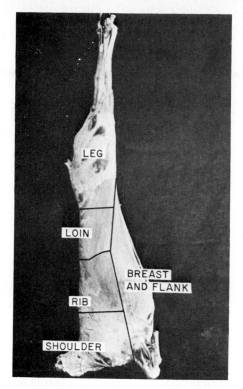

Figure 2-3. Lamb carcass showing the location of the wholesale cuts of lamb.

clear plate are not now considered as wholesale cuts but, instead, they go into fat trimmings to be rendered for lard. Today, the shoulder is often left intact rather than being first cut into Boston butt and picnic cuts. Four cuts of pork (ham, loin, shoulder, and bacon) are quite valuable. The fat cuts of fat back, clear plate, and jowl are much less valuable.

The wholesale cuts of lamb are shown in Figure 2-3. These include leg, loin, rib (also called the rack), shoulder, and breast and flank. Three cuts, leg, loin, and rib are valuable. The cuts of leg and loin, which comprise about 40% of the carcass weight, represent about 65% of the value of the carcass. Consumers prefer leg, loin and rib cuts of lamb but do not desire the other cuts; therefore these three cuts sell at a high price while other cuts are difficult to market.

The wholesale cuts of veal are shoulder, rib, loin, sirloin, leg (round), flank, breast, and shank. Veal has less flavor than beef, but is tender and nutritious. Veal is low in fat content; therefore, it is a desirable meat for weightwatchers.

The wholesale cuts are either made at the slaughterhouse or by wholesale jobbers, while retail cuts are made at retail outlets. Retail cuts are also made by processors who cut and package retail cuts prior to their delivery to retail stores.

Steaks are made from the loin, rib, and round wholesale cuts of beef. Roasts are made from the round, the rib, and other cuts (Figure 2-4). Some of the less desirable cuts from carcasses of high grades and all cuts from carcasses of low grades may be ground into hamburger or used in processed meats.

Figure 2-4. How the wholesale cuts of beef are prepared for retail sales of cuts and other preparations.

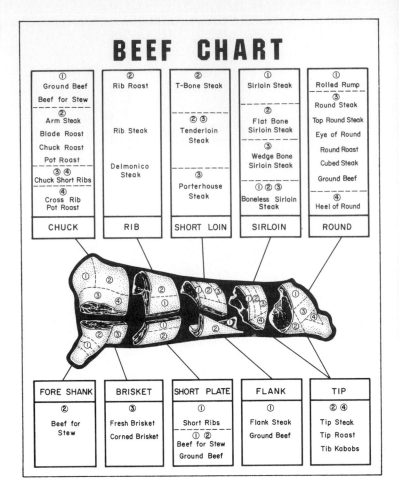

The leg of lamb is usually sold and consumed as such. The loin and rib of lamb are made into lamb chops (Figure 2-5). The less desirable cuts of lamb may be made into lamburger, lamb stew, or roasts. With the exception of leg roast, which may be sliced and served cold, lamb is most palatable when served hot. Also, a hot, rather than an iced, drink should be served with hot lamb because cold drinks cause the fat of lamb to solidify in the mouth, resulting in an unpleasant sensation.

The leg of a pork carcass may be cured into ham or consumed as fresh pork (Figure 2-6). The loin is usually eaten fresh as pork chops. The belly and side are cured and used as bacon. The shoulder may be used as fresh roasts or cured as picnic.

FACTORS AFFECTING THE VALUE OF MEAT

Several criteria are used to evaluate carcass meat. The color is important in marketing meat because dark meat usually has a shorter shelf life and pale meat appears anemic to the consumer in contrast to bright red meat. The amount of marbling (an interspersion of fat

LAMB CHART

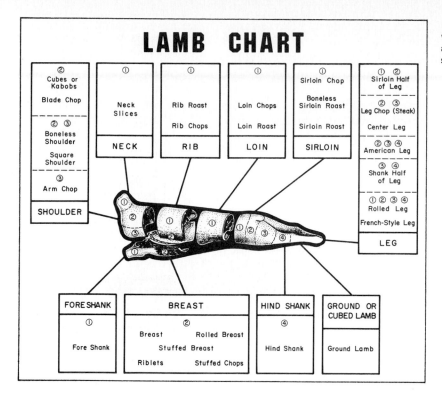

| ② Cubes or Kabobs
Blade Chop
----
② ③ Boneless Shoulder
Square Shoulder
----
③ Arm Chop

SHOULDER | ① Neck Slices

NECK | ① Rib Roast
Rib Chops

RIB | ① Loin Chops
Loin Roast

LOIN | ① Sirloin Chop
Boneless Sirloin Roast
Sirloin Roast

SIRLOIN | ① ② Sirloin Half of Leg
----
② ③ Leg Chop (Steak)
Center Leg
----
② ③ ④ American Leg
----
③ ④ Shank Half of Leg
----
① ② ③ ④ Rolled Leg
French-Style Leg

LEG |

| **FORE SHANK**
① Fore Shank | **BREAST**
② Breast   Rolled Breast
Stuffed Breast
Riblets   Stuffed Chops | **HIND SHANK**
④ Hind Shank | **GROUND OR CUBED LAMB**
Ground Lamb |

Figure 2-5. How the wholesale cuts of lamb are prepared for retail sales.

PORK CHART

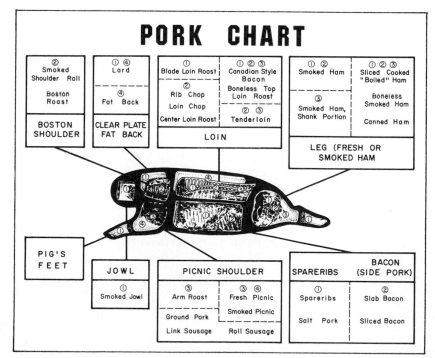

| ② Smoked Shoulder Roll
Boston Roast

BOSTON SHOULDER | ① ④ Lard
----
④ Fat Back

CLEAR PLATE FAT BACK | ① Blade Loin Roast
② Rib Chop
Loin Chop
Center Loin Roast | ① ② ③ Canadian Style Bacon
Boneless Top Loin Roast
② ③ Tenderloin | ① ② Smoked Ham
----
③ Smoked Ham, Shank Portion | ① ② ③ Sliced Cooked "Boiled" Ham
Boneless Smoked Ham
Canned Ham |
| | | **LOIN** | | **LEG (FRESH OR SMOKED HAM** | |

| **PIG'S FEET** | **JOWL**
① Smoked Jowl | **PICNIC SHOULDER**
③ Arm Roast
Ground Pork
Link Sausage | ③ ④ Fresh Picnic
Smoked Picnic
Roll Sausage | **SPARERIBS**
① Spareribs
Salt Pork | **BACON (SIDE PORK)**
② Slab Bacon
Sliced Bacon |

Figure 2-6. How the wholesale cuts of pork are prepared for retail sales.

between the muscle fibers) is important in giving meat more flavor and more juiciness. When one eats meat, the three important considerations are tenderness, flavor and juiciness. Tenderness is influenced by age—meat from young animals is more tender than that from old animals. It is also influenced by the amount of time a carcass hangs in the cooler and by the temperature of the cooler. Juiciness and flavor depend largely on the amount of marbling and to some extent on the age of the animal. Meat from very young animals and meat that is low in fatness (in particular, marbling) have a bland taste. Meat from mature and highly fattened animals has a highly desirable flavor. Thus, the consumer must to some extent decide between tender meat that lacks flavor and flavorful meat that lacks tenderness.

The amount of feed the animal has consumed influences the fatness of the carcass, particularly in pigs that are grown slowly and later allowed all the feed they will eat. Carcasses from such pigs are high in fat. Also, sex has a striking influence on fatness. For example, beef from bulls contains more lean and less fat than beef from steers, and beef from steers contains more lean and less fat than that from **heifers** (young female cows) when the animals are slaughtered at the same weights. Heifers tend to store more fat internally ("kidney fat") and subcutaneously (just under the skin) than either steers or bulls.

Exercise does not influence tenderness to any noticeable degree, but animals that are forced to exercise may tend to fatten at a slower rate than those that are rested.

HOW PRODUCERS ARE HELPING CONSUMERS

Producers of beef cattle, swine, and sheep are interested in producing animals that contain a high percentage of lean and a low percentage of excess fat and still have sufficient marbling to give good flavor. It is difficult to look at live animals and accurately determine which animals will make lean carcasses. However, some success in doing this has been achieved with beef cattle. If one looks for high development in the areas where fat is not generally stored in large amounts (the sides of the round) and less development in those areas where fat is stored in large amounts (**cod** of steers, or **udder** of heifers, and flank, over the tops of the shoulders, and on the brisket), one can reasonably evaluate beef cattle (Figure 2-7). A large ribeye area indicates leanness of the entire carcass (Figure 2-8).

Figure 2-7. A USDA Choice steer that made a No. 1 cutability carcass weighing approximately 600 lbs (see Figure 2-1). The ribeye area was 14 square inches with 0.3 inches of outside fat (see Figure 2-8).

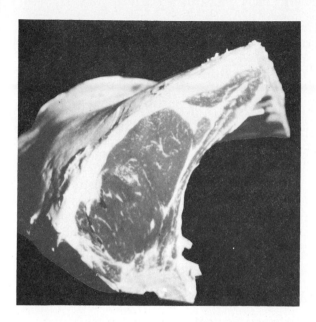

Figure 2-8. Ribeye from a U.S. No. 1 cutability carcass (see Figure 2-1). The lean ribeye was 14 square inches and there was only 0.3 inches of fat covering. There is plenty of marbling (white flecks in loin area) to give good flavor.

The **sonoscope**, an instrument that sends sound waves into the animal and indicates the point at which the waves are bounced back when they strike connective tissue or bone is used with varying degrees of success in determining the amount of lean and fat in live animals. When the waves from the sonoscope strike connective tissue between the fat and lean, some of the waves bounce back; when the waves strike bone, all of the waves bounce back. One can measure the depth of fat and lean by this technique, but the method requires a high degree of skill.

One can accurately determine fatness in live pigs by using a **probe** to measure fat thickness over the backs of pigs. An incision is made through the skin and a metal ruler is inserted through the fat. When the ruler reaches the lean tissue, it does not go deeper — thus the depth of fat can be measured directly. This probe has helped breeders select leaner swine for breeding over the last 30 years, with the result that pork products are leaner today than formerly. Visually appraising **carcass merit** (the value of the carcass for consumption) of sheep is difficult because of the sheep's wool covering. Usually the animal is felt over its ribs and back to judge what the carcass would show if the hide were removed.

If a carcass is weighed in air and then in water, one can determine its specific gravity. The determination of specific gravity gives figures from which one can calculate the relative percentages of lean and fat. This method has been used with success to determine carcass lean and fat, but it is difficult to apply it to live animals. In addition, cattle may have large quantities of gas in the rumen, which gives figures indicating a higher-than-actual percentage of fat in the body.

Carcasses are graded as Prime, Choice, or Good. There are some lower grades, but much of the meat of the lower grades is used in processed meats. The grades of meat and the terminology used were developed when consumers wanted fat meats. The fat was not eaten but the lean of fat animals was highly flavorful. Today, the consumer wants Choice meat that is not too fat. Animals and carcasses are now evaluated on the basis of **cutability** (the amount of trimmed lean retail cuts that a carcass yields). A USDA No. 1 yield grade beef or lamb carcass has a high lean cutability whereas a No. 4 carcass has a low lean cutability. Two Choice beef carcasses weighing 600 lbs each may differ in value by more than $200 because of differences in lean cutability.

In times past, animals were fed in the Midwest or West and shipped to big stockyards and slaughterhouses in such cities as Chicago and St. Louis. Slaughterhouses have recently been built in the areas of production because carcasses can be shipped at less expense than live animals. This change has been made possible by the development of refrigerated railcars and trucks.

Recent innovations in the meat industry include the processing and distribution of precooked meats and the development of vacuum packaging. When meats are processed into sausages and other prepared meats, other foods such as soybeans are often added. In the future one will likely see considerable amounts of soybean in steaks and roasts; whether these steaks and roasts can be made as palatable to the consumer as those without soybean is yet to be determined.

CONSUMER PROTECTION

Two protections are given the consumer when he or she purchases meat:
1. Meat is graded by the USDA Grading Service. The grade assures the consumer that, when he or she buys meat labeled "Choice," choice meat is indeed being purchased.
2. Inspections are made by the USDA Food Safety and Quality Service of the animals brought to slaughter, the slaughter-house, the processing facilities, and the carcasses. This assures the consumer that sanitation in the plant and the quality of health of the animals slaughtered meet federal and state government standards.

STUDY QUESTIONS

1. (True or False)
   a. As the percentage of fat in an animal increases, the percentage of lean decreases. Why or why not?
   b. Young animals produce carcasses having more flavor than older animals. Why or why not?
   c. Bull carcasses contain more fat and less lean than heifer carcasses of the same weight. Why or why not?
2. a. What are the three most valuable cuts of beef?
   b. What are the two most important characteristics contributing to desirability of beef?
   c. How is dressing percentage calculated?
   d. What qualities does the consumer prefer when he or she buys meat?
   e. From what wholesale cuts do we obtain each of the following: T-bone steak, lamb chops, rib roast, bacon?
3. Study the wholesale cuts of beef, pork, and lamb so that you can name each cut from a carcass drawing.
4. a. What are breeders doing to help satisfy the needs of the meat-purchasing consumer?
   b. What safeguards protect the consumer?
   c. What has the swine breeder done to provide more lean in pork carcasses?
   d. What effect does family income have on consumption of beef, pork, and lamb?
   e. What is meant by lean cutability, and how important is it in determining the value of a carcass?

SELECTED REFERENCES

American Meat Institute. 1980. A Statistical Summary About America's Largest Food Industry. *Meat Facts.* Washington, D.C. American Meat Institute.

American Meat Institute Foundation. 1960. *The Science of Meat and Meat Products.* San Francisco: W. H. Freeman and Co.

Boggs, D. L. and Merkel, R. A. 1979. *Live Animal, Carcass Evaluation and Selection Manual.* Dubuque, Iowa. Kendall-Hunt Publishing Co.

Forest, J. C., Aberle, E. D., Hedrick, H. B., and Merkel, R. A. 1975. *Principles of Meat Science.* San Francisco: W. H. Freeman and Co.

Milk Industry Foundation. 1977. *Milk Facts.* Washington, D.C. Milk Industry Foundation (in cooperation with U.S. Government and USDA).

National Dairy Council. 1979. Anonymous. *Nutrition and Vegetarianism.* Dairy Council Digest 50:1, 1-6. Rosemont, Illinois. National Dairy Council.

National Live Stock and Meat Board. 1973. *Lessons on Meat.* Chicago: National Live Stock and Meat Board.

Romans, J. R. and P. T. Ziegler. 1977. *The Meat We Eat.* 11th Edition. Danville, Illinois. The Interstate Printers and Publishers, Inc.

3

Milk and Milk Products

Milk, with its assortment of protein, fat, lactose (milk sugar), minerals, vitamins, pigments, enzymes, and water, is sometimes called nature's most perfect food. The world's people obtain most of their milk and milk products from cows, goats, sheep, and buffaloes, with horses, donkeys, reindeer, yak, and sows contributing to a lesser extent.

A versatile food that is in itself the basis of such products as cheese, ice cream, butter, and cottage cheese, milk is a major component of the human diet in many countries, including the United States. Table 3-1 compares the consumption of milk and milk products by Americans in 1974 and 1979 and reveals a shift toward nonfat and lowfat products. Consumers in other countries rely on certain milk products even more heavily than Americans do: for example, the American Milk Industry Foundation (1976) estimated the annual per capita consumption of fluid milk in Finland to be 564 lbs compared with 244 lbs in the United States; annual per capita consumption of butter in New Zealand, 32.8 lbs (compared with 4.7 lbs); and annual per capita consumption of cheese in France 33.3 lbs (compared with 14.5 lbs).

NUTRITIONAL IMPORTANCE OF MILK

Milk is particularly nutritious because its protein contains all the essential amino acids needed by humans. Lowfat milk is an especially important food in weight-control diets. The protein of milk is composed of **globulins**, **casein**, and **lactalbumin**. Casein is synthesized by the udder, globulins and albumins (the source of lactalbumin) are in the blood. Portions of the globulins of milk are structural parts of **antibodies**. Casein is the most abundant protein constituent of milk. It has many uses in addition to providing protein in the diet. Lactalbumin is part of the enzyme system that synthesizes lactose in the mammary gland. It is secreted into milk as a byproduct and becomes part of milk protein.

The first milk a female produces after the young is born is called **colostrum**. It contains many antibodies that give the newborn protection from harmful microorganisms that

Table 3-1. Annual per capita consumption of dairy products in the United States (1979)

Product	Consumption 1974		Consumption 1979	
	lb	(kg)	lb	(kg)
Fluid whole milk	200.0	90.8	157.0	71.3
Lowfat fluid milk	78.6	35.7	95.0	43.1
Fluid cream	5.7	2.6	5.6	2.5
Butter	4.3	1.9	4.6	2.0
Cottage cheese	4.7	2.1	4.7	2.1
American and other cheese	14.2	6.4	17.9	8.1
Evaporated and condensed milk	5.1	2.3	5.0	2.3
Ice cream	17.7	8.0	17.9	8.1
Ice milk	7.7	3.5	7.6	3.4
Nonfat dry milk	3.8	1.7	3.3	1.5

Source: From USDA. 1979 Handbook of Agricultural Charts. Agriculture Handbook No. 491. U.S. Department of Agriculture, October 1975; and from USDA and Cooperative Services, Washington, D.C. 1980.

invade the body and cause illness. The newborn animal has yet to develop antibodies of its own because it has yet to be exposed to any disease-causing microorganisms. The gut wall of the newborn is quite porous and permits antibodies in colostrum to enter the body. Within a few hours the gut wall becomes less porous and the antibody content of the milk diminishes, but the antibodies that have been absorbed into the body of the newborn give it protection (passive immunity) until it can develop antibodies in its own body. People seldom consume colostrum from animals because of its undesirable appearance and odor, although a few use it for pudding.

Other constituents of milk include **lactose**, minerals such as calcium and phosphorus (both of which are important in bone growth and other body functions), and vitamins. About 75% of the calcium in the American diet is supplied by dairy products. Milk is, however, quite low in iron; therefore young animals consuming nothing but milk may develop anemia. Baby pigs produced in confinement are especially likely to develop anemia if no iron is supplied. Young children who consume large quantities of milk at the expense of meat should be given some source of iron. Milk contains several important vitamins such as vitamin A, which helps keep the intestinal tract and skin in proper repair, the vitamin B complex, and vitamins D and E. Vitamin D, along with calcium and phosphorus, is important in bone growth and repair. Vitamin D is added to most marketed milk. Milk is low in vitamin C, which prevents **scurvy** (a disease characterized by bleeding, spongy gums and loose teeth); therefore, young children who depend heavily on milk for food should be given juices from citrus fruits.

In addition to cheese, ice cream, and various iced milk drinks, many delectable and nutritious foods are prepared from milk. Milk may have a portion of the water it contains removed and sugar added to produce condensed milk, or it may be dried to produce either dried whole milk or skim milk. Dried, condensed milk may be reconstituted to provide milk to drink or it may be used in cooking with or without reconstitution. **Buttermilk** is produced when butter is made, or it can be cultured from milk by the use of proper bacteria. Cottage cheese is made by curdling the milk and removing most of the liquid (whey).

SPECIES DIFFERENCES IN MILK COMPOSITION

Milk from different animals exhibits rather large differences in percentages of certain constituents. For example, water buffaloes, ewes, and sows all produce milk that is relatively high in total solids, fat, and protein, and low in water content. Milk from mares is low in fat. Human milk is low in protein (Table 3-2).

Schmidt (1971) gives the composition of milk from reindeer as 36.7% total solids of which 22.5% is **milk fat**, 10.3% is protein, 2.5% is lactose (milk sugar), and 1.4% is minerals. The remaining 63.3% is water. The milk from reindeer is the highest in nutrient constituents of any mammal that might someday be used for milk production.

Milk Fat

Milk fat is removed from the fluids of milk (called skim milk) by a separator. Not all but most of the fat is removed from skim milk. The presence of fat in milk is of significance to the American diet today as consumers turn increasingly to lowfat foods. The total *amount* of fat in milk is positively related to the amount of milk produced but the *percentage* of milk fat produced is inversely related to the amount of milk. Holstein cows, for example, are large and produce relatively large amounts of milk and milk fat; the percentage of fat in the milk of Holsteins is, however, lower (3.5% to 4.0%) than the percentage in milk from Jersey and Guernsey cows (4.5% to 6.0%).

Milk fat from Jersey and Guernsey cows has a rich, yellow color due to a yellow pigment, **carotene**; that from Holstein cows has a pale, yellow color; and that from goats is white. Carotene can be split into two molecules of vitamin A. The milk fat of Holstein cows is pale yellow because some of its carotene has been split in this way, and the milk fat of goats is white because the complete conversion has occurred.

Besides pigmentation, goat's milk differs from cow's milk in the type of **curd** (solidified protein) formed by the actions of stomach acids and enzymes on the milk. The curd formed from goat's milk is easier for some people to digest than the curd from cow's milk. Thus, people who cannot convert carotene into vitamin A readily and those who cannot digest

Table 3-2. Average percentage composition of milk from selected mammals

Mammal	Water	Total solids	Fat	Protein	Lactose
Human	87.8	12.2	3.8	1.2	7.0
Cow	87.3	12.7	3.9	3.3	4.8
Goat	87.6	12.4	3.7	3.3	4.7
Water buffalo	76.8	23.2	12.5	6.0	3.8
Ewe	81.6	18.4	6.5	6.3	4.8
Sow	82.4	17.6	5.3	6.3	5.0
Mare	90.2	9.8	1.2	2.3	5.9

Source: From *Introduction to Livestock Production*, Second Edition, by H. H. Cole. W. H. Freeman and Company. Copyright 1966.

cow's milk easily can benefit from goat's milk. Although dairy goats produce less total milk and milk fat than cows, their efficiency per unit of body weight is about the same.

BACTERIAL ACTION

Besides humans, other organisms thrive on milk—bacteria. In fact, milk is one of the best culture media for bacteria. This situation has both good and bad features. Properly processed milk from clean and healthy cows can be kept for 10 to 14 days in refrigerated conditions. Most of the milk sold commercially in the United States is pasteurized and homogenized. **Pasteurization** is a process of exposing milk to a temperature that destroys all pathogenic bacteria but neither reduces the nutritional value of the milk nor causes it to curdle. Pasteurization is done by raising milk to a temperature of 161 °F (71.7 °C) for 15 seconds, after which it is quickly cooled. This short-time, high-temperature pasteurization fits well into a continuous operation of pasteurization and subsequent processing.

Several highly desirable foods, such as the wide variety of cheeses, can be made as a result of introducing the proper types of bacteria into milk. However, pathogenic bacteria also thrive in milk, so people who consume raw milk should be certain that it comes from healthy animals, is produced under sanitary conditions, and is processed by healthy people.

HOMOGENIZATION

Much of the fluid milk consumed in the United States is **homogenized** to prevent the milk fat from separating from the liquid portion and rising to the top. Homogenization is a process of making a stable emulsion of milk fat and milk serum. A cream line does not appear in homogenized milk nor does it form butter when churned. Homogenizers are of three types: high-pressure type, low-pressure, rotary type, and sonic vibrator. The high-pressure type forces milk through a small orifice, thereby causing the fat globules to break apart and remain dispersed in a stable emulsion. Exceedingly high pressures of 5,000 lbs per square inch may be used in this process. The low-pressure, rotary type homogenizer employs pressures of less than 500 lbs per square inch. The rotary action shears the fat globules apart. The sonic vibrator type of homogenizer subjects milk to high frequency vibrations that tear the fat globules apart.

Fat globules in raw milk average about 6 micrometers (μm) in diameter. Homogenization results in fat globules that average fewer than 2 μm in diameter. To visualize how small the fat globules in homogenized milk are, note that 25,000 μm equal approximately 1 in. Homogenized milk will likely deteriorate by becoming rancid more rapidly than nonhomogenized milk, due to the greater surface area of the fat globules upon which lipolytic enzymes can act. Rancidity of homogenized milk is forestalled by pasteurizing the milk prior to or immediately following homogenization, thus destroying the action of the lipolytic enzymes.

CONSUMER PROTECTION

In the United States, state and federal agencies protect the milk consumer by several safeguards, such as enforcement of sanitation requirements for production sites, checks on bacterial counts of commercially sold milk, and checks on alterations of milk that occur when water is added to give greater volume. The most important safeguards from a health standpoint are the requirements that bacterial counts be low and that milk be pasteurized. Milk sold through commercial outlets is certified to be from herds that are tested and found free from Bang's disease and tuberculosis. The consumer who buys milk and milk products can be assured of obtaining a safe, desirable, wholesome food.

STUDY QUESTIONS

1. Answer statements 1a through 1e as true or false and briefly explain your answer.
   a. The percentage of fat is inversely related to amount of milk produced. Why or why not?
   b. The most important solid in milk for human nutrition is milk fat. Why or why not?
   c. The yellow color of milk is a good indicator of its milk fat content. Why or why not?
   d. Milk is high in most minerals, but is very low in iron content. Why or why not?
   e. Dairy cows and dairy goats are approximately equal in efficiency of milk production. Why or why not?
2. a. What are the solids-not-fat of milk?
   b. What is the first milk given after a female's young are born called?
   c. What safeguards are provided by state and federal regulations to protect the consumer who purchases milk?
   d. What are five food products made from milk?
   e. What vitamins are normally present in milk?
3. a. Why do some people drink goat's milk in preference to cow's milk?
   b. Milk is a perfect medium for bacterial growth. What hazards does this situation present? What advantages?
   c. What constituents of milk vary the most among different milk-producing species?
   d. What benefits are gained by homogenizing milk?
   e. Is milk a complete and perfect food for human infants? Why or why not?

SELECTED REFERENCES

USDA. 1980. Dairy Situation. DS379. Economics, Statistics, and Cooperative Service. Washington, D.C. USDA.

Kon, S. K., and Cowie, A. T. 1961. *Milk: The Mammary Gland and Its Secretion.* Vol. 2. New York: Academic Press.

Macy, I. G., Kelly, H. J., and Sloan, R. E. 1953. *The Composition of Milks.* Washington, D.C.: National Academy of Science, National Research Council Publication 254.

Milk Industry Foundation. 1976. *Milk Facts.* Washington, D.C.: Milk Industry Foundation (in cooperation with U.S. Government and USDA).

Schmidt, G. H. 1971. *Biology of Lactation.* San Francisco: W. H. Freeman and Co.

USDA. 1980. *Handbook of Agricultural Charts.* Washington, D.C.: USDA Agriculture Handbook 491.

4

Hides, Wool, Mohair, and Furs

The skin of common mammals produces a hairy covering to which various terms are applied depending on the nature of the growth. For example, cattle, pigs, horses, and dairy goats are said to have hair; sheep have wool; mink and nonangora rabbits have fur; angora rabbits have angora; and angora goats have mohair. The hair from most mammals has little commercial value (it is used mostly in padding and cushions), the hair of mink and nonangora rabbits is either naturally beautiful or can be dyed to give attractive colors; therefore, it has considerable value as fur. Fur differs from the hair of cattle, horses and pigs in that fur is composed of fine short fibers and relatively long, coarse guard hairs, and hair has only guard hairs.

HOW HIDES ARE USED BY PEOPLE

The skins (called hides or pelts) of certain mammals are frequently dried and made into useful products through the process of tanning. Tanning is used to make leather from hides of cattle and horses, coat liners and rugs from hides of sheep and goats, and furs from hides of mink and rabbits. The hair is removed in making leather, whereas wool or fur remain with their hides. The hides from pigs usually are not removed, because the hair is removed by scalding and the skin remains on the pork carcass. Some packers, however do skin pigs, and the skins can be used in making leather.

The value of hides may be reduced greatly by branding, by nicking the hide while skinning, or by **warbles** (larvae of heel flies). Warbles emerge from the back region of the hides of cattle in spring, thereby making holes in the hide. Sheep hides may be damaged by outgrowths of grasses in the production of seeds (called beards) which can penetrate the skin and by external parasites called keds.

The furs made from rabbits, mink, foxes, and bison (American buffalo) may be classed as status clothing because they are usually costly and special in appearance. White rabbit furs can be dyed any color (including the pastels), but colored furs cannot. This is why color **mutations** have been so important in the mink business. Color mutations provide a wide array of fur colors in mink. At present, the mink industry in the United States is suffering economically because of competition from furs made in other countries, and high costs of producing mink.

HOW HAIR, WOOL, AND MOHAIR FIBERS GROW

The hair fiber grows from a follicle located in the outer layers of the skin. Growth occurs at the base of the follicle, where there is a supply of blood, and the cells produced are pushed outward. The cells die after they are removed from the blood supply, because they can no longer obtain nutrients or eliminate wastes. A schematic drawing of a wool follicle is presented in Figure 4-1. The wool fiber has an inner core, the *cortex*, and an outer sheath, the *cuticle*. The cuticle of wool fibers cause the fibers to cling together. The intermingling of wool fibers is known as *felting*. Wool fibers have waves called *crimp*. Crimp is caused by the presence of hard and soft cellular material in the cortex. The soft cortex is more elastic and is on the outer side of the crimp. The felting of wool is advantageous in that wool fibers can be entangled to make **woolens**, but it is also responsible for the shrinkage that occurs when wool becomes wet.

Hair does not exhibit any crimp because all the cells in the cortex are hard. Also, the inner core of hair is not solid as is the inner core of most wool fibers. Some fibers in wool of a low quality are large, lack crimp, and do not have a solid inner core. These fibers are called kemp and they reduce the value of the fleece.

Mohair fiber follicles develop in groups consisting of three primary follicles each. Secondary follicles, per each primary follicle develop later. Mohair fibers have very little crimp (less than one crimp for each inch of length). The mohair cortex is composed largely of so-called *ortho* cells in contrast to the cortex of wool which is composed of both ortho and

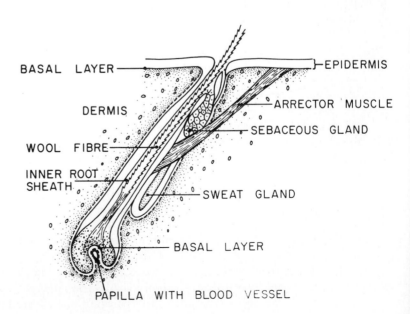

Figure 4-1. Schematic drawing of a wool follicle.

BASAL LAYER — EPIDERMIS

DERMIS — ARRECTOR MUSCLE

— SEBACEOUS GLAND

WOOL FIBRE

INNER ROOT SHEATH — SWEAT GLAND

— BASAL LAYER

PAPILLA WITH BLOOD VESSEL

para cells. Kemp and medullated fibers reduce the value of wool and mohair fleeces because such fibers do not dye well and they show in apparel made from fleeces containing such fibers.

PRODUCTION OF WOOL AND MOHAIR

Annual world wool production is approximately 5,396,600,000 lbs of which the United States produces 106,799,000 lbs or about 2.0%.

Growth of wool varies greatly among the breeds of sheep: fine-wool sheep grow 4 to 6 in. of wool per year, 1/4 to 1/2 blood breeds of sheep grow 5 to 10 in. of wool per year, while common to braid wool breeds grow 9 to 18 in. of wool per year. The weights of **fleeces** also vary greatly among the breeds of sheep. Even though fine-wool breeds have fleeces of short length, the weight of their fleeces varies from 9 to 12 lbs; 1/4 to 1/2 blood breeds of sheep produce fleeces weighing 7 to 15 lbs; and coarse wool (common and braid) breeds produce fleeces weighing 9 to 16 lbs.

Annual world mohair production is approximately 59 million lbs of which the United States produces between 29 and 30 million lbs or about half the world production. Texas produces about 97% of the mohair that is produced in the United States. Mohair growth is about 2 in. per month; therefore, goats are usually shorn twice per year with length of clip averaging 6 in.

HOW WOOL AND MOHAIR ARE USED

The fibers from sheep, angora rabbits and goats are used in making cloth and carpets. Cloth made from wool has both highly desirable and undesirable qualities. Wool has a pleasant warming quality and can even absorb considerable quantities of moisture while still providing warmth. Wool is also highly resistant to fire. Wool has a tendency to shrink when it becomes wet, however, and cloth from wool causes some people to itch.

Researchers in Australia and New Zealand have contributed greatly to the alteration of wool so that it will not shrink when washed. Recently the USDA Wool and Mohair Laboratory developed a process called WURLAN treatment, which makes woolen cloth machine washable. In this treatment the wool fibers are coated with a thin layer of resin. This resin layer is exceedingly thin (it adds only 1% to the weight of the wool); therefore, it does not alter wool in any significant way except to prevent the fibers from absorbing water.

Growers of sheep and wool in the United States have traditionally relied on tariffs to keep the price of wool at a profitable level rather than explore the possibilities of producing wool more efficiently or improving the quality of wool by technology. The result has been the development of synthetic fibers for making cloth. These synthetic fibers are now competing strongly with wool; they do not shrink when washed, have good wearing qualities, and do not irritate the skin of people who are bothered by wool. They lack the warming comfort of wool, however, and some are quite cold and clammy when wet.

Sheep producers need to realize that more financial progress can be made by improving rate and efficiency of meat production from sheep than by improving amount and quality of wool produced. At the present time, income from farm-flock sheep in the United States is about 10% to 15% from wool and 85% to 90% from lambs marketed. Range sheep yield a larger percentage of income from wool, but production of range sheep has declined greatly in the past 50 years.

The demand for black sheep has increased because their dark wool, which has an attractive natural appearance, does not require dyeing. Black color in sheep is a recessive hereditary trait (see Chapter 15). Therefore this trait will *breed true* (meaning that, for example, when two black sheep breed with each other, they produce only black offspring), making the production of black sheep simple.

Mohair differs from wool in that it lacks the cuticular development that results in felting. Therefore, clothing made from mohair has a harder finish. Both wool and mohair can be dyed any color that is desired. Much of the mohair that is produced in the United States goes to Europe and the United Kingdom for the making of desirable apparel.

HOME TANNING OF HIDES

The current interest that people are showing in providing homemade goods extends to tanning and quite satisfactory results can be obtained through a simple six-step procedure:
1. Wash the fresh pelt in a washing machine using cool water as one would wash a wool sweater.
2. After the pelt is washed, let it spin as one would a sweater, for a short time and at a slow speed.
3. Remove the pelt from the washing machine, place the pelt on a drying rack with the flesh side up, and let it dry.
4. Make a thick paste of baking soda and kerosene. Apply the paste heavily onto the flesh side of the pelt and leave it on for 30 days.
5. Rub the flesh side with pumice rock to clean and soften.
6. Color the pelt with an ordinary dye the same as one would dye woolen cloth.

FACTORS AFFECTING VALUE OF WOOL

The four most important factors affecting the value of wool are its fineness, length of fiber, amount of crimp, and amount of grease and dirt. Generally, fine wool has more grease than coarse wool. Wool is scoured to remove the grease before being processed into cloth. When a considerable amount of grease is present, a relatively large amount of dirt is also usually present because the grease makes dirt adhere to the wool. A considerable loss in weight occurs when grease and dirt are removed from wool. Long fibers can be made into **worsteds**, which hold a press well, wear well, and make a nice appearance. Short fibers must be made into woolens because they are too short to be combed and spun into strong yarn. Fine wool is more desirable than coarse wool because it can be made into attractive suits and dresses; coarse wool is used in rug making and upholstering. Generally, fine wool tends to be shorter than long wool and to have more crimp. The ideal wool is fairly long and very fine.

Wool is graded according to its fineness, either by the "blood" or the **"spinning count"** method. The "blood" system is based on the fineness of wool that one would expect from a sheep that is 1/2 or 1/4 Merino breeding. This method was developed many years ago when Merino sheep were the outstanding producers of wool. "One-half blood" equals the fineness of the wool of a sheep produced when fine-wool (Merino) sheep are mated to

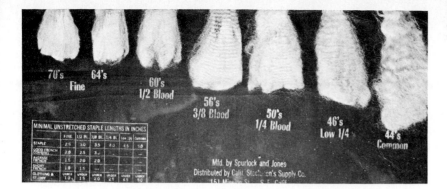

Figure 4-2. Wool samples of the major grades of wool.

nondescript sheep. "One-fourth blood" refers to the fineness of the wool of a sheep of 1/4 fine-wool breeding. The "spinning count" method is based on the theoretical number of **hanks** of yarn that can be spun from 1 lb of clean wool that has been combed to remove "tops" (short fibers). One hank equals 560 yards. Thus a grade of 50 by the "spinning count" method means that 28,000 yards (560 yd × 50 = 28,000 yd) of yarn can be spun. Fine wool provides more yarn per unit of weight than coarse wool. Both systems of grading, along with the breeds from which the grades of wool come, are presented in Table 4-1. Wool samples from each of the major grades are shown in Figure 4-2.

The value of a fleece (the wool from all parts of a sheep) can be influenced by anything that causes a severe stress on the sheep, because a break, or weak zone, is created in the fleece when wool ceases to grow. Being without food or water for a few days or having an illness with a high fever often causes a break in the fleece. Although black wool is now highly prized, black fibers or black tips on wool fibers are extremely objectionable in a white fleece. These fibers do not take on light-colored dyes and consequently stand out in a garment made from such wool dyed a light pastel color. Therefore, a fleece that contains a combination of black and white wool sells at a low price. Also, some sheep grow coarse, hairlike fibers intermingled with the wool fibers. These coarse fibers, called **kemp**, are not

Table 4-1. Market grades of wool

USDA grades based on spinning count	USDA grades based on blood system	Breeds according to average grades
80, 70, 64	Fine	Merino, Rambouillet
62, 60	1/2 blood	Targhee, Southdown
58, 56	3/8 blood	Montadale, Columbia, Shropshire, Hampshire, Corriedale
54, 50, 48	1/4 blood	Dorset, Suffolk, Cheviot, Oxford
46	Low 1/4 blood	Romney, Leicester
44	Common	Cotswold
49, 30	Braid	Lincoln

solid fibers; each fiber has a large medulla in its center. Cloth and wool from these coarse fibers are highly objectionable and thus are low in price. Kemp fibers in tanned sheep are not objectionable when the tanned hides are used as rugs but they may cause some discomfort if present in coat liners.

STUDY QUESTIONS

1. Define or explain:
   a. Mohair
   b. Felting
   c. Hank
   d. Spinning count
   e. Worsteds
2. a. What are some important ways that hides can be damaged?
   b. Why are black fibers detrimental to a white fleece?
   c. What are the bases for grading wool?
   d. From what animals is mohair obtained?
   e. What two characteristics of wool have the largest effect on its value?
3. What are the two most highly desirable characteristics of wool? What are the two major undesirable characteristics?
4. You should learn the grades of wool according to each of the grading systems and the breeds of sheep that produce each grade.
5. a. Where in the United States is most mohair produced?
   b. How does the United States compare in world production of wool and mohair?
   c. How often are goats shorn?

SELECTED REFERENCES

Lyne, A. G., and Short, B. F., eds. 1965. *Biology of the Skin and Hair Growth.* New York: Elsevier Scientific Publishing Co.

Onions, W. J. 1862. *Wool: An Introduction to Its Properties, Varieties, Uses, and Production.* London: Ernest Benn Brothers.

Pohle, E. M., Keller, H. R., Ray, H. D., Lineberry, C. T. and Reals, H. C. 1977. *Physical Properties of Grease Mohair and Related Mill Products.* (Spring and Fall Clips.) Marketing Research Report #954. Washington, D. C. Agricultural Marketing Service, USDA.

Terrill, C. E. 1971. Mohair. From *Encyclopedia of Science and Technology.* 3rd Edition. McGraw-Hill Book Co.

5

Market Classes and Grades of Livestock and Poultry

The production and movement of more than 80 billion pounds of highly perishable livestock and poultry products to the consumer in the United States is accomplished through a vast and complex marketing system. Marketing is the physical movement, transformation, and pricing of goods and services, and a number of buyers and sellers work to convey livestock and livestock products from the point of production to the point of consumption. Producers need to understand marketing if they are to produce products preferred by the consumer, decide intelligently between various marketing alternatives, understand how animals and products are priced, and eventually raise productive animals profitably.

Market classes and **grades** have been established to segregate animals and carcasses into uniform groups based on preferences of buyers and sellers. The USDA has established extensive classes and grades to make the marketing process simpler and more easily communicated. Use of USDA grades is voluntary. Some packers have their own private grades, but these grades are typically used in combination with the USDA grades. An understanding of market classes and grades helps producers recognize quantity and quality of the products they are supplying to consumers.

MARKET CLASSES AND GRADES OF RED MEAT ANIMALS

Slaughter Cattle

Slaughter cattle are separated into classes based primarily on age and sex. Age of the animal has a significant effect on tenderness, with younger animals typically producing more tender meat than older animals. The age classification for meat from cattle are veal, calf, and beef. Veal is from young calves 1 month to 3 months of age, with carcasses weighing less than 150 lbs. Calf is from animals ranging in age from 3 months to 10 months

Table 5-1. Official USDA grade standards for live slaughter cattle and their carcasses.

Class or kind	Quality grades (Highest to lowest)	Yield grades (Highest to lowest)
Beef		
Steer and Heifer	Prime, Choice, Good, Standard, Commercial, Utility, Cutter, Canner	1, 2, 3, 4, 5
Cow	Choice, Good, Standard, Commercial, Utility, Cutter, Canner	1, 2, 3, 4, 5
Bullock	Prime, Choice, Good, Standard, Utility	1, 2, 3, 4, 5
Bull	(no designated quality grades)	1, 2, 3, 4, 5
Veal	Prime, Choice, Good, Standard, Utility	not applicable
Calf	Prime, Choice, Good, Standard, Utility	not applicable

and carcass weights between 150 and 300 lbs. Beef comes from more mature cattle over 12 months of age having carcass weights higher than 300 lbs. Classes and grades established by the USDA for cattle are based on sex, quality grade, and yield grade, all of which are used in the classification of both live cattle and their carcasses (Table 5-1).

The sex classes for cattle are **heifer, cow, steer, bull,** and **bullock.** Occasionally the sex class of **stag** is used to refer to males that have been castrated after their secondary sex characteristics have developed. Sex classes separate cattle and carcasses into more uniform carcass weights, tenderness groups, and how carcasses are processed. Quality grades are intended to measure certain consumer palatability characteristics, whereas yield grades measure amount of fat, lean, and bone in the carcass. Slaughter steers representing some of the eight quality grades and five yield grades are shown in Figures 5-1 and 5-2.

Quality grades are based primarily on two factors: (1) maturity (physiological age) of the carcass, and (2) amount of **marbling.** Maturity is determined primarily by observing bone and cartilage structures. For example, soft, porous, red bones with a maximum amount of pearly-white cartilage, characterize A maturity whereas very little cartilage and hard, flinty bones characterize C, D, and E maturities. As maturity in carcasses increases from A to E, the meat becomes less tender.

Marbling is intramuscular fat or flecks of fat within the lean, and marbling is estimated from the exposed ribeye muscle between the twelfth and thirteenth ribs. Nine degrees of marbling, ranging from abundant to practically devoid, are designated in the USDA marbling standards. Various combinations of marbling and maturity that identify the carcass quality grades are shown in Figure 5-3. Note that C maturity starts at 48 months of age. Slaughter cows more than 4 years old will grade commercial, utility, cutter, or canner regardless of amount of marbling. Figure 5-4 shows ribeye sections of several different quality grades and several different degrees of marbling.

The **yield grades,** sometimes referred to as **cutability,** measure the quantity of boneless, closely trimmed retail cuts (BCTRC) from the round, loin, rib, and chuck. A numerical scale of 1 through 5 is used to rate yield grade, with 1 denoting the highest grade. Although marketing communications generally quote yield grades in whole numbers, these yield grades are often shown in tenths in research data and information on

SLAUGHTER STEERS
U.S. GRADES
(QUALITY)

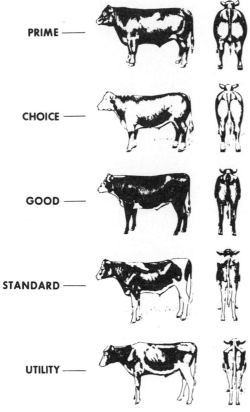

PRIME

CHOICE

GOOD

STANDARD

UTILITY

Figure 5-1. USDA quality grades (commercial, cutter, and canner omitted).

SLAUGHTER STEERS
U.S. GRADES
(YIELD)

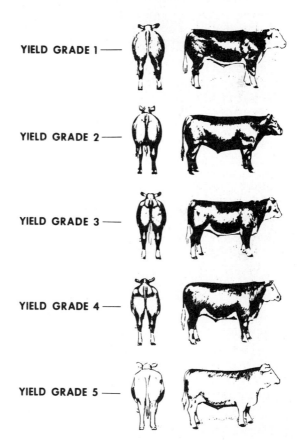

YIELD GRADE 1

YIELD GRADE 2

YIELD GRADE 3

YIELD GRADE 4

YIELD GRADE 5

Figure 5-2. USDA yield grades for slaughter cattle.

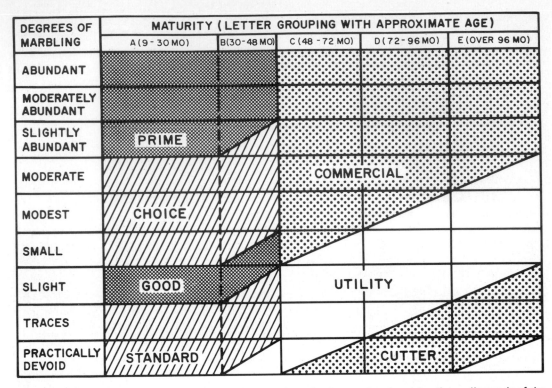

DEGREES OF MARBLING	MATURITY (LETTER GROUPING WITH APPROXIMATE AGE)				
	A (9 - 30 MO)	B (30-48 MO)	C (48 - 72 MO)	D (72-96 MO)	E (OVER 96 MO)
ABUNDANT					
MODERATELY ABUNDANT	PRIME		COMMERCIAL		
SLIGHTLY ABUNDANT					
MODERATE					
MODEST	CHOICE				
SMALL					
SLIGHT	GOOD		UTILITY		
TRACES	STANDARD				
PRACTICALLY DEVOID			CUTTER		

Figure 5-3. Marbling and the maturity of the carcass are the major factors that determine the quality grade of the beef carcass.

carcass contests. Table 5-2 shows the yield grades and their respective percentage of BCTRC. As an example, a carcass with a yield grade of 3.0 has a BCTRC of 50%; a 700-lb. carcass with this yield grade would produce 350 lbs of boneless, closely trimmed retail cuts from the round, loin, rib, and chuck.

Yield grades are determined from the following four carcass characteristics: (1) amount of fat measured in tenths of inches over the ribeye muscle, also known as the *longissimus dorsi* (Figure 5- 5), (2) kidney, pelvic, and heart fat (usually estimated as percent of carcass weight), (3) area of the ribeye muscle, measured in square inches (Figure 5-6), and (4) hot carcass weight. The latter measurement reflects amount of intramuscular fat. Generally, as the carcass increases in weight, amount of fat between the muscles increases also.

Measures of fatness in beef carcasses have the greatest effect in determining yield grade. A tentative yield grade can be determined by estimating or measuring the fat of the ribeye muscle. Figure 5-7 shows the five yield grades with varying amounts of fat over the ribeye muscle and the area of the ribeye.

Feeder Cattle

The USDA feeder grades for cattle are intended to predict feedlot weight gain and the slaughter weight end point of cattle fed to a desirable fat-to-lean composition. The USDA

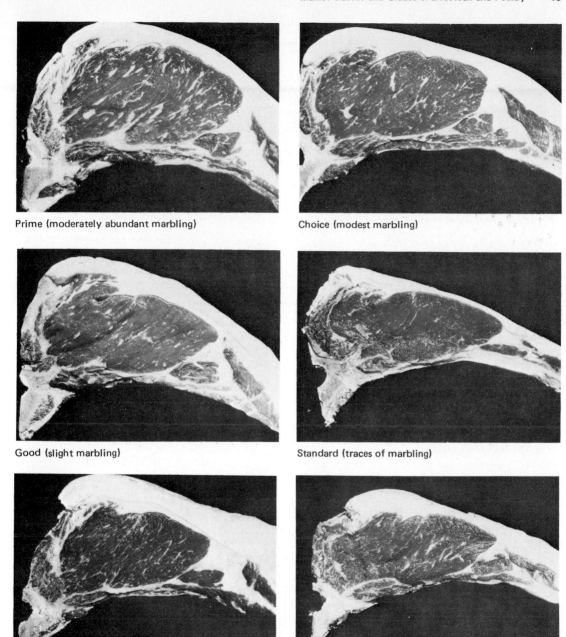

Prime (moderately abundant marbling)

Choice (modest marbling)

Good (slight marbling)

Standard (traces of marbling)

Commercial (moderately abundant marbling; however,
E maturity put it into this grade) See Figure 5-3.

Utility (modest marbling)

Figure 5-4. Exposed ribeye muscles (between twelfth and thirteenth ribs) showing various degrees of marbling of several beef carcass quality grades.

Table 5-2. Beef carcass yield grades and the yield of BCTRC
(boneless, closely trimmed retail cuts) from the round, loin, rib, and chuck

Yield grade	BCTRC %	Yield grade	BCTRC %	Yield grade	BCTRC %
1.0	54.6	2.8	50.5	4.6	46.4
1.2	54.2	3.0	50.0	4.8	45.9
1.4	53.7	3.2	49.6	5.0	45.4
1.6	53.3	3.4	49.1	5.2	45.0
1.8	52.8	3.6	48.7	5.4	44.5
2.0	52.3	3.8	48.2	5.6	44.1
2.2	51.9	4.0	47.7	5.8	43.6
2.4	51.4	4.2	47.3	—	—
2.6	51.0	4.4	46.8	—	—

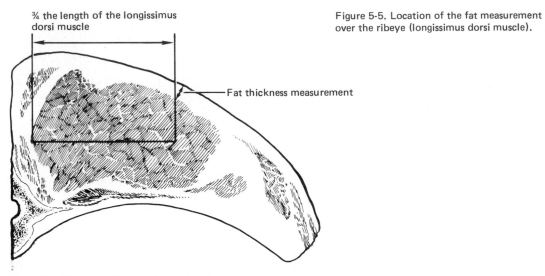

¾ the length of the longissimus dorsi muscle

Fat thickness measurement

Figure 5-5. Location of the fat measurement over the ribeye (longissimus dorsi muscle).

Figure 5-6. Plastic grid is placed over the ribeye muscle to measure the area. Each square represents 0.1 square inch.

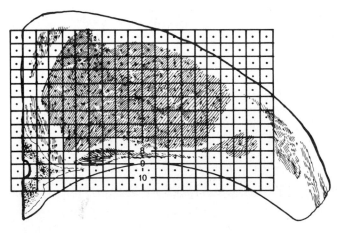

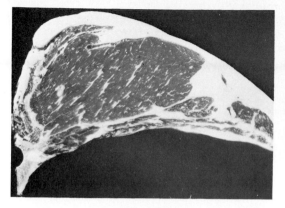

Yield Grade 1
(Fat 0.2 in. ribeye area 13.9 sq in.)

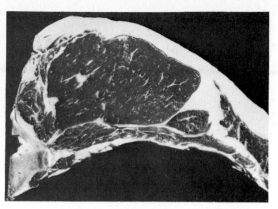

Yield Grade 2
(Fat 0.4 in. ribeye area 12.3 sq in.)

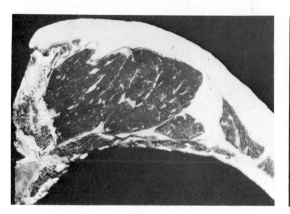

Yield Grade 3
(Fat 0.6 in., ribeye area 11.8 sq in.)

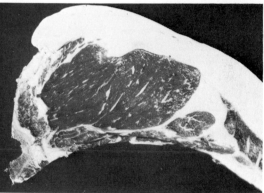

Yield Grade 4
(Fat 0.9 in., ribeye area 10.5 sq in.)

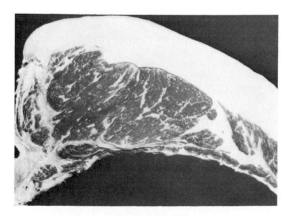

Yield Grade 5
(Fat 1.1 in., ribeye area 10.9 sq in.)

Figure 5-7. The five yield grades of beef shown at twelfth and thirteenth ribs.

cattle feeder grades were revised in 1979. The two criteria used to determine feeder grade are frame size and thickness. The three measures of both frame size and thickness are shown in Figures 5-8 and 5-9. Feeder cattle are given a USDA grade of inferior if the cattle are unhealthy or double muscled. These cattle would not gain satisfactorily in the feedlot.

a. Large frame

Figure 5-8. The three frame sizes of the USDA feeder grades for cattle.

b. Medium frame

No. 1

c. Small frame

No. 2

Figure 5-9. The three thickness standards of the USDA feeder grades for cattle.

No. 3

Although frame size and ability to gain weight in the feedlot are apparently related in the sense that large framed cattle usually gain fastest, frame size appears to predict more accurately carcass composition or yield grade at different slaughter weights than gaining ability. The USDA feeder grade specifications identify the different live weights from the three frame sizes when they reach the same thickness of fat over the ribeye (Table 5-3).

Slaughter Swine

Sex classes of swine are **barrow** and **gilt, sow, boar,** and **stag.** Boars and sows are older breeding animals, whereas gilts are younger females that have not produced any young. The barrow is the male pig castrated early in life, and the stag is the male pig castrated after it has developed certain boarlike characteristics. Because of relationships between sex and sex condition and the acceptability of prepared meats to the consumer, separate grade standards have been developed for barrow and gilt carcasses and for sow carcasses. There are no official grade standards for boar and stag carcasses.

The grades for barrow and gilt carcasses, which are not separated according to sex, are based on two general criteria: (1) quality characteristics of the lean, and (2) expected combined yields of the four lean cuts (ham, loin, blade shoulder, and picnic shoulder). The grade standards provide two general levels of quality of lean in the carcass: acceptable and unacceptable. Quality of the lean is assessed by observing the exposed surface of a cut muscle, usually at the tenth rib. Quality is evaluated on marbling, firmness, and color. Carcasses that have unacceptable lean quality or bellies too thin for suitable bacon production are graded U.S. Utility, as are soft and oily carcasses. Carcasses with acceptable lean quality are graded U.S. No. 1, U.S. No. 2, U.S. No. 3, or U.S. No. 4. These grades are based on expected yields of the four lean cuts as shown in Table 5-4.

Table 5-3. Slaughter weights of large-, medium-, and small-frame slaughter cattle at 0.50 in. of fat

	Slaughter weight	
Frame size	Steers (lbs)	Heifers (lbs)
Large	more than 1,200	more than 1,000
Medium	1,000-1,200	850-1,000
Small	less than 1,000	less than 850

Table 5-4. Expected Yields of the four lean cuts, based on percent of chilled carcass weight

Grade	Percent yield of four lean cuts
U.S. No. 1	53.0 and higher
U.S. No. 2	50.0 to 52.9
U.S. No. 3	47.0 to 49.0
U.S. No. 4	less than 47.0

Carcasses differ in their yields of the four lean cuts primarily because of differences in amount of fatness and muscling in relation to skeletal size. Average backfat thickness in relation to carcass length (or weight) together with a muscling evaluation is used to determine the numerical grade. Figure 5-10 shows where the measurements for average backfat thickness and carcass length are taken. Carcass length is measured from the forward edge of the first rib to the forward edge of the aitch (pelvic) bone. Average backfat thickness is an average of three measurements (which include the skin), made perpendicular to the outside surface opposite (1) the first rib, (2) the last rib, and (3) the last lumbar vertebra.

Figure 5-11 shows how backfat thickness, carcass length or weight, and muscling are combined to determine the USDA pork carcass grades. Figure 5-12 shows muscling scores

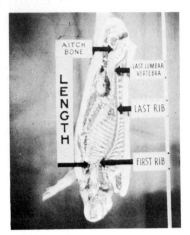

Figure 5-10. Pork carcass measurements for average backfat thickness and carcass length.

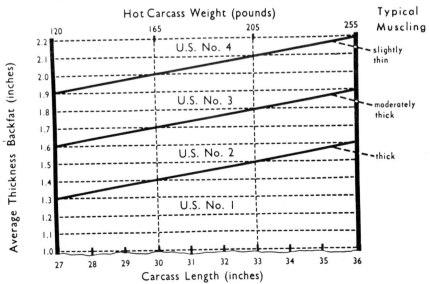

Figure 5-11. Relationship of average backfat thickness, carcass length or weight, and muscling in determining the carcass grade in swine.

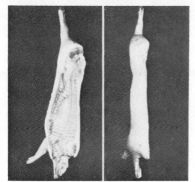

CARCASS LENGTH .... 30.4 inches
AVERAGE BACKFAT THICKNESS..... 1.3 inches
DEGREE OF MUSCLING .... thick

U.S. NO. 1

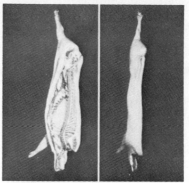

CARCASS LENGTH .... 29.5 inches
AVERAGE BACKFAT THICKNESS ..... 1.6 inches
DEGREE OF MUSCLING .... moderately thick

U.S. NO. 2

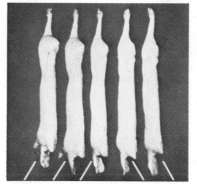

VERY   THICK   MODERATELY   SLIGHTLY   THIN
THICK   THICK   THIN

**DEGREES OF
MUSCLING**

CARCASS LENGTH ....30.0 inches
AVERAGE BACKFAT THICKNESS..... 1.9 inches
DEGREE OF MUSCLING ....slightly thin

U.S. NO. 3

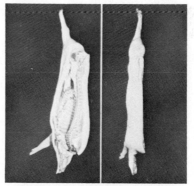

CARCASS LENGTH .... 28.5 inches
AVERAGE BACKFAT THICKNESS .. 2.2 inches
DEGREE OF MUSCLING ....thin

U.S. NO. 4

Figure 5-12. USDA grades for pork carcasses. Five of the six degrees of muscling used in determining the grades are given. Utility grade is omitted.

that are used to determine grades and also shows examples of each of the four pork carcass grades. Figure 5-13 shows pictures of live slaughter barrows representative of the five USDA carcass grades.

Feeder Pig Grades

The USDA grades of feeder pigs are U.S. No. 1, U.S. No. 2, U.S. No. 3, U.S. No. 4, and Utility (Figure 5-14). These grades correspond to USDA grades for market swine to be slaughtered at 200 to 260 lbs. Utility grade is for unthrifty or unhealthy feeder pigs. Thus feeder pig grades combine an evaluation for thriftiness and slaughter potential.

SLAUGHTER SHEEP

Slaughter sheep are classified by their sex and maturity, and the live sheep and their carcasses are also graded for quality grades and yield grades (Table 5-5).

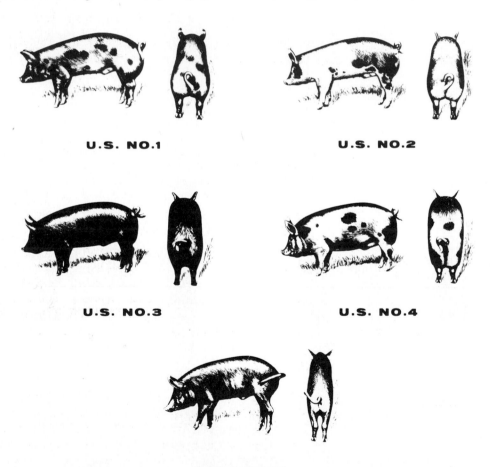

U.S. NO.1   **U.S. NO.2**

U.S. NO.3   **U.S. NO.4**

U.S. UTILITY

Figure 5-13. The USDA grades for live slaughter swine.

U.S. NO.1   **U.S. NO.2**

U.S. NO.3   **U.S. NO.4**

U.S. UTILITY

Figure 5-14. The USDA grades for feeder pigs.

Table 5-5. USDA Maturity groups, sex classes, and grades of slaughter sheep

Maturity group		Grade	
		Quality (highest to lowest)	Yield (highest to lowest)
Lambs	Ewe, wether, or ram	Prime, Choice, Good, Utility, Cull	1, 2, 3, 4, 5
Yearling mutton	Ewe, wether, or ram	Prime, Choice, Good, Utility, Cull	1, 2, 3, 4, 5
Mutton	Ewe, wether, or ram	Prime, Choice, Good, Utility, Cull	1, 2, 3, 4, 5

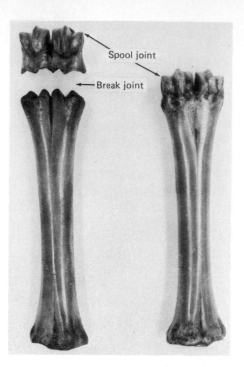

Figure 5-15. Break and spool joints. The cannon bone on the left exhibits the typical break joint in which the foot and pastern are removed at the cartilaginous junction. In the cannon bone on the right, the cartilaginous junction has ossified, making it necessary for the foot and pastern to be removed at the spool joint.

Lamb carcasses, ranging from approximately 2 to 14 months of age, always have the characteristic **break joint** on their shanks following the removal of their front legs (Figure 5-15). Mutton carcasses are distinguished from lamb carcasses by the appearance of the **spool joint** instead of the break joint (Figure 5-15). The break joint ossifies as the sheep mature. Yearling mutton carcasses which range from 12 to 25 months of age usually have the spool joint present but may occasionally have a break joint. Yearling mutton is also distinguished from lamb and mutton by color of the lean (intermediate between pinkish red of lamb and dark red of mutton) and shape of rib bones. Most American consumers prefer lamb to mutton because it has a milder flavor and is more tender.

Quality grades are determined from a composite evaluation of conformation, maturity, degree of feathering, flank streaking, and flank firmness and fullness. Conformation is an assessment of overall thickness of muscling in the lamb carcass. Maturity of lamb carcasses is determined by bone color and shape and muscle color. Feathering and flank streaking predict marbling since lamb carcasses are not usually ribbed to expose the marbling in the ribeye muscle. Feathering is intramuscular fat in muscles between the ribs; flank streaking is intramuscular fat observed on the surface of the flank muscle. Flank firmness and fullness is determined by taking hold of the flank muscle with the hand. Most lamb carcasses grade either Prime or Choice with few carcasses grading in the lower grades.

Lamb carcasses may be quality graded, yield graded, or both. Yield grades estimate boneless, closely trimmed retail cuts from the leg, loin, rack, and shoulder. The approximate percent retail cuts for selected yield grades are shown in Table 5-6.

Table 5-6. Lamb carcass yield grades and the percent retail cuts

Yield grade	Percent retail cuts	Yield grade	Percent retail cuts
1.0	49.0	3.5	44.6
1.5	48.2	4.0	43.6
2.0	47.2	4.5	42.8
2.5	46.3	5.0	41.8
3.0	45.4	5.5	41.0

Yield grades are determined from a composite evaluation of leg conformation score, percent kidney and pelvic fat, and fat thickness. Leg score is subjectively evaluated on a numerical scale from 1 to 15 by assessing thickness, fullness, and plumpness of muscling in the hind leg. Figure 5-16 shows examples of two different leg scores. Amount of kidney and pelvic fat is evaluated subjectively and expressed as a percent of the carcass weight. Fat thickness, in tenths of inches, is measured over the center of the ribeye muscle between the twelfth and thirteenth ribs. This measurement of amount of external fat is the most important yield grade factor since it is a good indicator of the amount of fat trimmed in

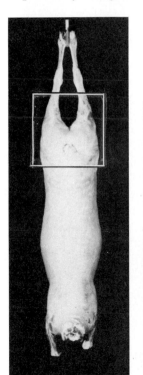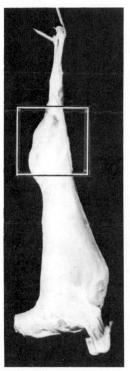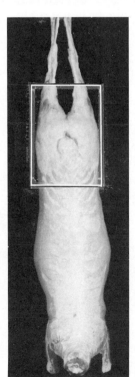

Leg score = Average Prime (score of 14)   Leg score = Low Choice (score of 10)

Figure 5-16. Comparison of two different lamb leg scores used as a partial determining factor for yield grade. Boxes define the area evaluated for thickness, fullness, and plumpness of muscling in the hind leg.

making retail cuts. Figure 5-17 shows cross sections of lamb carcasses representing the five yield grades and amount of fat thickness over the ribeye for each yield grade.

Feeder Lamb Grades

Choice and Prime slaughter lambs are produced in relatively large numbers from grass and their mother's milk. Lambs weighing less than 100 lbs at weaning time are considered feeder lambs. They require additional feeding to produce a more desirable carcass.

The USDA grades for feeder lambs are Prime, Choice, Good, Utility and Cull. These standards are not adhered to closely at the present time since most lambs, correctly finished for slaughter, are graded as Choice regardless of their breed or body shape. Most feeder

Fat .05 in.
Yield Grade 1.7

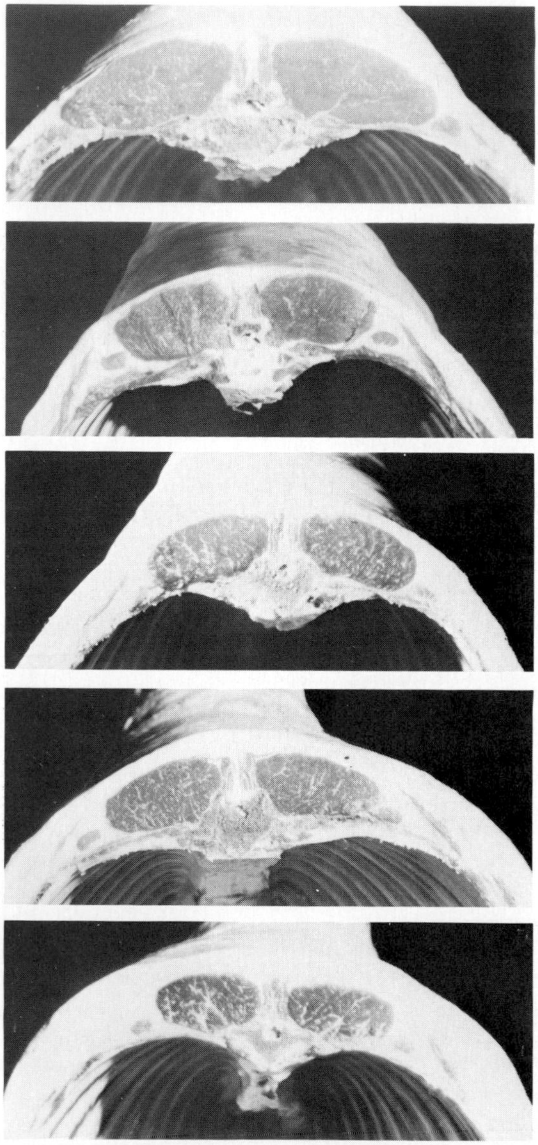

Figure 5-17. The five yield grades of lamb showing the progressive increases in the amount of fat over the loin eye at the twelfth and thirteenth ribs.

Fat .10 in.
Yield Grade 2.4

Fat .25 in.
Yield Grade 3.6

Fat .35 in.
Yield Grade 4.6

Fat .45 in.
Yield Grade 5.5

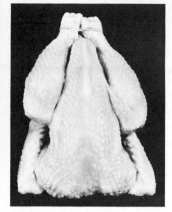

Broiler or fryer, A quality   Broiler or fryer, B quality

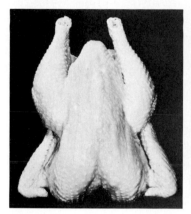

Hen or stewing chicken, A quality   Hen or stewing chicken, B quality

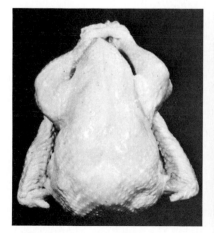
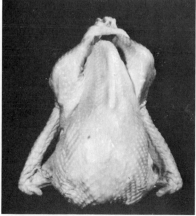

Young turkey, A quality   Young turkey, B quality

Figure 5-18. Classes and grades of ready-to-cook poultry.

lambs are classified by weight rather than grade. "Light weight" feeder lambs typically weigh 60 lbs to 75 lbs; "medium weight" 75 to 85 lbs; and "heavy weight" more than 85 lbs. Lambs may also be classified as natives (produced in midwestern and eastern farm states) or westerns (produced in the western United States).

POULTRY

The three classes of ready-to-cook poultry are broiler (fryer), hen (stewing chicken), and young turkey. Broiler and hen classifications are based on age, with the broiler being younger and more tender than the hen. The grades are A and B for each of the three classes, with A being the higher grade (Figure 5-18).

Carcasses of A quality are free of deformities that detract from their appearance or that affect normal distribution of flesh; they have a well-developed covering of flesh and a well-developed layer of fat in the skin; they are free of pinfeathers and diminutive feathers, exposed flesh on the breast and legs, and broken bones, and have no more than one disjointed bone, and they are practically free of discolorations of the skin and flesh and defects resulting from handling, freezing, or storage. Carcasses of B quality may have moderate deformities; they have a moderate covering of flesh, sufficient fat in the skin to prevent a distinct appearance of the flesh through the skin, and no more than an occasional pinfeather or diminutive feather; they may have moderate areas of exposed flesh and discoloration of the skin and flesh; they may have disjointed parts but no broken bones; and they may have moderate defects resulting from handling, freezing, or storage.

STUDY QUESTIONS

1. What is the difference between a market class and a market grade?
2. What is the difference between a quality grade and a yield grade?
3. Define the following terms:
   a. marbling
   b. wether
   c. barrow
   d. feathering
   e. percent lean cuts
4. What is the difference between a break joint and a spool joint and how are they used in sheep carcass evaluation?
5. What are the primary differences in A and B in ready-to-cook poultry grades?
6. Determine the difference in the BCTRC from a yield grade 2.0 and a yield grade 4.0 where the weight of the beef carcass is 600 lbs for each of the two yield grades.
7. Why cannot a six-year-old cow produce a carcass with a quality grade of Choice?
8. Discuss the similarity of beef yield grades, lamb yield grades and percent lean cuts in swine.

SELECTED REFERENCES

Boggs, D. L., and Merkel, R. A. 1979. *Live Animal Carcass Evaluation and Selection Manual.* Dubuque, Iowa: Kendall/Hunt Publishing Co.

Council for Agricultural Science and Technology (CAST). *Foods From Animals: Quantity, Quality, and Safety.* Report No. 82, March 1980.

McCoy, J. H. 1979. *Livestock and Marketing.* Westport, Connecticut: AVI Publishing Co.

_____ 1977. *Meat Evaluation Handbook.* National Live Stock and Meat Board, Chicago, Illinois.

6

Visual Evaluation of Slaughter Red Meat Animals

The goal of livestock production is generally to put a highly productive animal into the most merchandisable package that will command the highest market price. The productivity of breeding and slaughter meat animals is best identified by using meaningful performance records and effective visual appraisal. Opinions as to the relationship of form and function in the red meat animals (cattle, sheep, swine) differ widely among producers, who continually discuss the value of visual appraisal and measurements of body form and function in breeding programs, in determining market grades, and in defining the so-called "ideal types." It is important to separate the true relationships from opinion, in the area of animal form and function, which continues to be a controversial topic.

Type is defined as an ideal or standard of perfection combining all the characteristics which contribute to an animal's usefulness for a specific purpose. **Conformation** implies the same general meaning as type and refers to the form and shape of the animal. Both type and conformation describe an animal based on its external form and shape. Form and shape can be evaluated visually or measured more objectively with a tape or some other measuring device.

Red meat animals have three productive stages: (1) breeding (reproduction), (2) feeder (growth), and (3) slaughter (carcass or product). It has been well demonstrated that performance records are much more effective than visual appraisal in improving animal productivity and growth stages. Visual appraisal does have importance in these stages, primarily in identifying reproductive and skeletal soundness and health status. Visual appraisal has some importance in evaluating productive differences in the carcass stage of productivity. Both performance records (such as backfat probes and **sonoray** readings for meatiness) and visual appraisal are important in evaluating slaughter animals. Visual appraisal of the slaughter red meat animals can be used effectively to predict primarily carcass composition (fat, lean, and bone) when relatively large differences exist.

It should be recognized that most of the slaughter red meat animals are purchased at market time based on a visual appraisal of their apparent carcass merit.

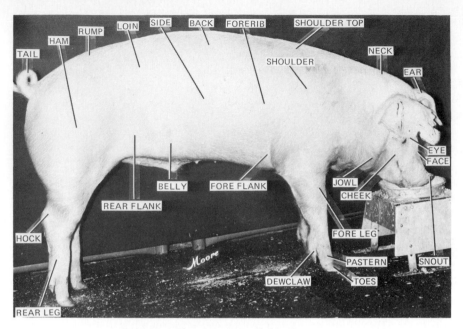

Figure 6-1. The external parts of swine.

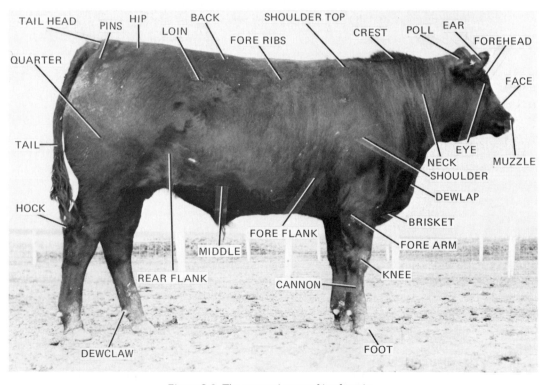

Figure 6-2. The external parts of beef cattle.

EXTERNAL BODY PARTS

Effective communication in many phases of the livestock industry requires a knowledge of the external body parts of the animal. The locations of the major body parts for swine, cattle, and sheep are shown in Figures 6-1, 6-2, and 6-3.

LOCATION OF THE WHOLESALE CUTS IN THE LIVE ANIMAL

The next step for the effective visual appraisal, after becoming familiar with the external parts of the animal, is to understand where the major meat cuts are located in the live animal. The wholesale and retail cuts of the carcasses of beef, sheep, and swine were previously identified in Chapter 2. A carcass is evaluated after the animal has been slaughtered, **eviscerated**, and split into two halves in the case of beef and swine. The lamb carcass remains as a whole carcass, not being split in half. Furthermore, when it is evaluated, the carcass is hanging by the hind leg from the rail in the packing plant. This makes it difficult to perceive how the carcass would appear as part of the live animal standing on all four legs. Figure 6-4 which shows the location of the wholesale cuts on the live animal, helps in understanding the major carcass component parts of the live animal.

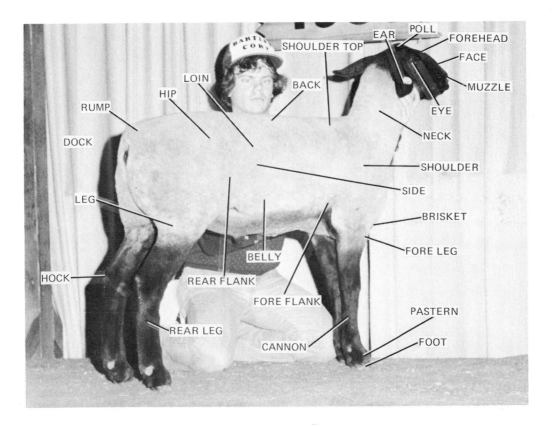

Figure 6-3. The external parts of sheep.

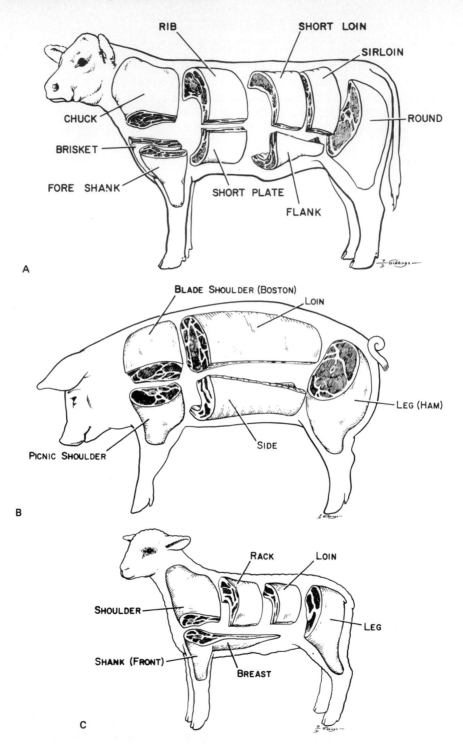

Figure 6-4. Location of the wholesale cuts on the live steer, pig, and lamb.

VISUAL PERSPECTIVE OF
CARCASS COMPOSITION OF THE LIVE ANIMAL

The carcass is composed of fat, lean (red meat), and bone. The meat industry goal is to produce large amounts of highly palatable lean and minimal amounts of fat and bone. These composition differences are reflected in the yield grades of beef and lamb and the lean percentages of swine discussed in Chapter 5.

Effective visual appraisal of carcass composition requires knowing the body areas of the live animal where fat deposits and muscle growth occur. Plates B and C show the fat and lean composition at several cross sections of a yield grade 2 steer and a yield grade 5 steer. The ribbon on the frozen carcasses shows the location of the cross-sections. Cross-section A takes off the bulge to the round, cross-section B is in front of the hip bone down through the flank, cross-section C is at the twelfth and thirteenth rib, and cross section D is at the point of the shoulder down through the brisket. Note the contrasts in fat (18% vs 44%) and lean (66% vs 43%) percentages. Both steers graded Choice and were slaughtered at approximately 1,100 lbs. The conformation characteristics of the two steers, which are primarily influenced by fat deposits and some muscling differences are contrasted in Figures 6-5 and 6-6. The conformation of these two steers is also contrasted to the conformation of a slaughter steer that is underfinished and thinly muscled (Figure 6-7). All of the body parts referred to in these figures can be identified in Figure 6-2 and Figure 6-4, with the exception of the twist. The **twist** is the distance from the top of the tail to where the hind legs separate as observed from a rear view.

Plates D and E show the cross sections of a yield grade 1 lamb and a yield grade 5 lamb. Cross section A is cut through the shoulder area, cross section B is at the back, and cross section C is through the leg area. Note that many of the fat deposits and muscle areas are similar to those in Figure 6-5.

Plates F and G show the cross section of a U.S. No. 1 pig and a U.S. No. 4 pig. Percent of lean cuts is used in pork carcass evaluation instead of yield grades. While the terminology is different, they both measure basically the same things. Percent lean cuts are measured by obtaining the total weight of the trimmed boston shoulder, picnic shoulder, loin, and ham (Figure 6-4) and dividing the total by the carcass weight. The lean cuts percentage for the No. 1 would be 53% or more as contrasted with less than 47% for the U.S. No. 4. The ribbon on the frozen carcass marked the location of the cross sections. Cross section A is through the shoulder area, B is through the middle of the back, and cross section C is through the ham area.

Even though the size and shape of slaughter cattle, sheep, and swine are different, these three species are remarkably similar in muscle structure and fat deposit areas. Therefore, what an individual can learn from the visual evaluation of one species can be applied to another species. Regardless of the species, an animal that shows a square appearance over the top of its back and appears blocky and deep from a side view has a large accumulation of fat. Fat accumulates first in the areas of the flank, brisket, dewlap, and throat (jowl), between the hind legs, and over the edge of the loin. Fat also fills in behind the shoulders and gives the animal a smooth appearance. Movement of the shoulder blade can be observed when lean cattle and swine walk. The wool covering of sheep can easily camouflage the fat covering. The amount of fat in sheep can be determined by pressing the fingers of the closed hand lightly over the last two ribs and over the spinal processes of the vertebrae. Sheep that have a thick padding of fat in these areas will produce poor yield grading carcasses.

Conformation characteristics

1. short, deep body (side view)
2. flat, wide top (rear view)
3. pear shaped (rear view)
4. deep in the **twist** (rear view)
5. deep in rear flank which makes a straight underline (side view)
6. uniform width or wider in middle of back (top view)
7. full dewlap and brisket (front view)
8. filled in behind the shoulders (side and rear view)

Figure 6-5. Conformation characteristics of slaughter steer typical of yield grade 5.

Conformation characteristics

1. relatively long body with moderate body depth (side view)
2. well turned (curved) top (rear view)
3. wide through center of round (rear view)
4. trim in the **twist** (rear view)
5. higher in rear flank than foreflank (side view)
6. wider in the rump than in the middle of the back (top view)
7. trim in dewlap and brisket (front view)
8. slightly dished behind the shoulders (side and rear view)

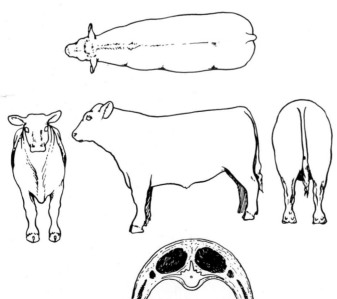

Figure 6-6. Conformation characteristics of slaughter steer typical of yield grade 2.

Underfinished

1. narrow, pleated brisket (front view)
2. relatively shallow body (side view)
3. prominent hip and rib bones (side and rear view)
4. markedly dished behind the shoulders (side and rear view)

Thinly muscled

1. narrow through center of round (rear view)
2. flat and narrow forearm (front and side view)

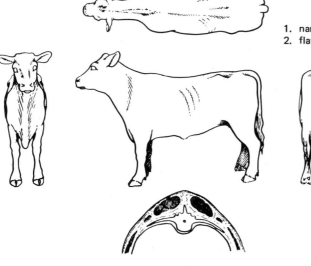

Figure 6-7. Conformation characteristics of slaughter steer that is underfinished (contrast with Figure 6-5) and thinly muscled (contrast with Figure 6-6).

Slaughter red meat animals having an oval turn to the top of their back while having thickness through the center part of their hind legs (as viewed from the rear) have a high proportion of lean to fat. It is important that slaughter animals have an adequate amount of fat because thin animals typically do not produce a highly palatable consumer product.

Accuracy in visually appraising slaughter red meat animals is obtained by making visual estimates of yield grades and percent lean cuts and their component parts, and then comparing the visual estimates with the carcass measurements. Accurate visual appraisal can be used as one tool in producing red meat animals with a more desirable carcass composition of lean to fat.

STUDY QUESTIONS

1. Why is it important to know the body parts of livestock?
2. Name the wholesale cuts of beef, pork, and lamb carcasses and know where they are located on the live animal.
3. Discuss the similarities and differences of the carcass cross-sections for the three red meat species.
4. What value is visual appraisal in evaluating the productivity of the slaughter red meat animals?
5. Where are the major fat deposit areas in beef cattle, sheep, and swine?
6. What body areas of beef, sheep, and swine should be visually appraised for muscling differences?

SELECTED REFERENCES

Beeson, M. W., Hunsley, R. E., and Nordby, J. E. 1970. *Livestock Judging and Evaluation.* Danville, Illinois: The Interstate Printers and Publishers, Inc.

Boggs, D. L. and Merkel, R. A. 1979. *Live Animal Carcass Evaluation and Selection Manual.* Dubuque, Iowa: Kendall/Hunt Publishing Co.

Crouse, J. E., Dikeman, M. E., and Allen, D. M. 1974. Prediction of beef carcass composition and quality by live-animal traits. *Journal of Animal Science* 38:264.

Kauffman, R. G., Grummer, R. H., Smith, R. E., Long, R. A., and Shook, G. 1973. Does live-animal and carcass shape influence gross composition? *Journal of Animal Science* 37:1112.

COLOR PLATE SECTION

Carcass Composition

What should be the proper ratio of fat to lean in an animal carcass is a controversial issue that has seesawed back and forth in the minds of producers, feeders, and packers since the industry began. At the turn of the century, fat-type animals such as the lard hog, baby beef, and fat lamb were the cornerposts of that era's livestock production system. Today, however, vegetable oils have replaced lard, lean beef has replaced the fatted calf, and even the lamb has been stretched longer and leaner through modern production techniques.

Today's consumer demands lean meat, and today's grading system reflects those demands.

By visually appraising a hanging carcass, one can quickly identify the relationship of lean to fat. Appraisal of the live animal, however, can be a more difficult and discriminating challenge. The following set of photographs graphically demonstrates a lean/fat ratio both on individual animals and as a comparison between different yield grades. The pictures are courtesy of Iowa State University, where the entire carcasses were frozen. This process allowed technicians to remove different layers of the carcass such as the hide, fat, and muscling, as well as to section entire carcasses for easy comparison purposes.

The pictures are organized by color plates, with individual captions for each photo presented here by plate and photo number.

Plate A compares body conformation of different yield grades in both live animals and their carcasses.

   1 - Thickness of muscling in two medium-framed feeder steers. The steer on the left has more thickness of lean meat in the round than the steer on the right.
   2 & 3 - Pounds of fat trimmed from one-half of the body of a yield grade 4 steer and a yield grade 2 steer. Black stripes are hide and fat which remain at key locations on the body.

4 - Rear view showing how the fat increases in thickness from the middle of the back to the edge of the loin.

5 - Two yearling Hereford bulls of approximately the same weight. Breeding cattle can be visually appraised for fat and lean compositional differences. The bull on the left would sire slaughter steers having yield grade 4 or 5 carcasses, while the bull on the right would sire yield grade 1 or 2 steers. This assumes that the bulls would be bred to similar frame-size cows, and the steers slaughtered at approximately 1,150 pounds. Compare to plates B and C.

Plates B & C compare fat to lean composition on a yield grade 2 steer (plate B) and a yield grade 5 steer (plate C).

1 - Live animal—side view

2 - Numbered ribbons indicate where following cross sections were made

3, 4, 5, & 6 - Cross sections from rump, hip, mid-carcass, and shoulder

7 - The percentages of fat, lean, and bone found in this carcass.

Plate D & E compare the fat to lean composition of lamb carcasses on yield grade 1 lamb (plate D) and yield grade 5 lamb (plate E).

1 - Side view

2 - Rear view

3 - Numbered ribbons indicate where following cross sections were made

4, 5, & 6 - Shows cross sections from shoulder, mid-carcass and rump

Plate F & G compare fat to lean composition of a U.S. No. 1 pig (plate F) with a U.S. No. 4 pig (plate G).

1 - Live animal—side view

2 - Rear view

3 - Numbered ribbons indicate where following cross sections were made

4, 5, & 6 - Shows cross sections from shoulder, mid-carcass and rump

Plate H shows the composition and appearance of a steer carcass at various stages of removal of hide, fat, and muscle.

1 - Live steer with hair clipped from one side

2 - Frozen steer with hide removed and fat exposed

3 - Fat removed from one-half of steer's body

4 - Rear view with fat removed from the left side. Note cod and twist fat on the right side.

5 - Skeleton of the beef animal after all the muscle has been removed.

1

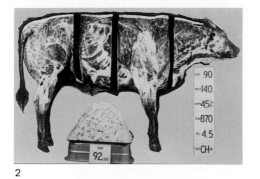

2

3

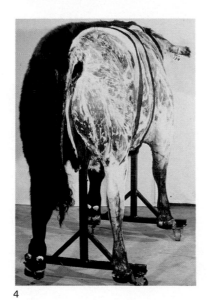

5

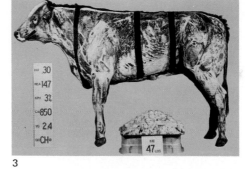

4

Plate A

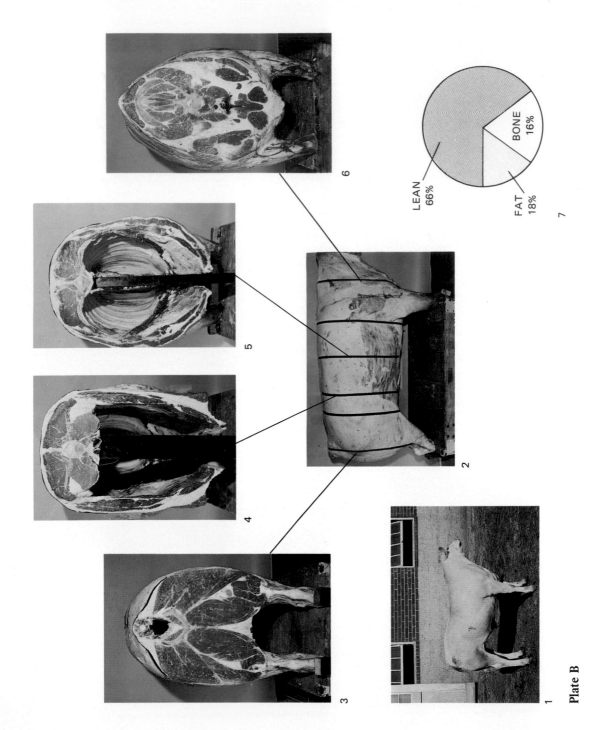

LEAN 66%

FAT 18%

BONE 16%

7

6

5

4

2

3

1

Plate B

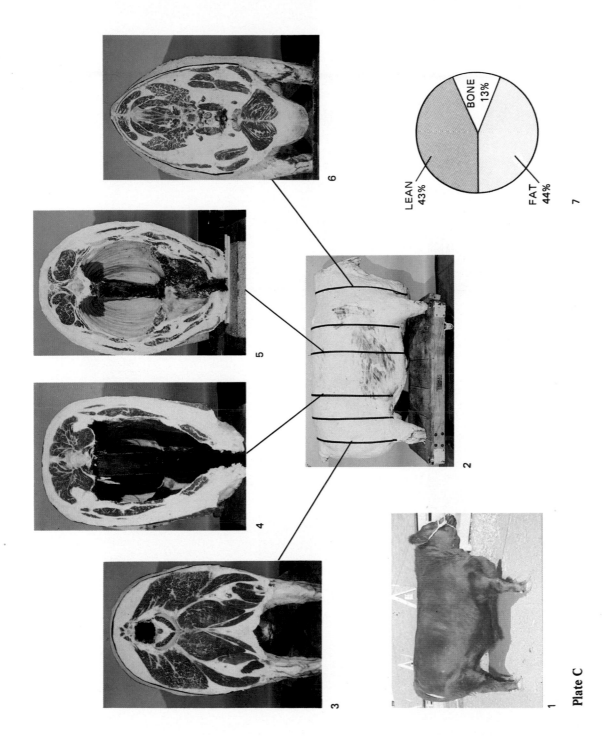

LEAN
43%

BONE
13%

FAT
44%

7

6

5

4

3

2

1

Plate C

Plate D

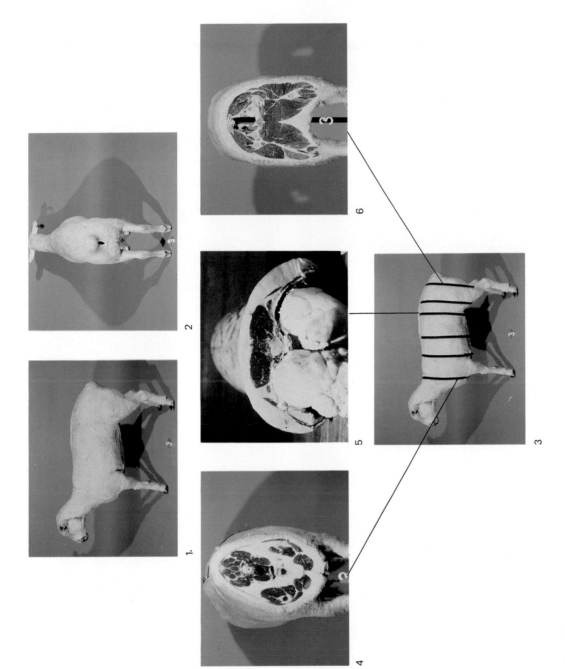

Plate E

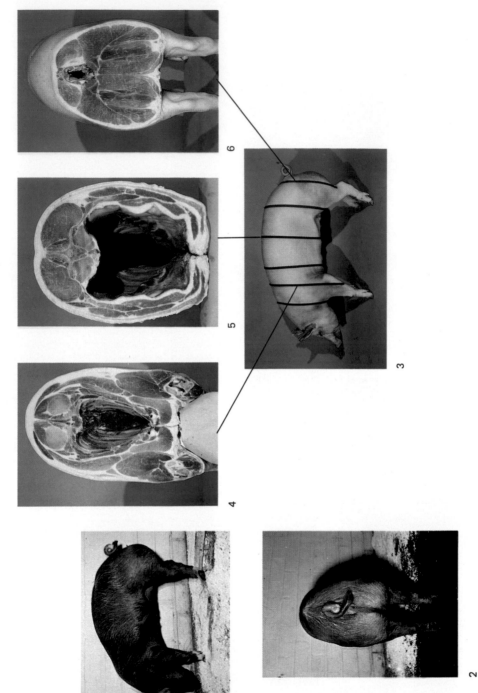

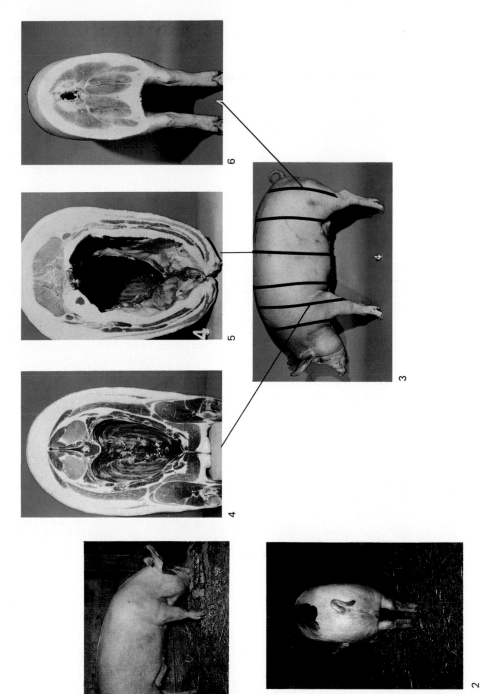

Plate G

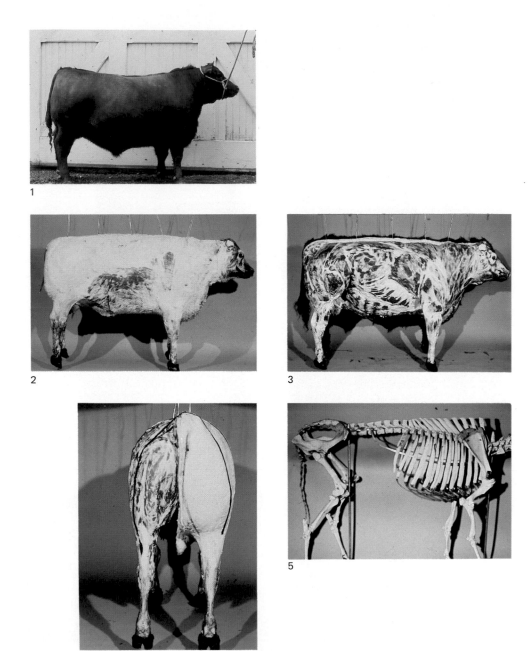

1

2

3

4

5

Plate H

7

Reproduction

Reproductive efficiency in farm animals, as measured by number of calves or lambs per 100 breeding females or numbers of pigs per litter, is considered to be the trait of greatest economic importance in farm animal production. It is essential to understand the reproductive process in the creation of new animal life because it is the focal point of overall animal productivity. Producers who manage animals for high reproductive rates must understand the production of viable sex cells, estrous cycles, mating, pregnancy, and birth.

FEMALE ORGANS OF REPRODUCTION AND THEIR FUNCTION

Figures 7-1 and 7-2 show the reproductive organs of the cow and sow. The anatomy of the various female species is similar although there are a few obvious differences.

The organs of reproduction of the typical female farm **mammal** consist, first of all, of a pair of **ovaries**, which are suspended by ligaments just back of the kidneys, and a pair of open-ended tubes, the oviducts (also called the Fallopian or ovarian tubes), which lead directly into the uterus (womb). The uterus itself has two horns, or branches, that merge together at the lower part into one structure in our farm mammals, so that the lower opening, or exit, from the uterus is a canal. This canal, called the cervix, is surrounded by muscles. Its surface is fairly smooth in the mare and the sow, but is folded in the cow and ewe. The cervix opens into the vagina, a relatively large canal or passageway that leads posteriorly to the external parts, which are the vulva and its clitoris. The urinary bladder empties into the vagina through the urethral opening.

Ovaries

The ovaries produce eggs (female sex cells, also called **ova**) and hormones (see Figure 7-3). Each egg, or **ovum**, is generated in a recently formed follicle within the ovary (Figure 7-4). Some tiny follicles develop and ultimately attain maximum size, about 10 mm

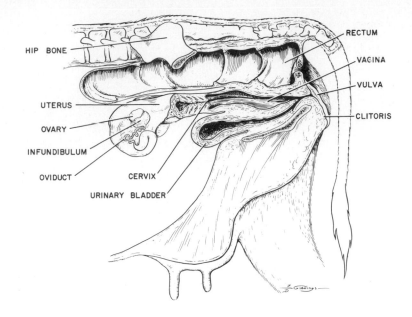

Figure 7-1. Reproductive organs of the cow.

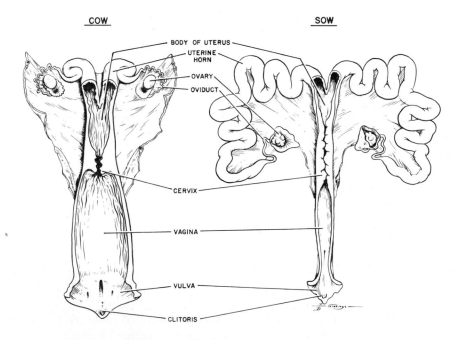

Figure 7-2. A dorsal view of the reproductive organs of the cow and the sow. The most noticeable difference is the longer uterine horns of the sow compared to the cow.

Figure 7-3. Bull sperm and cow egg, each magnified 300 X. Note the comparative sizes. Each is a single cell and contains one-half the chromosome number typical of other body cells.

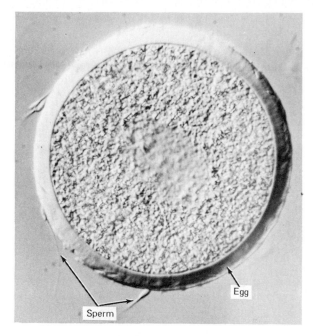

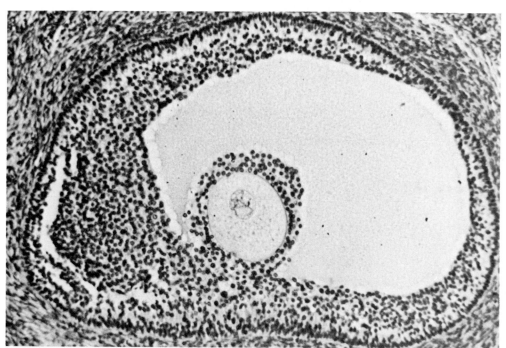

Figure 7-4. The large structure, outlined with a circle of dark cells, is a follicle located on a cow's ovary (magnified 265 X). The smaller circle, near the center, is the egg. The large light-gray area is the fluid which fills the follicle. When the follicle ruptures, the fluid will wash the egg into the oviduct.

in diameter, after having migrated from deep in the ovary to the surface of the ovary. These mature (Graafian) follicles rupture, thus freeing the ovum (**ovulation**). Many of the tiny follicles grow to various stages, cease growth, deteriorate, and are absorbed (Figure 7-5).

After the egg escapes from the mature follicle, the follicle is turned into a "yellow body", or corpus luteum, which becomes a vitally important structure if pregnancy occurs.

Oviducts

The oviducts receive the ova immediately after the ova leave the ovaries. The ova are tiny, 200 μm or less in diameter (200 μm = 1/5 mm), which is approximately the size of a dot made by a sharp pencil. The ovarian tubes also receive the sperm after the female is inseminated (naturally or artificially) and therefore are the sites where ova and sperm meet and where fertilization takes place. The open end (funnel, infundibulum) of each oviduct is quite spacious so that the fertilized ova traverse it in a short time. For the ova to travel down the remaining two-thirds of the oviduct, however, 3 to 5 days are required in cows and ewes and probably about the same amount of time in other farm animals. From the oviduct, the newly developing embryos pass to the uterus and soon attach to it.

Uterus

The uterus varies in shape from the type that has long, slender left and right horns, as in the sow, to the type that is mostly a fused body with very short branches, as in the mare. In the sow, the embryos develop in the uterine horn, while in the mare, the embryo develops in the body of the uterus. Each embryo develops into a fetus and remains in the uterus until parturition (birth). It is through the uterus that sperm must pass en route to the upper ends of the oviducts, there to meet the ova.

The lower outlet of the uterus is the cervix, a muscular organ that constitutes a formidable gateway from the uterus to the vagina, and from the vagina to the uterus. Like the rest of the reproductive tract, the cervix is lined with mucosal cells. These cells change very much as the animal goes from one period of estrus to another, and they change in pregnancy, also. The cervical passage changes from one that is tightly closed or sealed in pregnancy, or nearly closed when the animal is not in **estrus** (heat), to a relatively open, very moist canal at the height of estrus.

Vagina

The vagina serves as the organ of copulation at mating and as the birth canal at parturition. Its mucosal surface changes during the estrous cycle from very moist when the animal is ready for mating to almost dry, even sticky, between periods of heat. The tract from the urinary bladder joins the posterior ventral vagina; from this juncture to the exterior vulva, the vagina serves the double role of a passage for the reproductive and urinary systems.

Clitoris

A highly sensitive organ, the clitoris, is located ventrally and at the lower tip of the vagina. The clitoris is the homologue of the penis in the male (that is, it came from the same

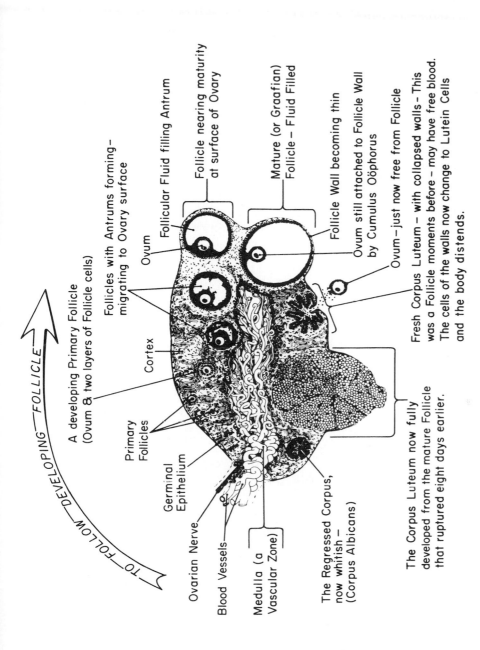

TO FOLLOW DEVELOPING FOLLICLE

A developing Primary Follicle
(Ovum & two layers of Follicle cells)

Follicles with Antrums forming –
migrating to Ovary surface

Ovum

Follicular Fluid filling Antrum

Follicle nearing maturity
at surface of Ovary

Mature (or Graafian)
Follicle – Fluid Filled

Follicle Wall becoming thin

Ovum still attached to Follicle Wall
by Cumulus Oöphorus

Ovum – just now free from Follicle

Fresh Corpus Luteum – with collapsed walls – This
was a Follicle moments before – may have free blood.
The cells of the walls now change to Lutein Cells
and the body distends.

Cortex

Primary
Follicles

Germinal
Epithelium

Ovarian Nerve

Blood Vessels

Medulla (a
Vascular Zone)

The Regressed Corpus,
now whitish –
(Corpus Albicans)

The Corpus Luteum now fully
developed from the mature Follicle
that ruptured eight days earlier.

MAMMALIAN OVARY

Figure 7-5. A cross-section of the bovine ovary showing how a follicle develops to full size and then ruptures, allowing the egg to escape. The follicle then becomes a "yellow body" (corpus luteum) which is actually orange-colored in cattle. The corpus luteum degenerates in time and disappears. Of course, many follicles cease development, stop growing, and disappear without ever reaching the mature stage.

embryonic source as the penis). There is some research that indicates that clitoral stimulation or massage will increase conception, but it has not been well verified.

Reproduction in Poultry Females

The hen differs from the farm mammals in that the young are not suckled. The egg is layed outside the body, and there are no well-defined estrous cycles or pregnancy. Since eggs are an important source of human food, hens have been selected and managed to lay eggs consistently through the year.

The anatomy of the reproductive tract of the hen is shown in Figures 7-6 and 7-7. At hatching time the female chick has two ovaries and two oviducts. The right ovary and oviduct do not develop. Therefore the sexually mature hen has only a well developed left ovary and oviduct. The ovary appears as a cluster of tiny gray eggs or yolks in front of the left kidney and attached to the back of the hen. As the hen reaches sexual maturity some of

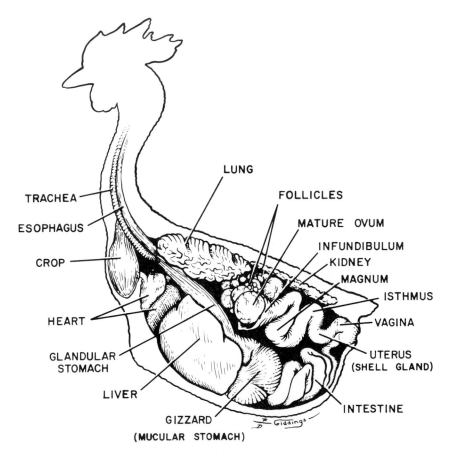

Figure 7-6. Reproductive organs of the hen in relation to other body organs. The single ovary and oviduct are on the hen's left side; an underdeveloped ovary and oviduct are sometimes found on the right side, having degenerated in the developing embryo.

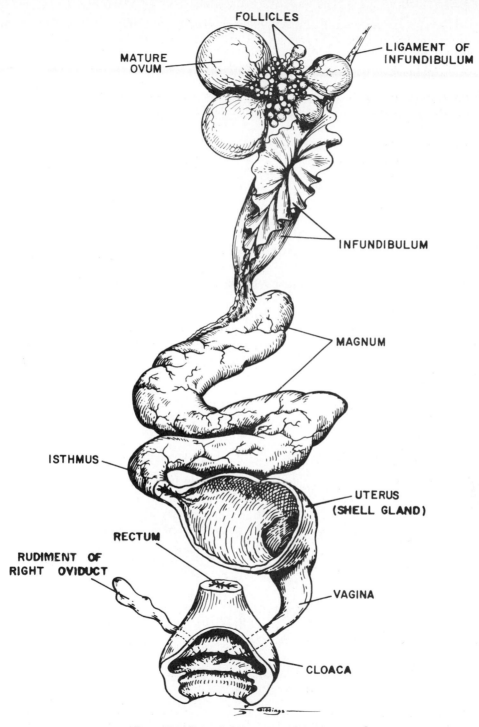

Figure 7-7. Reproductive organs of the hen.

the ova develop into mature yolks (yellow part of the laid egg). The oviduct is a long, glandular tube leading from the ovary to the cloaca (common opening for reproductive and digestive tracts). The oviduct is divided into five parts: the infundibulum, which receives the yolk; the magnum, which secretes the thick albumen or white of the egg; the isthmus, which adds the shell membranes; the uterus or shell gland, which secretes the thin white, the shell, and the shell pigment; and the vagina. The total time for the egg to pass through the oviduct is approximately 24 hours.

MALE ORGANS OF REPRODUCTION AND THEIR FUNCTION

Figures 7-8 and 7-9 show the reproductive organs of the bull and boar. The organs of reproduction of the typical male farm mammal consist, first of all, of the two testicles, which are bean-shaped organs held in the scrotum. Male sex cells (called sperm or spermatozoa) are formed in the tiny seminiferous tubules of the testicles. The sperm from each testicle then pass through very small tubes into an epididymis, which is a tube that is held in a covering applied closely to the exterior of the testicle. Each epididymal tube leads to a larger tube, the deferent duct (also called the **vas deferens** or ductus deferens). The two deferent ducts converge from the left and right sides of the body to connect with the urethral canal at its upper end, very near to where the urinary bladder opens into the urethra. The urethra is the large canal that leads through the penis to the outside of the body. The penis has a triple role: (1) that of a passage for urine, (2) that of a passage for the products originating in the other organs of reproduction, and (3) that of an organ of copulation. The seminal vesicles and the prostate gland are also found at the base of the

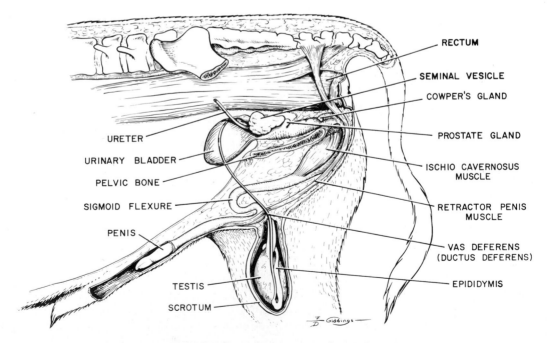

Figure 7-8. Reproductive organs of the bull.

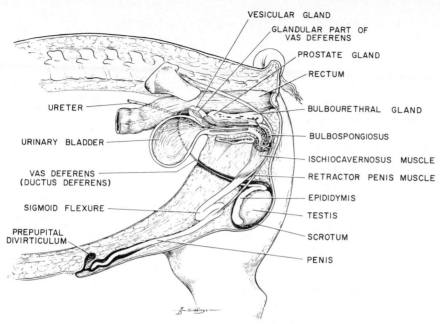

Figure 7-9. Reproductive organs of the boar.

urinary bladder. The left and right parts of the seminal vesicles, which lie against the urinary bladder, consist of glandular tissue that supplies a secretion that moves through the exit tube of each seminal vesicle into the urethra. The prostate gland is composed of a group of some 12 or more glandular tubes, each of which empties into the urethra, also near the opening from the urinary bladder and dorsal to (above) it. Another gland, the bulbourethral (Cowper's) gland, which also empties its secretion into the urethral canal, is approximtely 40 centimeters (cm) posterior to (behind) the prostate. Details of the structure of the spermatozoa of the bull are shown in Figures 7-10, 7-11, and 7-12.

Testicles

The testicles produce (1) the sperm cells that fertilize the eggs, or ova, of the female, and (2) a hormone called testosterone that conditions the male so that his appearance and behavior are masculine. Thus, if both testicles are removed (as is done in castration of males), the individual loses his sperm factory and is left sterile. Also, without testosterone his masculine appearance fades and he approaches the status of a neuter, that is, an individual whose appearance is somewhere between that of a male and that of a female. To illustrate, the effect of castration of the bull calf is to make him a steer, that is, a sterile individual lacking prominent horn development, lacking the crest, or powerful neck, of the bull, lacking to a degree the bull's propensity to grow to his full potential, and having a weaker voice than if he had been left intact. If the calf is castrated while he is immature, such organs of reproduction as the deferent ducts, seminal vesicles, prostate, and bulbourethral glands all but cease further development. If castration is done in the mature animal, the remaining genital organs tend to shrink in size and in function.

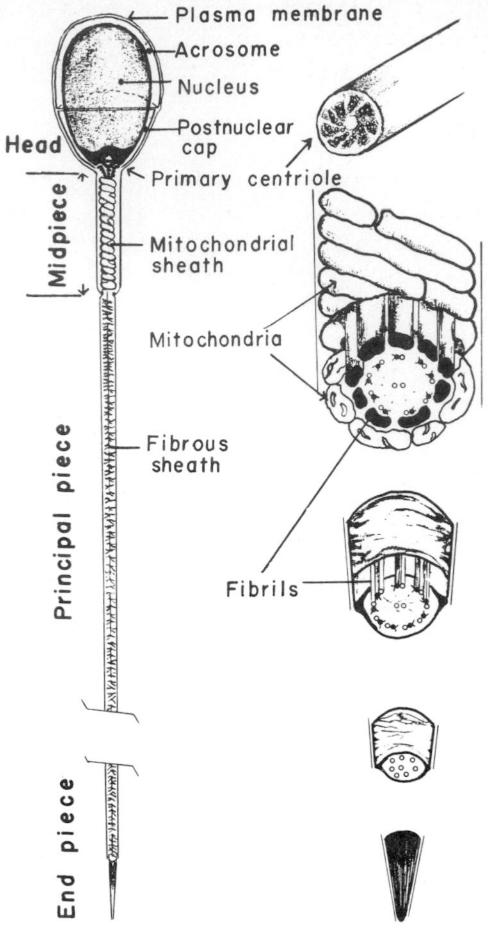

Figure 7-10. A diagrammatic sketch of structure of bull sperm.

Plasma membrane
Acrosome
Nucleus
Postnuclear cap
Head
Midpiece
Primary centriole
Mitochondrial sheath
Mitochondria
Principal piece
Fibrous sheath
Fibrils
End piece

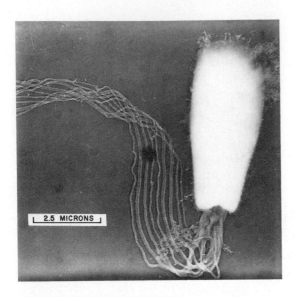

2.5 MICRONS

Figure 7-11. A spermatozoa (sperm cell) from a bull (7000 ✕) viewed through the electron microscope after treatment with 0.15 N NaOH at 25 °C for 16 hours. Note how the covering membrane of the neck region of the sperm has been removed, exposing the nine fibrils of the axial filament. Three filaments are larger than the rest.

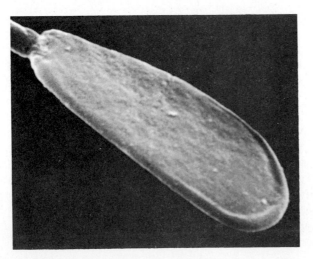

Figure 7-12. Bull sperm (12,000 ✕) viewed with the scanning electron microscope showing the depth of the sperm head covered by the raised acrosome.

Within each testicle, the sperm cells are generated in the seminiferous tubules and the testosterone is apparently produced in the cells between the tubules, which are called interstitial cells or cells of Leydig (Figure 7-13).

Epididymis

The epididymis affords the opportunity for the sperm cells, which enter it from the testicle, to mature. In passing through this very long tube (30 to 35 m in the bull, longer in the boar and stallion), the sperm acquire more and more capacity to fertilize ova. Sperm taken from the part of the epididymis nearest the testicle are not likely to be able to fertilize ova, whereas those taken from areas farther along this long, winding tube increasingly show the capacity to fertilize.

In the sexually mature male animal, sperm reside in the epididymis in vast numbers. In time the sperm age and degenerate and are absorbed in the part of the epididymis farthest from the testicle unless they have been moved on into the deferent ducts and have been ejaculated.

Scrotum

The scrotum is a two-lobed sac that contains and protects the two testicles—but it does more. It is the scrotum that regulates the temperature of the testicles, maintaining them at a temperature lower than body temperature—1.6 °C to 3.9 °C lower in the bull and 5 °C to 7 °C lower in the ram and goat. When the environmental temperature is low the tunica dartos muscle of the scrotum contracts, drawing the testicles toward the body and its warmth; when the environmental temperature is high, this muscle relaxes, permitting the testicles to drop away from the body and its warmth. This heat-regulating mechanism of the scrotum begins at about the time of puberty when the testicular hormone starts to function.

When the environment is so hot that the testicles cannot cool sufficiently, the formation of sperm is impeded and a temporary condition of lowered fertility is produced.

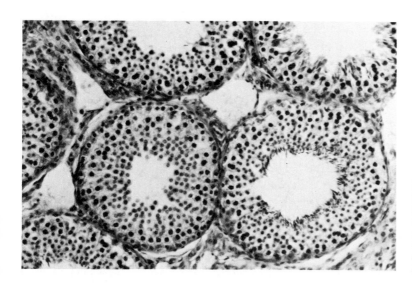

Figure 7-13. A cross-section through the seminiferous tubules of the testis of the bull (magnified 240 ×). The tubule in the lower right-hand corner demonstrates the more advanced stages of spermatogenesis as the spermatids are formed near the lumen (opening) of the tubule.

Providing shade, keeping the males in the shade during the heat of the day, providing cold water to drink and at the same time providing cool air to breathe over the cold water, or even providing air conditioning are ways to manage and prevent this temporary sterility.

Vas Deferens

The **vas deferens** are essentially transportation tubes that carry the sperm-containing fluid from each epididymis to the urethra. The vas deferens join the urethra near its origin as the urethra leaves the urinary bladder. In the mature bull, the vas deferens is about 3 mm in diameter except in its upper part where it widens into a reservoir, or ampulla, about 10 to 17 cm long and 1 cm wide. The ampulla of the vas deferens is profusely supplied with nerves from the pelvic plexus of the sympathetic nervous system.

Under the excitement of anticipated mating, the secretion loaded with spermatozoa from each epididymis is propelled into each vas deferens and accumulates in the ampulla of the deferent duct. This brief accumulation of **semen** in the ampulla is an essential part of sexual arousal. The sperm reside briefly in the ampulla until the moment of ejaculation, when the contents of each ampulla are pressed out into the urethra, and then through the urethra and the penis en route to their deposition in the female tract.

The ampulla is found in the bull, stallion, goat, and ram—species that ejaculate rapidly. It is not present in the boar or dog, animals in which ejaculation normally takes several minutes (8 to 12 minutes is typical in swine). In such animals, sperm in numbers travel all the way from the epididymis through the entire length of the vas deferens and the urethra. On close observation of the boar at the time of mating, one can see the muscles over the scrotum quivering rhythmically as some of the contents of each epididymis are propelled into vas deferens and on into the urethra. This slow ejaculation of the boar contrasts to the sudden expulsion of the contents of the ampulla of the vas deferens at the height of the mating reaction, or orgasm, in the bull, stallion, ram, and goat.

Urethra

The urethra is a large, muscular canal extending from the urinary bladder. The urethra runs posteriorly through the pelvic girdle and curves downward and forward through the full length of the penis. Very near the junction of the bladder and urethra, the tubes from the seminal vesicles and the tubes from the prostate gland join this large canal. The bulbourethral gland joins the urethra at the posterior floor of the pelvis. The urethra is lined with many tiny glands whose watery secretion is clear and high in mucoproteins.

Seminal Vesicles

The seminal vesicles are sizable organs which lie over the neck of the bladder and on either side of the pelvic urethra and which open into the urethra near the place where the deferent ducts open into it. In the bull the seminal vesicles are about 10 cm long and 2.5 cm in diameter. The boar's seminal vesicles are large, thin-walled, pyramid-shaped glands in which the apex points posteriorly. The combined weight of the two glands varies from 150 to 850 g and the contents from 38 to more than 500 g.

To the semen (the fluid that contains the sperm), the seminal vesicles contribute ascorbic acid, citric acid, inorganic phosphorus, acid-soluble phosphorus, and the bulk of the seminal fructose, and in the case of the boar, much ergothionine. Seminal vesicles are prominent in the bull, stallion, ram, and goat, but are absent in the dog and cat.

Prostate Gland

The prostate gland also lies near the neck of the bladder and surrounds the urethra. In the bull the main part of the prostate is about 37 mm across (left to right) and 12 mm in diameter. Additional prostate tissue is scattered among the muscles of the pelvic urethra. The prostate empties into the urethra through a small opening in the muscular wall of the urethra. It supplies antagglutin and minerals to semen plasma.

Bulbourethral Glands

The bulbourethral glands (sometimes referred to as Cowper's glands) are located on either side of the pelvic urethra, just posterior to the urethra-penis where the urethra-penis dips downward in its curve. The bulbourethral glands are covered by fibrous tissue. Except for the boar, they are small in farm animals. In the bull, they are about 25 by 12 mm, but in the boar they are comparable in shape and size to a banana (15 by 3 to 5 cm). The secretion from the bulbourethral glands is thick and viscous, very slippery, or lubricating, and whitish in color. It has a high sialoprotein content. This sialoprotein is involved in the formation of the gelatinous fraction of the semen in the boar and probably in the stallion.

The seminal vesicles, prostate, and bulbourethral glands are known as the **accessory sex glands**. Their primary functions are to add volume and nutrition to the sperm-rich fluid coming from the epididymis. The semen characteristics of some of the farm animals are shown in Table 7-1.

Penis

The penis is the organ of copulation. It also provides a passageway for the escape of urine. It is a muscular organ characterized especially by its spongy, erectile tissue that fills with blood under considerable pressure during periods of sexual arousal, making the penis rigid and erect.

Table 7-1. Semen characteristics

Male animal	Semen characteristic			
	Volume per ejaculate (ml)	Composition of ejaculate	Sperm concentration per ml $\times 10^9$	Total sperm per ejaculate $\times 10^9$
Bull (cattle)	3-10	single fraction	0.8-1.2	4-18
Ram (sheep)	0.5-2.0	single fraction	2-3	1-4
Boar (swine)	150-250	fractionated	0.2-0.3	30-60
Stallion (horse)	40-100	fractionated	0.15-0.40	8-50
Buck (goat)	0.5-2.5	—	2.0-3.5	1-8
Dog (dog)	1.0-5.0	—	2-7	4-14
Buck (rabbit)	0.5-6.5	—	0.3-1.0	1.5-6.5
Tom (turkey)	0.1-0.7	—	8-30	1-20
Cock (chicken)	0.1-1.5	—	0.4-1.5	0.05-2.0

The penis of the bull is about 1 m in length and 3 cm in diameter, tapering to the free end. In the bull, boar, and ram the penis is S-shaped when relaxed. This S-curve or Sigmoid flexure, is eliminated when the penis is erect. The S-curve is restored after copulation when the relaxing penis is drawn back into its sheath by a pair of retractor muscles. The stallion penis has no S-curve; it is enlarged by engorgement of blood in the erectile tissue.

The free end of the penis is termed the glans penis. The opening in the ram penis is at the end of a hairlike appendage that extends about 20 to 30 mm beyond the larger penis proper. This appendage also becomes erect and whirls around in a circular fashion when free. It does not regularly penetrate the ewe's cervix, as some investigators claim it does. Only a small portion of the penis of the bull, boar, ram, and goat extends beyond its sheath during erection. The full extension awaits the thrust after entry into the vagina has been made. The stallion and ass usually extend the penis completely before entry into the vagina.

It is emphasized here that all these accessory male sex organs depend on testosterone for their tone and normal condition. This dependence is especially apparent when the testicles are taken away (at castration); the usefulness of the accessory sex organs is then diminished or even terminated. Boars at least retain their fertility without seminal vesicles and bulbourethral glands.

Reproduction in Male Poultry

The reproductive tract of male poultry is shown in Figure 7-14. There are several differences when compared to the reproductive tracts of the farm mammals previously described. The testes of male poultry are contained in the body cavity. Each vas deferens

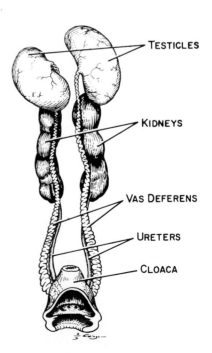

Figure 7-14. Male poultry reproductive tract (ventral view).

opens into small papillae which are located in the cloacal wall. The male fowl has no penis but does have a rudimentary organ of copulation. The sperm are transferred from the papillae to the rudimentary copulatory organ which transfers the sperm to the oviduct of the hen during the mating process.

WHAT MAKES THE TESTICLES AND OVARIES OPERATE

Testicles

The testicles produce their hormones under stimuli coming to them from the anterior pituitary body (AP) situated at the base of the brain. The AP elaborates two hormones important to male performance, the luteinizing hormone (LH) and the follicle-stimulating hormone (FSH). These two hormones are known as gonadotropic hormones because they stimulate the gonads (ovary and testicle). LH produces its effect on the interstitial tissue of the testicle (probably the interstitial cells), causing this tissue to produce the male hormone, testosterone. FSH stimulates cells in the seminiferous tubules to develop into functional spermatozoa. There is also some evidence that FSH also has an influence on testosterone production.

Some species that respond to changes in length of day exhibit more seasonal fluctuation in reproductive activities than others. These influences are on the neurophysiological mechanism, the hypothalamus. The hypothalamus is the floor and part of the wall of the third ventricle of the brain and secretes releasing factors through portal vesicles that affect the anterior pituitary and its production of FSH and LH.

Ovaries

Ovarian hormones in the sexually mature female own the rhythmicity of their production to hormones that originate in the anterior pituitary body and to the interplay between gonad-stimulating hormones produced there and ovarian hormones whose level and potency vary as the estrous cycle progresses. FSH (the same hormone as FSH in the male) circulates through the bloodstream and affects the responsive follicle cells of the ovary, which respond by secreting the so-called estrogens or estrous-producing hormones, estradiol and estrone. When the amount of estradiol and estrone in the blood reaches a sufficient level, the pituitary is caused to reduce its production of FSH. With this drop in FSH production, the production of estradiol and estrone subsides. When the estrous-producing hormones have been lost from the body and their depressing effect on the anterior pituitary has been spent, the anterior pituitary again steps up its production of follicle-stimulating hormone and the cycle is repeated.

Luteal cells, the successors of the follicle cells, secrete a hormone called progesterone, which plays a role in the female reproductive cycle. Progesterone helps cause the occurrence and recurrence of the desire to mate, that is, the estrous cycle. Synchronized with this cycle is the important and essential phenomenon of ovulation. Ovulation occurs in the cow after estrus. It occurs in the sow, ewe, goat, and mare toward the latter part of, but nevertheless during, estrus. These species all ovulate spontaneously—that is, ovulation takes place whether copulation occurs or not. By contrast, copulation (or some such stimulation) is necessary to trigger ovulation in such animals as the rabbit, cat, ferret, and mink. Ovulation in these animals takes place at a fairly consistent time after mating.

Ovulation is controlled by hormones. The follicle of the ovary grows, matures, fills with fluid, and then softens some few hours before rupture. The follicle ruptures due to the interaction of hormones rather than bursting as a result of pressure inside.

The FSH of the anterior pituitary accounts for the increase in size of the ovarian follicle and for the increase in the amount of estradiol and estrone, which are products of the follicle cells. Now another part of the anterior pituitary elaborates the LH, which alters the follicle cells and granulosa cells of the ovary, changing them into luteal cells. These luteal cells are in turn stimulated by luteotrophic hormone (LTH) from the anterior pituitary to produce progesterone.

When a certain delicate balance between high amounts of estrous-producing hormones and a minimal amount of progesterone is attained, the membranes over part of the large follicle thin out, separate, and break, thus allowing the follicular fluid to escape and carry the ovum into the open end of the oviduct. Thus, ovulation is accomplished. After the ovum escapes from the follicle, the hole through which it escaped is sealed over and the cells that formed the ruptured follicle are luteinized by LH. In the course of 7 to 10 days what was formerly an egg-containing follicle develops into the corpus luteum, a luteal body of about the same size and shape as the mature follicle. The egg, having escaped and migrated down the oviduct, is of course absent from the corpus luteum. The luteal cells of the corpus luteum produce a quantity of progesterone sufficient to depress the part of the anterior pituitary that produces FSH until the luteal cells reach maximum development (in nonpregnant animals), cease their development, and (in 3 weeks) lose their potency and disappear.

If pregnancy occurs, the corpus luteum continues to flourish, persists in its progesterone production, and prevents further estrous-producing cycles. Thus, no more heat occurs until after pregnancy has terminated. The process of follicle development, ovulation, and corpus luteum development and regression is shown in Figure 7-5. Table 7-2 shows the length of estrus, estrous cycles, and time of ovulation for the different farm animals.

The changing length of day is a potent factor that influences the estrous cycle, the onset of pregnancy, and the seasonal fluctuations in male fertility. Changing day length acts both directly and indirectly on the animal—it acts directly on the hypothalmus by influencing the secretion of hormones and indirectly by affecting plant growth, thus altering the level of quality of nutrition available. In cattle and horses, increasing length of day is associated with increased reproductive activity in both males and females. In sheep, the breeding season reaches its height in the autumn as the hours of daylight shorten. Of course, individuals of both sexes vary in their intensity and level of fertility. When selection has resulted in improvements in the traits associated with reproduction, individuals exhibit higher levels of fertility (that is, more intense expression of estrus, occurrence of estrus over more months of the year, or occurrence of spermatogenesis at a high level over more months of the year) than do unimproved breeds.

PREGNANCY

When the sperm and the egg unite (fertilization), conception occurs and pregnancy (gestation) is initiated. The fertilized egg initiates a series of cell divisions (Figure 7-15). In farm mammals, the new organism migrates through the oviduct to the uterus in 3 to 4 days.

Table 7-2. Duration and frequency of heat and time of ovulation

| Female animal | Duration of heat | | Length of cycle | | Approximate time of ovulation |
	Average	Range	Average (days)	Range	
Heifer Cow (cattle)	16 hrs	10-27 hrs	21	19-23	30 hours after begining of heat
Ewe (sheep)	30 hrs	20-42 hrs	17	14-19	26 hours after beginning of heat
Mare (horse)	6 days	1-37 days	21	10-37	1 day before the end of heat
Gilt Sow (swine)	44 hrs	1½-4 days	21	19-23	30-38 hours after beginning of heat
Doe (goat)	39 hrs	20-80 hrs	17	12-27	On second day of heat
Doe (rabbit)[b]		Constant estrus			8-10 hours after mating
Queen (cat)	5 days	4-7	10	8-14	24 hours after mating
Bitch (dog)[b]	9 days	4-13	[a]	[a]	24-48 hours after heat
Jill (mink)	2 days	Seasonal breeding (March)			40-50 hours after mating

[a]The dog has no cycle. There are generally two heats per year—in the fall and in the spring.

[b]The dog and rabbit may exhibit a pseudo or false pregnancy after mating has occured.

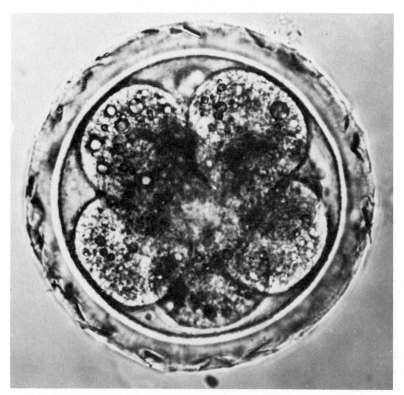

Figure 7-15. A bovine embryo in the six-cell stage of development (magnified 620 X).

By then it has developed to the 16- or 32- cell stage. The chorionic and amniotic membranes develop around this new embryo and attach it to the uterus. The embryo (and later the fetus) gets nutriton and discharges wastes through these membranes. This period of attachment (thirtieth to thirty-fifth day of pregnancy in cattle) is critical. Unless the environment is sufficiently favorable, the embryo dies. Embryonic mortality causes a significant economic loss in farm animals, especially swine, in which multiple ovulations and embryos are typical of the species. It is vital that management protect the female early in pregnancy by providing feed of sufficient quality and by minimizing stress. The female should enter breeding season in a thrifty (vigorous), gaining condition and maintain this condition throughout the first weeks of pregnancy.

The embryonic stage in the life of an individual is defined as that period in which the body parts differentiate to the extent that the essential organs are formed. This period lasts 45 days in cattle.

When the embryonic stage is completed, the young organism is called a fetus. The fetal period, which lasts until birth, is mainly a time of growth. The duration of pregnancy in cattle is shown in Table 7-3. The length of pregnancy varies chiefly with the breed and the age of the mother.

PARTURITION

Parturition (birth) marks the termination of pregnancy (Figure 7-16). The extra-embryonic membranes which had been formed around the embryo in early pregnancy, are shed at this time and are known as afterbirth. These membranes are attached to the uterus during pregnancy by the organ known as the placenta which is responsible for the transfer of nutrients and wastes between mother and fetus. The extra-embryonic membranes and the placenta develop, in some species, the capacity to produce hormones, especially estrogens and progesterone; in the mare, progesterone is produced in the "endometrial cups" in the lining of the uterus. A balance between estrogens and progesterones is attained in which the estrogens predominate in quantity. The uterine muscles become sensitive to the hormone oxytocin, which is produced by the posterior pituitary. Under the stimulus of oxytocin, the weak, rhythmic contractions of the uterus that prevail through most of pregnancy become pronounced and cause labor pains, and the parturition process is under way. Parturition is a synchronized process. The cervix, until now tightly closed,

Table 7-3. Gestation length and number of offspring born

	Gestation length	Usual number of offspring born
Cow (cattle)	283	1
Ewe (sheep)	147	1-3
Mare (horse)	346	1
Sow (swine)	114	6-14
Doe (goat)	150	2-3
Doe (rabbit)	31	4-8
Jill (mink)	50	4
Queen (cat)	52	4
Bitch (dog)	60	7

Figure 7-16. Parturition in the ewe. **(A)** The water bag (a fluid-filled bag) becomes visible. **(B)** The head and front legs of the lamb appear. **(C)** The lamb is forced out by the uterine contractions of the ewe. **(D)** In a few minutes after birth, the lamb is nursing the ewe.

relaxes. The relaxation of the cervix, along with the pressure generated by the uterine muscles on the contents of the uterus, permits the passage of the mature fetus into the vagina and on to the exterior. Another hormone, relaxin, is thought to aid in parturition. Relaxin, which originates in the corpus luteum or in the placenta, helps to relax cartilage and ligaments in the pelvic region to help parturition occur.

At the beginning of parturition, the offspring typically assumes a position that will offer the least resistance as it passes into the pelvic area and through the birth canal. Fetuses of the cow, mare, and ewe assume similar positions in which the front feet are extended with the head between them (Figure 7-17). Fetal piglets do not orient themselves in any one direction, which does not appear to affect the ease of birth. Calves, lambs, or

Figure 7-17. Normal and some abnormal presentations of the calf at parturition.

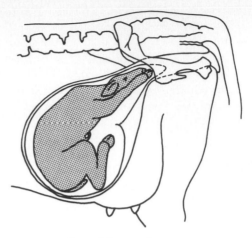

Normal Presentation

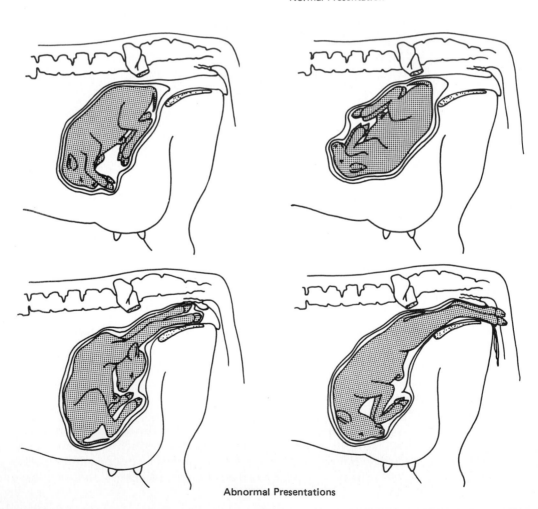

Abnormal Presentations

foals may occasionally present themselves in a number of abnormal positions (Figure 7-17). In many of these situations, assistance needs to be given at parturition. Otherwise, the offspring may die or be born dead, and in some instances, the mother may die as well. An abnormally small pelvis opening or an abnormally large fetus can cause some mild to severe parturition problems.

Producers can manage their herds and flocks for high reproductive rates. Management decisions that cause males and females, selected for breeding, to reach puberty at early

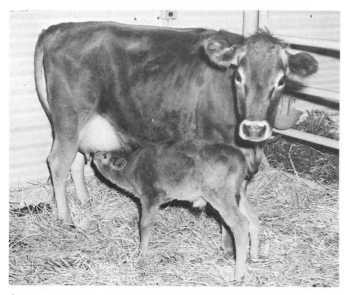

A

B

Figure 7-18. Live offspring are born to each breeding female in the herd or flock when the intricate mechanisms of reproduction function properly. (A) Dairy cow and calf. (B) Sow and litter of pigs. (C) mare and foal. (D) Ewe and lambs. (E) Chicks in the process of hatching. (F) Beef cow and calf.

C

D

E

F

Figure 7-18/continued.

ages, have high conception rates, and have minimum difficulty at parturition are most critical. Keeping the animals healthy, providing adequate levels of nutrition, selection of genetically superior animals, and producer attention at parturition are some of the more critical management inputs to minimize reproductive loss. Live offspring born to each breeding female is the key endpoint to successful farm animal reproduction (Figure 7-18).

STUDY QUESTIONS

1. How important is reproductive efficiency to the profitability of farm animal production?
2. What organs of the reproductive systems do the sperm traverse from the time that they are freed to move about until they reach the ova?
3. What organs of the male contribute to the composition of the semen?
4. How does castration affect the anatomy and function of the reproductive organs of the bull?
5. Describe the epididymis of a farm mammal and explain how it contributes to fertility.
6. Trace the development of the cow's egg from its early beginning to the stage when it can be fertilized.
7. Do all eggs of our farm mammals develop and mature? If not, what happens to them?
8. Do all sperm of our farm mammals develop and fertilize eggs? If not, what happens to them?
9. How does the uterus of the cow contrast with that of the sow? With that of the mare?
10. Which male farm animal has the largest semen volume per ejaculate?
11. How can one explain the fact that an unbred sow, ewe, cow, or mare comes back into heat over and over again?
12. After an ovarian follicle ruptures and the egg escapes, what happens to this follicle?
13. Describe ovulation (what happens and why).
14. What in the environment influences the estrous cycle?
15. Is the volume of semen per ejaculate highly related to the size of the animal? Give examples.
16. Why is it important for the producer to check the females at parturition time?
17. How does the scrotum function to keep wide temperature fluctuations from affecting the spermatazoa?

SELECTED REFERENCES

Battaglia, R. A., and Mayrose, V. B. 1981. *Handbook of Livestock Management Techniques.* Minneapolis: Burgess Publishing Co.

Bone, J. F. 1975. *Animal Anatomy and Physiology.* 4th ed. Corvallis: Oregon State University Book Stores.

Cole, H. H. and Cupps, P. T. 1977. *Reproduction in Domestic Animals.* New York: Academic Press.

Hafez, E. S. E., ed. 1980. *Reproduction in Farm Animals.* Philadelphia: Lea and Febiger.

Nalbandov, A. V. 1976. *Reproductive Physiology of Mammals and Birds.* San Francisco: W. H. Freeman and Co.

Pickett, B. W., Voss, J. L., Squires, E. L., and Amann, R. P. 1981. *Management of the Stallion for Maximum Reproductive Efficiency.* Fort Collins: Colorado State University, Animal Reproduction Laboratory General Series 1005.

Sorenson, A. M., Jr. 1979. *Animal Reproduction: Principles and Practices.* New York: McGraw-Hill Book Co.

Wu, A. S. H., and McKenzie, F. F. 1955. Microstructure of Spermatozoa After Denudation as Revealed by the Electron Microscope. *Journal of Animal Science* 14: 1151-66.

8
Artificial Insemination, Estrus Synchronization, and Embryo Transfer

In the process of artificial insemination (**AI**), semen is deposited in the female reproductive tract by artificial techniques rather than by natural mating. AI was first successfully accomplished in the dog in 1780 and in horses and cattle in the early 1900s. AI techniques are also available for use in sheep, goats, swine, poultry, laboratory animals, and bees.

The primary advantage of AI is that it permits extensive use of outstanding sires to maximize genetic improvement. For example, a bull may sire 30 to 50 calves naturally per year over a productive lifetime of 3 to 8 years. In an artificial insemination program, a bull can produce 200 to 400 units of semen per ejaculate, with four ejaculates typically collected per week. If the semen is frozen and stored for later use, hundreds of thousands of calves can be produced by a single sire (one calf per 1.5 units of semen) and many of these offspring can be produced long after the sire is dead. AI also can be used to control reproductive diseases and sires can be used that have been injured or are dangerous when used naturally.

SEMEN COLLECTING AND PROCESSING

There are several different methods of collecting semen: the most common method is the artificial vagina (Figure 8-1), which is constructed to be similar to an actual vagina. The artificial vagina is used commonly to collect semen from bulls, stallions, rams, and buck goats and rabbits. The semen is collected by having the male mount an estrous female or training him to mount another animal or object (Figures 8-2 and 8-3). When the male mounts, his penis is directed into the artificial vagina by the person collecting the semen and the semen accumulates in the collection tube. Semen from the boar and dog is not typically collected with an artificial vagina. It is collected by applying pressure with a gloved hand, after grasping the extended penis, as the animal mounts another animal or object.

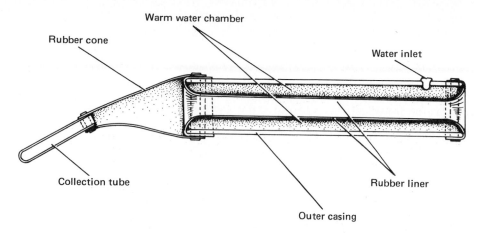

Figure 8-1. Longitudinal section of an artificial vagina.

Figure 8-2. Using the artificial vagina to collect semen from the stallion. The handler of the stallion is holding the front leg to prevent the horse from striking the collector.

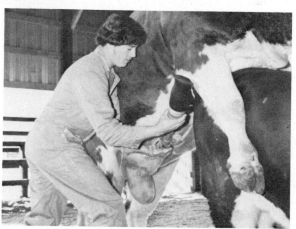

Figure 8-3. Collecting semen from a bull using the artificial vagina.

Semen can also be collected by using an electroejaculator, in which a probe is inserted into the rectum and an electrical stimulation causes ejaculation. This is used most commonly in bulls and rams that are not easily trained to use the artificial vagina and from which semen is collected infrequently.

If semen is collected too frequently, the number of sperm per ejaculate decreases. Semen from a bull is typically collected twice a day for two days a week (Figure 8-4). Semen from rams can be collected several times a day for several weeks, while bucks (goats) must be ejaculated less frequently. Boars and stallions give large numbers of sperm per ejaculate so semen from them is usually collected every other day at most. Semen is collected from tom turkeys two to three times per week. The simplest and most common technique for collecting sperm from toms is the abdominal massage. This technique usually requires two persons: the "collector" helps hold the bird and operates the semen collection apparatus while the "milker" stimulates a flow of semen by massaging the male's abdomen (Figure 8-5).

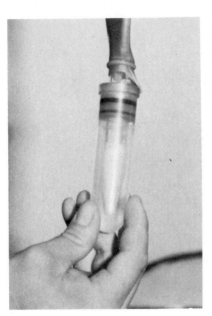

Figure 8-4. Approximately 4 ml of bull semen in the collection tube attached to the artificial vagina. Note that the collection tube is immersed in water to control the temperature. The livability of the sperm is decreased when they are subjected to sudden temperature changes.

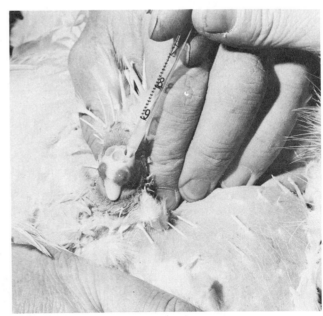

Figure 8-5. Collection of semen from a turkey. Two individuals are usually involved in the collection process. One holds the turkey while the other individual manually stimulates semen into the cloaca while drawing semen into the collection tube.

After the semen is collected, it is evaluated for volume, sperm concentration, motility of the sperm, and sperm abnormalities (Figure 8-6). The semen is usually mixed with an extender which dilutes the ejaculate to a greater volume. This greater volume allows a single ejaculate to be processed into several units of semen, where one unit of semen is used for each female inseminated. This extender is usually composed of nutrients such as milk and egg yolk, a citrate buffer, antibiotics, and glycerol. The amount of extender used is based on the projected number of viable sperm available in each unit of extended semen. For example, each unit of semen for insemination in cattle should contain 10 million motile, normal spermatozoa.

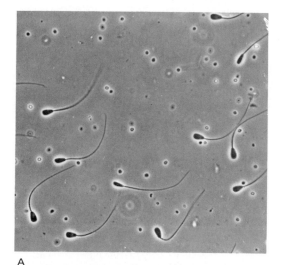

A

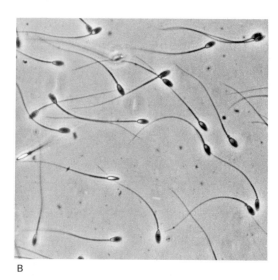

B

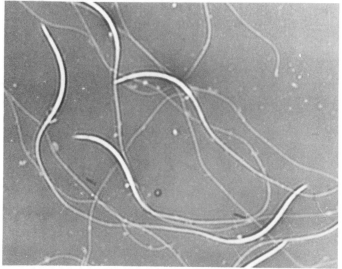

C

Figure 8-6. **(A)** Normal bull semen. **(B)** Normal stallion semen. **(C)** Normal semen from domestic fowl as viewed under the microscope.

Figure 8-7. Semen is typically stored and frozen in ampules or straws. Three ampules appear at the top of the figure, with three straws of bull semen below. The animal's name, registration number, and location of collection are printed on the ampules and straws.

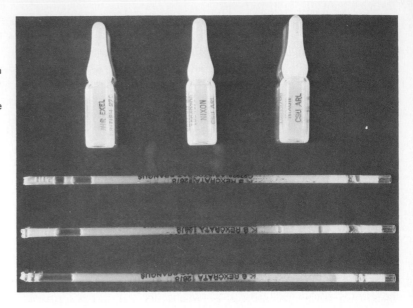

Some semen is used fresh; however, it can only be stored this way for a day or two. Most semen is frozen in liquid nitrogen and stored in ampules (glass vials), plastic straws (Figure 8-7), or pellets. The semen can be stored in this manner for an indefinite period of time and still retain its fertilization capacity. Most cattle are inseminated with frozen semen that has been thawed, while ram semen must be used as fresh semen. The fertilizing capacity of frozen semen from boars and stallions is only modestly satisfactory; however, improvement has been noted in the past few years. Turkey semen cannot be frozen satisfactorily. For maximum fertilization capacity, it should be used within 30 minutes of collection. The semen cannot be extended, therefore 12 toms are usually kept for each 100 hens in the flock.

INSEMINATION OF THE FEMALE

Prior to insemination, the frozen semen is thawed. Semen that is thawed should not be refrozen and used again because conception rates will be reduced.

High conception rates using AI depend on the females cycling and ovulating, detecting estrus, using semen that has been properly collected, extended and frozen, thawing and handling the semen satisfactorily at the time of insemination, insemination techniques, and avoiding extremes in stress and excitement to the animal being inseminated.

DETECTING ESTRUS

Estrus must be detected accurately because it signals the time of ovulation and determines the proper timing of insemination. The best indication of estrus is the condition called "standing heat" in which the female stands still when mounted by a male or another female.

Cows are typically checked for estrus twice daily, in the morning and evening. They are usually observed for 30 minutes to detect standing heat. Other observable signs are restlessness, attempting to mount other cows, and a clear mucous discharge from the vagina. Some producers use sterilized bulls or hormone-treated cows as "heat checkers" in the herd. These animals are sometimes equipped with a head harness that greases or paints marks on the back of the cow when the cow is mounted.

Estrus in sheep or goats is checked using sterilized males equipped with a brisket marking harness. Gilts and sows in heat assume a rigid stance with ears erect when hands are placed firmly on their backs. The vulva is usually red and swollen. The presence of a boar and the resulting sounds and odors can help detect heat in swine as the females in estrus will attempt to locate a boar. Signs of estrus in the mare are elevation of the tail, contractions of the vulva (winking), spreading of the legs, and frequent urination.

PROPER TIMING OF INSEMINATION

The length of estrus and ovulation time is quite variable in farm animals. This variability poses difficulty in determining the best time for insemination. An additional challenge is that the sperm are short-lived when put into the female reproductive tract. Also, estrus is sometimes expressed without ovulation occurring; sows, for example, typically show estrus 3 to 5 days after **farrowing**, but ovulation does not occur. Sows should not be bred at this time either artificially or naturally.

Insemination time should be as close to ovulation time as possible. Cows found in estrus in the morning are usually inseminated the evening of the same day, and cows in heat in the evening are inseminated the following morning. Insemination, therefore, should occur toward the end or after estrus has been expressed in the cow. Ewes are usually inseminated in the second half of estrus and goats are inseminated 10 to 12 hours after the beginning of estrus. Sows ovulate from 30 to 38 hours after the beginning of estrus, so insemination is recommended at the end of the first day or at the beginning of the second day of estrus. Sometimes the sows are inseminated both days, which improves the conception rate.

Insemination of dairy cows occurs while the cow is standing in a stall or stanchion. Beef cows are penned and inseminated in a chute that restrains the animal. The most common insemination technique in cattle involves the inseminator having one arm in the rectum to manipulate the insemination tube through the cervix (Figure 8-8). The insemination tube is passed just through the cervix and the semen is deposited into the body of the uterus. The insemination procedure for sheep and goats is similar to cattle, however, a speculum (a tube approximately 1.5 in. in diameter and 6 in. long) allows the inseminator to observe the cervix in sheep and goats. The inseminating tube is passed through the speculum into the cervix where the semen is deposited into the uterus or cervix.

The sow is usually inseminated without being restrained. The inseminating tube is easily directed into the cervix because the vagina tapers into the cervix. The semen is expelled into the body of the uterus. The mare is hobbled or adequately restrained prior to insemination. The vulva area is washed and the tail is wrapped or put into a plastic bag. The plastic-covered arm of the inseminator is inserted into the vagina and the index finger is inserted into the cervix. The insemination tube is passed through the cervix and the semen is deposited into the uterus.

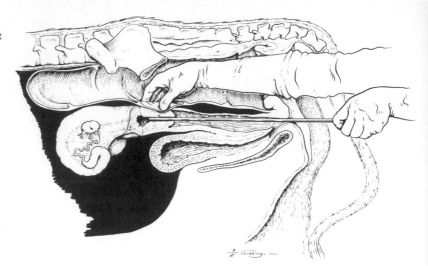

Figure 8-8. Artificial insemination of the cow. Note that the insemination tube has been manipulated through the cervix. Inseminator's forefinger is used to determine when the insemination rod has entered into the uterus.

The turkey hen is inseminated by first applying pressure to the abdominal area to cause the eversion of the oviduct (Figure 8-9). The insemination tube is inserted into the oviduct approximately 2 in., and the semen released. The first insemination is made when 5 to 10% of the flock have started laying eggs. A second insemination a week later assures a high level of fertility. Thereafter, insemination is done at two-week intervals. Fertility in the turkey usually persists at a high level from 2 to 3 weeks after insemination. This is possible because sperm are stored in special glands in the hen, where they are nourished and retain their fertilizing capacity. Most of these glands are located near the junction of the uterus and vagina.

EXTENT OF ARTIFICIAL INSEMINATION

The number of farm animals inseminated each year is not well documented. It is estimated that in the United States more than 60% of the dairy cows (approximately 8 million head) and 5% of the beef cows (a total of approximately 2 million head) are inseminated each year.

It is estimated that approximately 15,000 sows are artificially bred in the United States each year, which is less than 1% of the total sows bred in this country. Satisfactory techniques to freeze boar semen were accomplished in 1971, but frozen semen yields smaller litter sizes than fresh semen. More than 400,000 sows are artificially inseminated each year in certain European countries, which represents 20 to 30% of the sows bred in these countries.

AI in horses is still limited because of difficulty in providing extended semen storage. AI in sheep and goats in the United States is limited because the herds and flocks are dispersed over wide areas and the cost per unit of semen is high.

Little AI is done in chickens; however, AI is rather extensive in turkeys. It is especially important in the broad-breasted turkey, which, because of the size of its breast has difficulty mating naturally.

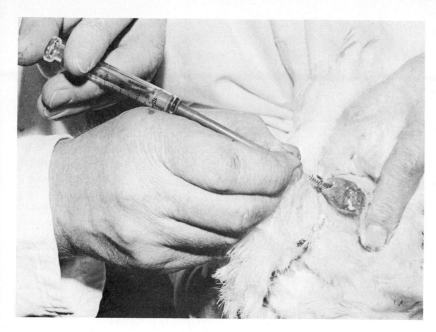

Figure 8-9. Insemination of a hen turkey is accomplished by everting the opening of the oviduct through the opening of the cloaca. The tube is inserted one to two inches, pressure on the oviduct is relaxed, and the proper amount of semen is deposited as the syringe is slowly withdrawn.

ESTRUS SYNCHRONIZATION

Estrus synchronization is controlling or manipulating the estrous cycle so that females in a herd or flock express estrus at approximately the same time. Estrus synchronization is a useful part of an AI program because checking heat and breeding animals under range conditions is time consuming and expensive. Also, estrus synchronization is a successful tool in making embryo transfer programs successful.

Considerable research has been done on estrus synchronization over the past few decades, where several different products have been evaluated. However, none of the products have been available commercially because they have not been cleared by the Food and Drug Administration. Also, some of the earlier research work demonstrated that conception rates were lowered when estrus synchronization products were used.

In 1979, a **prostaglandin** was cleared for use in cattle. Prostaglandins are naturally occurring fatty acids that appear to have important functions in several of the body systems. The prostaglandin that has a marked effect on the reproductive system is prostaglandin F_2 alpha ($PGF_2\alpha$).

In Chapter 7, it was pointed out that the corpus luteum (CL) controls the estrous cycle in the cow by secreting the hormone progesterone. Progesterone prevents the cow from expressing heat and ovulation. The prostaglandin destroys the CL, destroying thus the source of progesterone. About three days after the injection of prostaglandin, the cow will be in heat. For prostaglandin to be effective, the cow must have a functional CL. It is ineffective in heifers that have not reached puberty or in noncycling mature cows. Also, prostaglandin is ineffective if the CL is immature or has already started to regress. Prostaglandin is, then, only effective in heifers and cows that are in days 5 to 18 of their estrous cycle. Because of this relationship, prostaglandin is given on either a one-injection or two-injection system.

One-Injection System

This system requires one prostaglandin injection and an 11-day AI breeding season. The first 5 days are a conventional AI program of heat detection and insemination. After 5 days, a calculation can be made to determine what percentage of the females in the herd have been in heat. If a smaller-than-expected percentage are cycling, the cost of the prostaglandin may not justify the anticipated benefit. If the decision is to proceed with the injection, the remaining animals not previously bred are injected on day 6. Then the AI breeding season continues 5 more days for a total of 11 days.

Field trial results in herds where a high percent of the females cycling show a 50% to 60% pregnancy rate at the end of 11 days of AI breeding for those receiving prostaglandin versus a 30% to 40% pregnancy rate for those cows not receiving it. The amount of prostaglandin required per calf that is produced by AI, will be between 1.2 and 2.0 units.

Two-Injection System

In this system all cows are injected with prostaglandin at two different times. Counting the first injection as day 1, the second injection is administered on day 11, 12, or 13. All cows capable of responding to the drug should be in heat during the first five days after the second injection of prostaglandin. Heat detection and insemination can occur each of these five days or a fixed-time insemination can be performed 76 to 80 hours after the second injection.

Field trial results in herds in which a high percent of the females are cycling show a 35% to 55% pregnancy rate at the end of the fifth day or 76 to 80 hour one-time insemination versus 10% to 12% in the cows not receiving prostaglandin. The two-injection system will require 4 to 7 units of prostaglandin per AI calf.

Prostaglandin is not a wonder drug. It will only work in well-managed herds where a high percentage of the females are cycling. Biologically, it has been well demonstrated that prostaglandin can synchronize estrus. However, producers must weigh the cost against the economic benefit. Caution should be exercised in adminstering prostaglandin to pregnant cows as it may cause abortion. Estrus synchronization in cattle may not be advisable in areas where inadequate protection can be given to young calves during severe blizzards. Also, herd health programs must be excellent to prevent high losses from calf scours and other diseases that become more serious where large numbers of newborn calves are grouped together.

In a sense, there are some natural occurrences of estrus synchronization. The weaning process in swine is an example because the sow will typically show heat 3 to 8 days after the pigs are weaned. Estrus is suppressed through the suckling influence, so when the pigs are removed the sow will show heat.

EMBRYO TRANSFER

Embryo transfer is sometimes referred to as ova transplant or embryo transplant. In this procedure, an embryo in its early stage of development is removed from its own mother's (the donor's) reproductive tract and transferred to another female's (the recipient's) reproductive tract.

The first successful embryo transfer was accomplished in 1890. In the past several decades, successful embryo transfers have been reported in sheep, goats, swine, cattle, and horses.

In recent years commercial embryo transfer companies have been established in the United States and several foreign countries. Most of the commercial work is done with beef and dairy cattle. It is estimated that approximately 20,000 pregnancies from embryo transfers in cattle occurred in North America in 1980. This compares to 20 done in 1972.

Superovulation is the production of a greater-than-normal number of eggs. Females that are donors of eggs for embryo transfer are injected with hormones to stimulate increased egg production. In cattle the average donor produces approximately 7 transferable embryos per superovulation treatment; however, a range of 0 to 33 embryos can be expected (Figure 8-10). This procedure gives embryo transfer its greatest advantage, that is, increasing the number of offspring that a superior female can produce. The key to the justification of embryo transfer is the identification of genetically superior females. Embryo transfer is usually confined to seedstock herds where genetically superior females can be more easily identified and where the high costs can be justified.

Until recently embryo transfer had to be done surgically, but now nonsurgical techniques can be used. The procedure after superovulation is to breed the donor 12 and 24 hours after she comes into heat with two doses of semen each time. The fertilized eggs (embryos) are recovered about a week later by flushing the uterus. This is essentially the reverse of the artificial insemination process and does not require surgery. The embryos are located and evaluated using a microscope. Nonsurgical transfer of an embryo into a recipient is done with an artificial insemination straw gun. Conception rates are usually higher if the transfer is done surgically.

Embryos usually are transferred shortly after being collected, so recipient females need to be in the same stage (within 24 hours) of the estrous cycle for successful transfer to occur. Large numbers of females must be kept for this purpose or estrus synchronization of a smaller number of females is necessary. Embryos can now be frozen in liquid nitrogen and remain dormant for months or years. Conception rates are lower when frozen embryos are

Figure 8-10. Ten Holstein embryo transfer calves resulting from one superovulation and transfer from the Holstein cow in the background. The ten recipient cows are shown on the left-hand side of the fence.

used than when fresh embryos are used, but research is bringing about improvement. In cattle, approximately 50% to 60% of the frozen embryos will be normal after thawing. Thirty to 40% of these normal embryos will result in confirmed pregnancies at 60 to 90 days. This compares to a pregnancy rate of 55% to 65% from fresh embryos transferred the same day of collection.

Recent advances in embryo transfer research has permitted the embryo to be mechanically divided so that identical twins can be produced from a single embryo. Perhaps in the future an embryo of a desired mating and sex could be selected from an inventory of frozen embryos. This embryo would then be thawed and transferred nonsurgically in the cow when she comes into heat. This is the way the reproduction technology is presently developing in the industry.

STUDY QUESTIONS

1. What is the primary advantage of artificial insemination?
2. What is prostaglandin?
3. How does prostaglandin affect: a. noncycling cow b. cycling cow in day 2 of her cycle c. cycling cow in day 12 of her cycle d. pregnant cow?
4. What are some of the different methods of collecting semen from farm animals?
5. What are the main factors affecting conception rates in an artificial insemination program?
6. What is unique about sperm survival in the reproductive tract of a hen turkey compared to other farm animal females?
7. Why is artificial insemination more widely practiced in dairy cattle?
8. Define embryo transfer.
9. What is the main advantage for a producer using embryo transfer?
10. How do artificial insemination, estrus synchronization, and embryo transfer complement each other?

SELECTED REFERENCES

Ernst, R. A., Ogasawara, F. X., Rooney, W. F., Schroeder, J. P., and Ferebee, D. C. 1970. *Artificial Insemination of Turkeys.* University of California Agricultural Extension Publication AXT-338.

Herman, H. A., and Madden, F. W. 1974. *The Artificial Insemination of Dairy and Beef Cattle (Including Techniques for Goats, Sheep, Horses and Swine).* Columbia, Missouri: Lucas Brothers Publishers.

Lake, P. E., and Steward, J. M. 1978. *Artificial Insemination in Poultry.* Ministry of Agriculture, Fisheries, and Food, No. 213.

Perry, E. J., ed. 1968. *The Artificial Insemination of Farm Animals.* New Brunswick, New Jersey: Rutgers University Press.

Pickett, B. W., and Back, D. G. 1973. *Procedures, Collection, Evlauation and Insemination of Stallion Semen.* Fort Collins: Colorado State University Experiment Station Animal Reproduction Laboratory, General Series 935.

Salisbury, G. W., VanDemark, N. L., and Odge, J. R. 1978. *Physiology of Reproduction and Artificial Insemination of Cattle.* 2nd Edition. San Francisco: W. H. Freeman and Co.

Seidel, G. E. Jr., Seidel, S. M., and Bowen, R. A. 1978. *Bovine Embryo Transfer Procedures.* Colorado State University Experiment Station Animal Reproduction Laboratory, General Series 975.

Seidel, G. E. Jr. 1981. Superovulation and embryo transfer in cattle. *Science,* 211:351.

_____. 1979. *Synchronization of Beef Cattle with Prostaglandin.* American Breeders Service, DeForest, Wisconsin.

9

Growth and Maturation

Growth is a complex phenomenon and any definition of it is likely to be inadequate. Here we will define growth as the amount of protein synthesized over the amount that is lost. By this definition, growth is accomplished through cell multiplication, through increases in cell size, or through a combination of both. Fattening, bone deposition, and water storage in the tissues, each of which can be increased or decreased at any time in postnatal life, is not considered as growth. However, it is extremely difficult to actually measure what percentage of an animal is stored fat, water, and bone. The increase in weight that occurs with time in young animals is considered to be growth because rapidly gaining animals do not store large quantities of fat and the percentage of bone in an animal is low.

MEASURING GROWTH

Growth is a curved-line function regardless of how it is expressed mathematically. If one considers growth as an increase in mass (weight), the growth curve is somewhat S-shaped in that it starts at a slow rate, accelerates at a very rapid rate, and then slows markedly. The S-shaped growth curve is illustrated in Figure 9-1. It has been found that the shape of this growth curve applies to all farm mammals even though actual weights in different species are quite different.

The general mathematical description of growth as an increase in weight with time is:

$$\text{Growth} = \frac{w}{t}, \text{ in which } w = \text{weight and } t = \text{time.}$$

The mathematical description of the rate at which an animal gains weight during a particular period of time is:

$$\text{Rate of gain} = \frac{\Delta w}{t}, \quad \begin{array}{l} \text{in which } \Delta w = \text{the change in weight that occurred} \\ \text{in the time period involved } (\Delta = \text{``change in'' and} \\ t = \text{time}). \end{array}$$

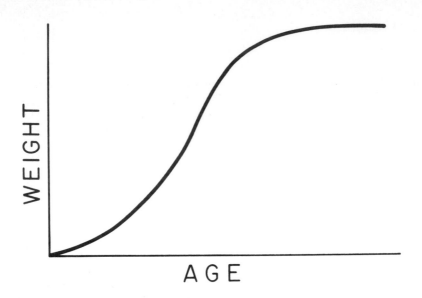

Figure 9-1. Growth curve showing S-shaped function of increase in mass (weight) with increasing age until maturity.

For cattle in the postweaning period, weight is expressed mathematically as a straight line because the cattle gain weight at a constant rate throughout this period. However, the rate of gain declines as the animal increases in age and approaches **puberty,** and growth ceases when the animal reaches maturity. If one wishes to describe growth as weight increase per unit of body weight as time changes, the equation is:

$$\text{Growth per unit of body weight} = \frac{\Delta w/w}{t}, \text{ in which } \Delta w = \text{change in weight, and } t = \text{time.}$$

The rate of increase in weight per unit of body weight always declines as time changes.

FACTORS AFFECTING GROWTH

Growth rate is influenced by both external and internal factors. The most important external factor is nutrition; animals that do not receive a sufficient quantity of food and those fed an improper diet do poorly. Two important internal factors are **inheritance** and endocrine, or hormonal, secretions. (It is likely that endocrine function is also under genetic control but endocrine function and inheritance will be discussed separately.) Postweaning rate of gain in farm animals is under strong genetic control, but the inheritance of growth is not simple. It appears that the maximum potential for growth of an animal is inherited, but whether the animal approaches this maximum depends greatly on its environment. Even with the best rations that scientists have been able to develop, some animals grow at a much slower rate than others. **Endocrine glands** are ductless glands which produce hormones that are secreted into the blood. A **hormone** is a chemical secretion from an endocrine gland that is carried by the bloodstream to other parts of the body where it exerts a specific effect. Although most of the endocrine glands contribute directly or indirectly to growth, the most important glands affecting growth are the pituitary, thyroid, ovaries, testicles, and adrenals.

Pituitary Gland

The pituitary gland is located under the brain, posterior to the optic chiasma. It produces several hormones, the most important to growth being **somatotropin** (growth hormone). Growth hormone stimulates nitrogen retention and results in true growth—more protein synthesis than protein loss. Administration of growth hormone increases growth over a short period of time, but it cannot be continued with success. The growth hormone is a protein, and repeated injections of it cause the animal to develop **antihormone** substances and antibodies. Thus, the injection of growth hormone not only becomes ineffective but may also cause anaphylactic shock (the antibodies developed as a result of injecting a foreign protein cause the injected protein to be coagulated, and this coagulation may result in death).

Thyroid Gland

The thyroid gland consists of two lobes, one on each side of the trachea (windpipe), which are connected by an isthmus. The thyroid secretes the hormone thyroxine, which controls body metabolism. Thyroxine differs from those of some hormones in that the effects of either an insufficiency or an excess is harmful. Insufficient thyroxine in early life results in disproportionate dwarfism, with greater development of the head and shoulders than of the posterior part of the body. In later life, insufficient thyroxine results in a harsh coat of dry hair, lethargy, and storage of large amounts of subcutaneous fat. An excess of thyroxine causes the animal to grow less rapidly than normal, because the animal is literally burning itself up due to an excessive rate of metabolism. In this situation the catabolic, or destructive, metabolic action becomes greater than the anabolic, or constructive, metabolic action.

The function of the thyroid gland is influenced by thyrotrophic hormone (which is secreted by the pituitary gland), by iodine intake in the diet, and by environmental temperature. If the diet is low in iodine, the thyroid gland cannot produce sufficient thyroxine. When denied sufficient iodine, the thyroid continues to be stimulated by the pituitary, increases in size, and develops into a **goiter** which is a gross enlargement in the throat (Figure 9-2). Goiters can be prevented by supplying iodized salt.

Ovaries

The ovaries produce progesterone and estrogens. One of the effects of progesterone is to increase protein retention or synthesis; it is, therefore, a growth stimulant. The growth-stimulating effect of **progesterone** may provide one explanation of why pregnancy is a stimulant to the female. Estrogenic effects vary with species. For example, the administration of estrogens reduces growth in rats, increases fattening in chickens, and increases growth and decreases fattening in cattle and sheep.

The administration of a synthetic estrogen-like substance, diethylstilbestrol (DES), to feedlot steers and wether (castrated male) lambs results in increased gains, in decreased

Figure 9-2. Sheep with goiter.

feed required per unit of gain, and in carcasses that are relatively high in lean and low in fat. Diethylstilbestrol has been administered either in the feed or as an implant in the ear so that any portion remaining in the body at slaughter is discarded. DES has been banned for use either as an implant or as mixed in the feed by the Federal Drug Administration. It is unfortunate that such action has been taken because levels of estrogenic activity in meat from steers given DES is lower than the amount in meat from untreated heifers. Also, most of the estrogenic activity that has been detected in treated steers is confined to the liver.

Synovex®, a combination of substances containing estrogen and progesterone, has been used in ruminant animals to stimulate rate and efficiency of gains. Ralgro®, a substance obtained from fungus, stimulates rate and efficiency of gains in a manner similar to DES but reportedly has no estrogenic properties. Rumensin (also called Monensin) is an antibiotic against **coccidia** (organisms that cause the disease called **coccidiosis),** but it has also been found to alter the ratio of volatile fatty acids in the rumen in such a way as to lead to increased efficiency. Ralgro® and Rumensin will come into widespread use since DES has been banned.

Testicles

The **testicles** produce **testosterone** and other androgenic (masculinizing) substances. **Androgens** stimulate growth, reduce feed required per unit of gain, and help form carcasses that are relatively high in lean and low in fat. Androgens have a much greater stimulating effect on females than on castrated males. It is interesting to note that in most farm animals, including cattle, sheep, goats, fowl, and pigs, males grow more rapidly and reach larger adult size than females. (Rabbits are an exception—female rabbits reach considerably heavier weights at maturity than males.) **Castration,** which deprives the animal of its testosterone, results in reduced growth in young male calves, lambs, pigs, and horses. Castrated animals are easier to handle, and in pigs castration prevents the unsavory flavor of meat that otherwise develops in boars. The price paid for intact males for slaughter is lower than for castrated animals of comparable quality. Spaying (castration) of female farm animals does not appear to influence growth. Spaying of heifers and implanting Synovex H* in them gives increased growth.

Adrenal Glands

The adrenal glands are located anteriorly (to the front) and medially (to the middle) to the kidneys. Each consists of two parts, the medulla and the cortex. The medulla, or center part, of the adrenal gland produces the hormone adrenalin. The cortex, or outer part, secretes many steroids. Some of these steroids regulate electrolysis and water metabolism; others influence protein, **carbohydrate,** and fat metabolism. Administration of cortisone (one of the steroids produced by the adrenal cortex) to cattle and sheep causes an increase in body fat. Animals having hyperadrenal activity often become obese.

PHASES OF GROWTH IN FARM ANIMALS

Growth has so far been discussed in a general way. Actually, three phases of growth occur in farm mammals and each phase has specific influences affecting it. The three growth phases are (1) **prenatal**, (2) preweaning (but postnatal), and (3) postweaning.

Prenatal Growth

Growth begins shortly after the egg is fertilized and continues until the animal reaches its mature size.

When the egg is fertilized by the sperm, a **zygote** is created. Shortly after formation of the zygote, cell divisions occur. The zygote does not initially increase in overall size by these cell divisions but, instead, the size of individual cells decreases. Soon the cells start to differentiate such that three basic layers, the **ectoderm** (outer layer); **mesoderm** (middle layer), and **endoderm** (inner layer) are formed. The ectoderm develops into the nervous system, skin, hair, wool, hooves, and certain endocrine glands. Mesoderm cells develop into muscle tissue (smooth, striated or voluntary and cardiac [heart]), circulatory organs (heart, blood, and lymph vessels), and connective tissue (bone, cartilage, ligaments, and tendons). The endoderm develops into the liver, digestive glands, certain endocrine glands, and the inner linings of the digestive system. The early stages of prenatal development collectively is called the embryonic stage and the late period is called the fetal stage. Evidence indicates that heredity of the individual strongly influences both prenatal and postnatal growth. Birth weights of calves when reciprocal crosses are made among Angus, Hereford, Holstein and Brown Swiss breeds are closely related to weight at 12 months of age.

Evidence also indicates that nutrition obtained early in prenatal growth affects growth rate. Twin lambs are usually smaller at birth than single lambs. The average number of **cotyledons** (attachment areas of placenta to uterus) for each twin lamb is about 70% that for each single lamb. Nutrient supply to the young in the **uterus** and waste removal are provided by diffusion of materials to and from the young at the cotyledon sites.

In animals that produce several young at one time **(multiparous),** prenatal growth is influenced by the number of fetuses present in the uterus of the pregnant female. High numbers may result in insufficient nourishment to the individual fetuses, causing small young to be born. In animals that produce only one young **(monoparous),** the size and age of the mother influences prenatal growth of the young. Young females and small females that are mature usually have smaller offspring than older and larger females because of differences in uterine environment, such as the size of the uterus. In addition, the birth weights of calves of beef cattle vary significantly with the sire, indicating that inheritance plays a significant role in prenatal growth.

Preweaning Growth

The growth of young animals in the preweaning (nursing) period is highly influenced by the quality and quantity of the milk provided by the mother. When very large litters of pigs are produced by a sow, she may be unable to produce sufficient milk for optimal growth for all of them. In sheep, twin lambs usually grow less rapidly than single lambs. Research has shown that twin lambs grow slower due to the lack of sufficient milk for optimal growth for both. Although a ewe nursing twins gives more milk than one nursing a single lamb, she does not produce double the quantity.

The preweaning weights of bull calves that were weighed each 14 days from birth until they reached a weaning weight of at least 425 lbs vary much more for those requiring a long time to reach the given weight than for those reaching the weight in a short time (Figure 9-3).

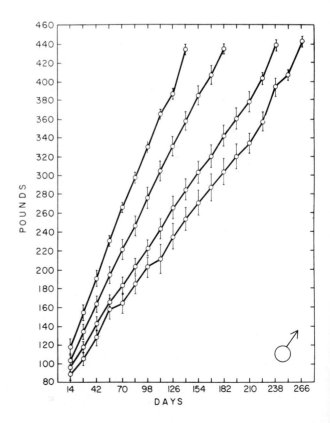

Figure 9-3. Preweaning weights of rapidly and slowly gaining male calves plotted by two-week periods showing the standard deviation (a measure of the amount of variation) at each weighing period. The sample sizes for the data lines plotted from left (fastest gains) to right (slowest gains) are respectively, 30, 50, 50, and 30.

It appears that some calves grow continuously at a rapid rate, some grow continuously at a slow rate, and some gain more rapidly and slowly at various times in the preweaning period. Growth during the preweaning period is computed as follows:

$$\text{Preweaning rate of gain} = \frac{\text{weight at weaning} - \text{weight at birth}}{\text{days from birth to weaning}}$$

Beef cattle producers often object to the use of rate of gain to weaning because the rate is expressed in decimals and small figures. Methods have been developed for computing weight at the age of 205 days, which is the average age at **weaning** (when milk from the mother is replaced by other foods) of range calves that were used to develop this method of adjusting weight of calves at weaning to a common age:

$$\text{Weight at 205 days} = \frac{\text{weight at weaning} - \text{weight at birth}}{\text{days from birth to weaning}} \times 205 + \text{weight at birth}$$

If birth weight has not been recorded, an average birth weight of 70 lbs is often used for the computation. The figures expressing 205-day weights are rounded to whole numbers, and some people find them more usable than figures expressing weight gains per day.

The amount of milk available to the offspring affects preweaning growth. Young females, aged females, and females of small mature size within a breed usually produce less milk than large, mature females. When feed is not available in sufficient quality or quantity, females produce less milk and thereby reduce preweaning growth of their young. When females that are capable of producing a satisfactory amount of milk are given favorable feeding conditions in which to nurse calves, intact male calves usually grow more rapidly than castrated male calves and castrated male calves grow more rapidly than female calves during the nursing period. When feeding conditions are unfavorable, little difference is noted in preweaning growth of male and female young.

Postweaning Growth

Postweaning growth of cattle is growth that occurs between weaning and slaughter weight of 1,000 to 1,100 lbs. Rate of gain during the postweaning period is computed as follows:

$$\text{Postweaning rate of gain} = \frac{\text{final weight} - \text{weight at weaning}}{\text{days from weaning to final weight}}$$

This equation can also be used for computing postweaning gains of pigs and lambs. Any differences of postweaning rate of gain among farm animals are under considerable genetic control, providing that the animals are not subjected to variations in food supply such as exist when they are nursing. Figures have been presented ranging from 40 to 70% for the heritability of rate of gain for beef cattle.

Several other factors may influence postweaning rate of gain. Animals that were under austere feed conditions during the nursing period because their dam failed to provide sufficient milk for them tend to compensate when they are weaned and are given better feed conditions, provided that the earlier austere conditions were not severe enough to cause permanent damage. By contrast, young that were nursed by a dam that gave an abundance of milk may find feed conditions following weaning less satisfactory than what they enjoyed while they were with their dams. Such young tend to grow less rapidly during the initial phase following weaning. Although animals tend to compensate for previous austerity, they do not overtake, in total weight, those young that had better feed conditions previously at a given age. Thus, one cannot expect to use **compensatory** gains as a means of marketing more weight of animals in a given time period. Animals that gain more rapidly in the feedlot require less feed to make a unit of gain than slowly gaining animals. They also store more lean and less fat in their bodies. Thus, rapidly gaining animals resemble younger animals physiologically whereas slowly gaining animals resemble older animals.

When intact male and female animals reach puberty and begin to develop sexually, their growth rates decline. For example, beef bulls such as Herefords reach puberty at about 15 months, at which time they weigh 1,200 lbs or more. They continue to grow for another 25 months, and may eventually reach 1,800 to 2,700 lbs. Growth rate declines continually from puberty until maturity is reached, at which time growth ceases entirely.

Males grow more rapidly than females during the postweaning period even though they eat no more feed per unit of body weight than females; consequently, males make their gains with much less feed per unit of gain than females. Rapidly gaining bulls, for example, may gain 3.5 to 4.0 lbs per day and may require 5.0 to 5.5 lbs of feed per lb of gain, whereas rapidly gaining heifers may gain 2.7 to 3.0 lbs per day and may require 6.5 to 8.0 lbs of feed per lb of gain.

The various tissues and organs grow rapidly at different times of life. The vital organs (brain and central nervous system, heart, lungs and digestive system) are the first organs to develop and are well developed at birth. After these organs begin developing, they are followed in order by bone, muscle and fat tissue. The development of the various kinds of tissues may be discussed either according to chronological age or according to physiological age. Chronological age is age in days from conception or from birth, while physiological age is the age at which certain physiological events, such as puberty occur. Type and sex influence the growth curve of bone, muscle, and fat. Bone and muscle growth occurs to a later chronological age in large-type animals (animals that reach a large size at maturity) than in small-type animals (animals that mature at a small size). Thus, small-type animals complete muscle growth and start storing fat at younger ages and smaller weights than large-type animals. Females also complete muscle growth and store fat at a younger age and a smaller weight than males.

Animals that produce muscle tissues for a longer period of time and start fattening relatively late have a higher lean-to-fat ratio than early-maturing animals do at the same slaughter weight. Therefore, when late-maturing and early-maturing animals are at the same weight, the late-maturing animals yield carcasses with a high lean cutability grade but with a low-quality grade because of lack of fat; particularly marbling, and the early-maturing animals yield carcasses with a low lean cutability grade but with a high-quality grade.

Young animals gain at about the same rate as old animals during the postweaning period, but they eat less feed; therefore, they are more efficient in use of feed. There are some explanations for this difference; small (young) animals have less body weight to maintain than large (old) animals. Also, large (old) animals tend to store more fat than small (young) animals. In general, animals that grow rapidly during the postweaning period tend to store more lean and less fat as body tissues. In most of the work done in which growth rate was increased genetically or by hormone administration, an increase in lean and a decrease in fat storage occurred. In animals of the same age, rapidly and slowly gaining animals may both have the same feed requirement per unit of gain but rapidly gaining animals will be larger. In animals of the same weight, rapidly gaining animals will use less feed per unit of gain and their bodies will contain more lean and less fat than slowly gaining animals.

In all meat producing animals, rapid growth is closely associated with efficiency of converting feed into body weight gains. Statistical methods have been developed to measure the association of one trait with that of another. The **coefficient of determination** measures the variation of the dependent variable that is attributable to variation in the independent variable. Our assumption is that rate of gain is the independent variable and that efficiency of feed use is the dependent variable. The coefficient of determination is computed by squaring the **correlation coefficient** between two traits, which, in this case, is 0.8 for the correlation between rate of gain and feed efficiency. When this correlation is squared, one obtains a value of 0.64 for the coefficient of determination. Our interpretation is that 64% (0.64 x 100) of the variation in feed efficiency is attributable to variation in rate of gain.

MATURITY

After an animal reaches maturity, it may change in weight by a considerable amount simply by increasing or decreasing the amount of fat or water that it stores. The increase in weight of a mature animal due to fattening is not growth because no net increase in body protein occurs. In fact, animals tend to lose body protein as they grow older. The loss of body protein in an animal is one of the phenomena in the aging process.

As most mammals reach maturity, a closure of the **epiphyseal-diaphyseal** junction occurs. This junction is between the shaft and the joint end of the long bones. Growth of long bones takes place in the cartilage at this junction, but after it is closed by bone formation (ossification), no further growth can take place. The epiphyseal-diaphyseal junction appears to close in all farm mammals. In the rat, as a contrasting example, the epiphyseal-diaphyseal junction never closes, consequently, one can initiate growth in mature (even old) rats by injecting growth hormones. Growth cannot be initiated by injecting growth hormones into mature cattle, sheep, or swine.

After animals mature, they may add or withdraw fat from their bodies, consequently, they may change in weight. They also can withdraw or replenish minerals from the bones. They do not normally add protein to their bodies.

As female animals age, they may lose their teeth and their productivity (reproductive and lactating ability) goes down. The age at which productivity declines varies widely. Most farm animals are sold for slaughter when productivity starts to decline. Most beef cows are marketed at 10 to 11 years of age because that is the age when many of them are beginning a decline in productivity. However, a few cows maintain high productivity and cows more than 20 years old sometimes produce good calves.

Female sheep are usually marketed at 7 to 8 years of age because their front teeth are usually gone at this age which interferes with their grazing. Female swine are often marketed at 2 to 3 years of age but this is not due to a decline in productivity. Large (old) sows consume much feed in relation to their pig production; therefore, it is more economical to use young females for producing pigs. In addition, sows at 2 to 3 years of age sell at a price that is near that of butcher hogs; therefore, the farmer does not hesitate in keeping young females for breeding and selling the mature sows.

STUDY QUESTIONS

1. a. Define growth.
    b. If growth is expressed as weight according to age, what type of growth curve is obtained?
    c. What is the appearance of the curve that expresses gain in weight per unit of time during the rapidly gaining period of growth?
    d. If the growth of an animal is expressed as change in weight per unit of body weight per unit of time, what is the appearance of the growth curve?
    e. After animals become mature, they may gain considerably in weight. Is this growth?
2. a. If growth is explained on a cellular basis, how is growth defined?
    b. What hormones affect growth? Where are these hormones produced?
    c. Males usually grow more rapidly than females. What effect does this relatively rapid rate of growth have on percentages of lean and fat in their carcasses? What effect does it have on the efficiency of gains?
    d. Can growth in farm animals be increased over long periods by injecting growth hormone? Why?
    e. Can growth in mature farm animals be initiated by injecting growth hormone? Why?
3. a. Diethylstilbestrol (a synthetic estrogen), when fed or injected, causes ruminants to react one way, chickens to react another way, and rats to react a third way. What are the reactions of each?
    b. What factors influence prenatal growth of farm animals?
    c. What factors greatly influence preweaning growth of farm animals? Which is the most important?
    d. In which of the postnatal periods (preweaning or postweaning) is growth most under genetic control?
    e. What is the relationship of rate of growth to percentages of lean and fat of the body?
4. What is the relationship of type to the storage of body lean and fat?
5. What is the order of growth of vital organs, fat, bone, and muscle tissues?
6. What is the relationship between rate of gain and efficiency of feed use in making gains? Why do you think this relationship exists?

SELECTED REFERENCES

Boggs, D. L. and Merkel, R. A. 1979. *Live Animal Carcass Evaluation and Selection Manual.* Dubuque, Iowa: Kendall-Hunt Publishing Co.

Brody, S. 1945. *Bioenergetics and Growth.* New York: Reinhold Publishing Corp.

Casida, L. E., Andrews, F. N., Bogart, R., Clegg, M. T., and Nalbandov, A. V. 1959. *Hormonal Relationships and Applications in the Production of Meats, Milk and Eggs.* Washington, D.C.: National Academy of Science Publication 74.

Hafez, E. S. E. and Dyer, I. A., eds. 1969. *Animal Growth and Nutrition.* Philadelphia: Lea and Febiger.

Hammond, J., ed., 1955. *Progress in the Physiology of Farm Animals.* Vol. 2. London: Butterworths Scientific Publications.

10

Lactation

Lactation, (the production of milk by the mammary glands) is a distinguishing characteristic of mammals whose young at first feed solely on milk from their mothers. Even after they start to eat other foods, the young continue to nurse until they are weaned (no longer able to nurse from their mothers). The length of time that they consume only milk from their mothers is governed by their state of development at birth. For example, guinea pigs are well developed at birth and start eating solid food the day they are born. Some guinea pigs even survive on solid food alone if the mother dies when they are born. Rabbits, by contrast, are less developed at birth; they consume only milk from their mothers for about 2 weeks.

Milk production by all mammalian farm animals to provide nourishment for their young is vital for continuation of the species, but milk production by the animals that supply milk for human consumption is also important. The milk drunk in the United States comes from the dairy cow and goat; in other countries, however, the source of supply may be the donkey, water buffalo, sheep, yak, reindeer, or sow.

DEVELOPMENT OF THE MAMMARY GLAND

At birth, the young female mammal has the rudiments of an udder with the characteristic number of teats for the species—for example, 2 for sheep, horses and donkeys, 4 for cattle, and 10 to 18 for swine. Development of the udder is slight until puberty, at which time periods of estrus begin.

At puberty, the ovary of the female begins to function under the stimulation of **follicle-stimulating hormone** (FSH), which is produced by the pituitary gland. This hormone causes the formation of a **follicle** or follicles on the ovary and stimulates the formation of an egg in each follicle. The follicle produces hormones called **estrogens** (estradiol, estriol, and estrone), which are circulated through the blood. Estrogens also act on the pituitary gland, causing it to reduce its output of FSH and to initiate the production of **luteinizing**

hormone (LH). The luteinizing hormone acts on the follicle. When there is a certain balance of FSH and LH, the follicle ruptures and releases the egg. Luteinizing hormone continues to act on the tissue that was the follicle, causing it to develop into a **corpus luteum**.

The corpus luteum produces progesterone, a hormone that stimulates lobulo-alveolar development at the ends of the ducts that were previously developed under the stimulation of estrogen. Although development of the duct and alveoli is achieved shortly after puberty (Figure 10-1), the mammary gland does not yet secrete milk.

At the onset of the **estrous** periods, development of the udder progresses when the estrogren level is high and regresses when it is low. The development is always greater than the regression; therefore, the udder becomes more developed with succeeding estrous periods. When the animal becomes pregnant, estrogens and progesterone are produced by the placenta in much larger quantities than by the ovaries. In pregnancy the major development of ducts, ductules, and alveoli occurs until the lobulo-alveolar and duct systems resemble a bunch of grapes; the alveolar and ductule mass is comparable to the grape, the small ducts are comparable to small stems, and the main duct that empties into the gland cistern is comparable to the main grape stem that connects with the vine. Mammary growth continues during early lactation and in the first three to four lactations.

Estrogens and progesterone were formerly thought to act directly on the mammary gland because injections of estrogens and progesterone into young females, or into females from which the ovaries have been removed, cause mammary gland development and even milk secretion. After the pituitary and adrenal glands are removed, however, injections of estrogens and progesterone do not stimulate the development of the mammary gland. Based on many studies, it is evident that estrogens act both directly and indirectly through the pituitary. Two hormones produced by the pituitary gland, lactogen (prolactin) and growth hormone, apparently are necessary both for proper udder development and, once the mammary gland is completely developed, for milk secretion.

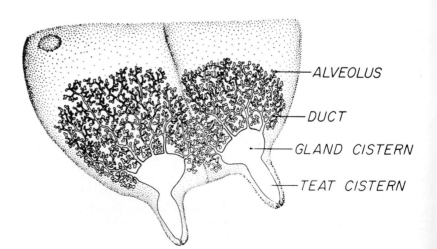

Figure 10-1. Diagrammatic structure of the udder of a cow at puberty. Blood vessels and connective tissue are not shown in this drawing.

ALVEOLUS

DUCT

GLAND CISTERN

TEAT CISTERN

INITIATION OF LACTATION

Speculation exists as to how lactation is initiated at the time the young are born. It appears that a sudden decrease in progesterone output triggers an increased output of prolactin from the pituitary and cortisol from the adrenal glands, and that these hormones bring about milk secretion. In some animals, a regression of the corpus luteum occurs prior to delivery of the young. In these species, milk secretion may occur prior to delivery of the young.

Secretion of Milk

Milk is made and secreted by the single layer of cells in the alveoli. The precursor of milk is the plasma from the blood; therefore, the udder contains a well-developed system of blood vessels. Large blood vessels can be seen leading to the udder in a cow that is producing much milk (Figure 10-2). The udder is large and has considerable weight, which is supported by well-developed connective tissue and a strong middle suspensory ligament attached to the pelvic arch.

Milk fat is secreted as small droplets. It is made from triglycerides, ketones, acetates, and glucose from the blood. Lactose is synthesized from glucose of the blood. Gamma casein, blood serum albumen, and immune globulins are absorbed from the blood by the udder. The other proteins of milk are synthesized by the udder cells from amino acids. Milk secretion requires great enzyme activity.

Milk Letdown

Female farm animals typically give milk in response to pressure exerted on the udder or teat by a suckling offspring or by a human or a mechanical milker in the absence of their

Figure 10-2. Jersey cow showing an udder with good attachments and well-developed circulation. The dairy cow is a unique factory. She can use hay, silage, grain, and pasture as raw products both for the manufacture of a highly nutritious food and as a source of energy for the manufacture of food.

young, but mares and ewes must have their young nearby to do so. An **intravenous** (within a vein) injection of oxytocin can be used to stimulate mares and ewes to give milk in the absence of their offspring.

The normal routine for milking dairy cows includes (1) stimulating the udder by washing it with warm water; (2) obtaining some milk in a cup to determine the quality of the milk; (3) applying suction cups to the teats within a minute after the udder has been stimulated; (4) removing suction cups from teats as soon as the cow has been milked; and (5) dipping the cow's teats in an approved teat dip to prevent infection.

FACTORS AFFECTING LEVEL OF PRODUCTION

The amount of milk produced is influenced strongly by the amount of stimulus given the mammary gland. If milk is not removed from the mammary gland, production ceases and involution of the mammary gland results, and if only part of the milk is removed, milk production is reduced. When, however, milk is removed at frequent intervals, the mammary gland responds with increased production. For example, milking farm animals three times daily increases milk production by 15% to 20% over the amount obtained when milking is done twice daily. Thus, cows that are milked several times each day or ewes that are raising twins produce greater amounts of milk than they would otherwise.

Inheritance determines the potential for milk production but feed and management determine whether or not this potential is attained. The best feed and care will not make a high-producing cow out of one that is genetically a low producer. Likewise, a cow with high genetic potential will not produce at a high level unless she receives proper feed and care. Production is also influenced by the health of the animal. For example, **mastitis** (an inflammation of the udder) in dairy cows can reduce production by 30% or more. Proper management of dairy cows, particularly adherence to routine milking and feeding schedules, will contribute to a high level of production.

In all meat-producing farm animals, the level of milk production is significant because milk from these females provides much of the required nutrients for optimal growth of the young. If beef cows, for example, are inherently heavy milkers, their milk-producing levels will be established by the ability of the young to consume milk. Normally, if one does not milk such a cow, she will adjust her production to what the calf will consume. However, if the cow is milked early in lactation, her milk flow may be far greater in amount than the calf can consume. This situation makes it necessary to either continue milking the cow until the calf becomes large enough to consume the milk, or to put another calf onto the cow.

Anything that causes a female to reduce her production of milk also usually causes some regression of the mammary gland, thereby preventing resumption of full production. If the young calf lacks vigor, it does not consume all the milk that its mother can supply, and her level of production is lowered accordingly. This is one reason why inbred calves grow more slowly and crossbred calves more rapidly during the nursing period. Crossbred calves are quite vigorous at birth and can stimulate a high level of milk production by their dams, whereas inbred calves are less vigorous and cause their dams to produce milk at a comparatively low rate.

Cows with bull calves produce more milk than cows with heifer calves. Great amounts of nutrients are needed to supply the requirements of lactating females for high levels of milk production, because much energy is required for milk secretion and because milk contains large quantities of nutrients. The need for adequate nutrition is much greater for lactation than for gestation; for example, a cow producing 100 lbs of milk daily must produce 4.0 lbs of butterfat, 3.3 lbs of protein, and almost 5.0 lbs of lactose in that milk. If nutrition is inadequate in quality or quantity, a cow that has the inherent capacity to produce 100 lbs of milk daily will draw nutrients from her own body as the body attempts to sustain a high level of milk production. Withdrawal of nutrients reduces the body stores. This withdrawal usually occurs to a limited extent in high producers even when they are well fed. Cows reduce their milk production in response to inadequate amounts or quality of feed, but usually do so only after the loss of nutrients is sufficiently severe to cause a loss in body weight. In gestation, a cow becomes relatively efficient in digesting her feed, so feed restrictions in gestation are less harmful than they might be at other times.

Goats and calves, as well as other mammals, are born with rudimentary mammary systems. Mammary glands grow and develop under the influence of estrogens, progesterone, prolactin, growth hormone and the adrenal corticoid hormones. Progesterone must be absent and prolactin and adrenal corticoids must be high to initiate lactation. Maintenance of lactation requires growth hormone, prolactin, thyroxine and adrenal corticoid hormones. Increases in milk production have been obtained by administration of thyroid and growth hormones; however, thyroxine increases metabolic rate, thereby creating a need for more feed. The extra feed may be as costly as the added value of the extra milk obtained.

STUDY QUESTIONS

1. a. What hormones influence the development of the alveolar and duct systems of the udder?
    b. What hormones influence initiation and maintenance of lactation?
    c. What happens to the lobulo-alveolar system if milk is not removed from the udder?
    d. Where is milk produced in the mammary gland?
    e. What causes milk to be let down?
2. a. What influence does frequent removal of milk from the mammary gland have on the amount of milk produced?
    b. Why do females produce more milk when raising cross-bred offspring than they produce when raising inbred offspring?
    c. What effect does litter size of swine have on milk produced by the sow?
    d. What is the relationship of pregnancy to the development of the mammary gland?
    e. If feed conditions are extremely austere when a female should be in the peak of her production, will she return to a high level of production if feed conditions become optimal?

SELECTED REFERENCES

Bath, D. L., Dickinson, F. N., Tucker, H. A. and Appleman, R. D. 1978. *Dairy Cattle: Principles, Practices, Problems, Profits.* Philadelphia, Pennsylvania: Lea and Febiger.

Kon, S. K., and Cowie, A. T. 1961. *Milk: The Mammary Gland and Its Secretions.* Vol. II. New York: Academic Press.

Larson, B. L., and Smith, V. R. 1974. *Lactation: A Comprehensive Treatise.* Volumes I, II and III. New York: Academic Press.

Reynolds, M., and Folley, S. J. 1969. *Lactogenesis: The Initiation of Milk Secretion at Parturition.* Philadephia: University of Pennsylvania Press.

Turner, C. W. 1952. *The Mammary Gland, I. The Anatomy of the Udder of Cattle and Domestic Animals.* Columbia, Missouri: Lucas Brothers Publishers.

_____ 1969. *Harvesting Your Milk Crop.* Chicago: Babson Institute Press.

11

Adaptation to the Environment

The Animal Kingdom is divided into two types with respect to body temperature, namely, those that are **poikilothermic**, or cold-blooded, and those that are **homeothermic**, or warm blooded. The body temperature of poikilothermic animals varies with and is approximately equal to environmental temperature. The body temperature of homeothermic animals is maintained at a uniform temperature that is characteristic of the species. Thus, the body temperature of homeothermic animals may be much above that of the environment in cold periods and somewhat below that of the environment when the weather is hot.

The body temperature of homeotherms usually varies from 98 to 105 °F (22.4 to 26.3 °C) and death will occur if environmental temperatures should become excessively high or low for any appreciable length of time unless the animal is capable of adapting. Because all farm animals are homeotherms, this chapter considers only homeothermic animals.

HEAT PRODUCTION IN ANIMALS

The routine body processes and physiological processes of farm animals, such as digestion and **assimilation**, rumen fermentation by protozoa and bacteria, cellular activity, muscular activity, growth, and milk production, produce a great amount of heat. Thus, most farm animals past the stages directly following birth, have more difficulty dissipating heat than keeping warm; for example, animals fed all they will eat (full fed) have difficulty eliminating heat unless the environmental temperature is 70 °F (21-22 °C) or lower.

The metabolic rate of very young animals is low and remains so until they obtain food and generate heat through the work of digestion and assimilation. Animals born in cold weather are especially vulnerable to chilling and possible resultant death because they are wet when they are born and evaporation of the moisture on them increases the cooling effect of the temperature. Animals born in cold conditions should be wiped dry, encouraged to nurse almost immediately and, as necessary, given an artificial source of heat.

ANIMALS DISSIPATE HEAT

Animals are commonly cooled by **convection** (air currents) if the environmental temperature is below that of the body. Convection increases evaporation from the surface of the body and thus helps cool an animal even if environmental temperature is higher than body temperature. Evaporation is more pronounced if the air is dry than if it contains considerable moisture, so evaporation is greatest when humidity is low. Heat may also be lost through conduction (movement through a conducting medium). For example, an animal lying on concrete or iron, both of which are good heat conductors, will lose more heat than an animal lying on wood, shavings, or sawdust. Some animals such as horses perspire, and this perspiration provides moisture on the body surface. Such animals are cooled greatly by evaporation when the air is dry. Other animals, such as sheep, hogs, dogs, and some cattle, may lack the ability to sweat except at the nose, and their greatest source of cooling through evaporation is their lungs. They pant to inhale dry air and exhale moist air in warm weather or after exercise.

The amount of body surface obviously affects the rate at which an animal can dissipate heat. Brahman cattle, which have a large skin surface per unit of body weight, are more capable of withstanding hot climates than cattle that have less body surface per unit of weight.

The amount and kind of fibers growing from the surface of animals influence heat loss, and the fibers may act as insulation. The hair of horses and cattle grows longer as the weather cools in the autumn and winter, and is shed in the spring or early summer as the temperature of the environment increases. Rather than growing a thick coat of hair, pigs lay down a thick layer of fat immediately under the skin that prevents excessive heat loss.

ADJUSTING TO ENVIRONMENTAL CHANGES

As the seasons change, two major kinds of changes occur in the environment: changes in temperature and changes in length of daylight. As summer approaches, temperature and length of daylight increase, thus increasing the amount of heat available to the animal. As autumn approaches, length of daylight and temperature decrease. Hormones enable the animal to respond physiologically to these seasonal changes.

The thyroid gland, whose lobes are located on both sides of the trachea, regulates the metabolic activity of the animal. When the environmental temperature is cool, the air inhaled tends to cool the thyroid gland. The thyroid responds by increasing its production of the hormone thyroxine; the thyroxine in turn stimulates metabolic activity and heat generated. The growing and shedding of insulating coats are slow changes that are also influenced by hormones. Sudden warm periods before shedding occurs in spring and sudden cool periods before a warm coat has been produced in autumn can have severe effects on an animal. These sudden temperature changes are the greatest predisposers for the onset of respiratory diseases such as pneumonia, influenza, and shipping fever.

MACROCLIMATE AND MICROCLIMATE

To cope with the stresses of the **macroclimate**, animals can create or use an effective **microclimate** for comfort and survival. Buildings that house people are examples of microclimates. Animals are sometimes also provided microclimates in the form of shelters to keep out rain and snow or windbreaks to prevent excessive air convection and heat. When the environmental temperature is high, animals can be provided with cooling shade, a sprinkler system to provide moisture to the body surface, and air currents generated by fans to abet evaporation of sweat. The combined use of sprinklers and shade is particularly helpful in dry climates.

TEMPERATURE ZONES OF COMFORT AND STRESS

The comfort zone, or **zone of thermoneutrality**, is 60 °F to 65 °F (15.5 °C to 18.3 °C) for farm animals; in this range of temperature, heat production and heat loss from the body are about the same. When the temperature is below 50 °F (10 °C) animals increase their feed intake, rate of exercise, and rate of heartbeat, and reduce blood flow to the surface and to the extremities. Mammals generate heat by shivering. Fowl fluff their feathers to create a large space of dead air about themselves to prevent rapid heat transfer from their bodies. Some animals may hunch to expose less body surface. Cattle congregate in an area that provides protection from blizzards and they huddle closely such that each animal receives heat from other animals. During a severe blizzard, some animals are killed from being trampled and some die from suffocation.

To avoid excessive heat loss, animals maintain the temperature of their extremities at a level below that of the rest of the body through a unique mechanism called countercurrent blood flow action. Blood in arteries coming from the core of the body is relatively warm and blood in veins in the extremities is relatively cool. The blood in veins in the extremities cools the warmer arterial blood so that the extremities are kept at a cool temperature. With the extremities thus kept at a relatively cool temperature, loss of heat to the environment is reduced.

Between 65 °F - 80 °F (18.3 °C - 26.7 °C) dilation of blood vessels near the skin and in the extremities occurs so that the surface of the animal becomes warmer, water consumption increases, respiration increases, and, in animals that can sweat, perspiration increases. Above 80 °F (26.7 °C) animals that have the capacity to sweat keep their body surfaces wet with sweat so that evaporation can help cool them. When the environmental temperature exceeds 90 °F (32.2 °C), farm animals suffer. Hogs may die from such heat. Animals may become less active, reducing the amount of heat they generate, and lie down in the shade, reducing their exposure to the sun. Animals typically increase their consumption of water and excretion of urine. If the water consumed is cooler than the temperature of the animal, it can help considerably to cool the body.

Animals must sometimes adapt to food or water shortages. Fat-tailed sheep and Brahman cattle store large quantities of fat that can be used as a source of energy if food becomes scarce over a short duration. Animals that eat lush growths of forages may only need to drink water once each day or once every other day, whereas animals that are eating dry feeds need water at frequent intervals.

When animals do work, the amount of heat that they produce increases. Animals on full feed produce a great deal of heat in performing digestion and assimilation. The dairy cow that is producing a large quantity of milk may produce two or three times the heat that is produced by a nonlactating cow of the same size. She must consume large amounts of food to produce large quantities of milk and this additional work produces heat. All animals generating large quantities of heat are affected more adversely by hot weather than by cold weather.

INABILITY OF ANIMALS TO COPE WITH HEAT AT TIMES

Animals are sometimes unable to control their body temperatures in conditions of extreme heat or cold. When the body temperature exceeds normal because the animal cannot dissipate its heat, a condition known as fever results. Fever is often caused by systemic infectious diseases and is often most severe when environmental temperatures are extremely high or low. Keeping the ill animal comfortable while medication is given will assist recovery. When the temperature becomes excessively high, some animals (notably pigs) may lose normal control of their senses and do things that aggravate the situation. If nothing is done to prevent them from doing so, pigs that get too hot will often run up and down a fence line, squealing, until they collapse and die. Pigs seen doing this should be cooled with water and induced to lie down on damp soil.

INABILITY OF ANIMALS TO COPE WITH OTHER STRESSES

Some animals are unable to adjust to environmental stresses. This inability appears to be genetically controlled.

When cattle are grazed at high altitudes, 1% to 5% develop high mountain (brisket) disease. This disease usually occurs at altitudes of 7,000 feet or more. It is characterized as right heart failure (deterioration of the right side of the heart). It was considered that this disease was caused by chronic **hypoxia** (lack of oxygen), which leads to pulmonary **hypertension** (high blood pressure). As the animal becomes more affected, marked **subcutaneous edema** (filling of tissue with fluid) develops in the brisket area. Some recent studies at the United States Poisonous Plant Center at Logan, Utah, have shown that certain poisonous plants that grow only at high altitudes will cause brisket disease when fed to cattle maintained at low altitudes.

Some young cattle and sheep whose dams are fed plants in selenium deficient areas develop **white muscle disease**. The disease is characterized by **calcification** of the heart and voluntary muscles, giving the muscles a white appearance. It has been shown (Oldfield, 1971) that white muscle disease can be prevented by administering selenium to the mother during pregnancy and to the young at birth. In the absence of treatment, some of the animals are affected and some are not; therefore, resistance and susceptibility appear to be inherited.

STUDY QUESTIONS

1. a. By what mechanisms may an animal lose heat?
    b. By what mechanisms may an animal conserve heat?
    c. Why is a very young animal more influenced by cold temperatures than when it is older and larger?
    d. An animal on full feed has a problem eliminating heat. Why?
    e. What are the body processes that generate heat?
2. a. Define homeotherm and poikilotherm. To which of these classes do farm animals belong?
    b. What is the zone of thermoneutrality?
    c. When temperatures are below 50 °F, what does an animal do initially to keep warm? What can one do to assist an animal in keeping warm?
    d. What physiological changes occur in an animal when it is subjected for a long time to a low temperature?

e. What physiological changes occur in an animal when it is subjected for a long time to high altitudes?

3. a. How can animals be assisted to remain cool when the weather is hot?
    b. What effect does circulating the air about an animal that does not sweat have on cooling the animal?
    c. When an animal's body surface is wet, what effect do air currents have?
    d. Is heat conservation or heat elimination more important in animals that are producing large quantities of milk?
    e. What happens to an animal's temperature when disease-causing organisms invade the body?

4. a. What is brisket disease and where does it occur?
    b. Why does only a small percentage in a herd develop brisket disease?
    c. What is white muscle disease and how can it be prevented?

SELECTED REFERENCES

Hafez, E. S. E. 1968. *Adaptation of Domestic Animals.* Philadelphia: Lea and Febiger.

Hammond, J., ed. 1954. *Progress in the Physiology of Farm Animals.* Vol. 1. London: Butterworths Scientific Publications.

Oldfield, J. E. 1971. *Selenium Deficiency in Soils and its Effect on Animal Health.* Corvallis, Oregon: Oregon Agricultural Experiment Station Special Paper 332.

Phillips, R. W. 1949. *Breeding Livestock Adapted to Unfavorable Environments.* Rome: FAO Agricultural Studies No. 1. Food and Agriculture Organization of the United Nations.

Rhoad, A. O. 1955. *Breeding Beef Cattle for Unfavorable Environments.* Austin: University of Texas Press.

12

Digestion and Absorption of Feed

Animals obtain substances needed for all body functions from the feeds that they eat and the liquids that they drink. Before the body can absorb and use them, feeds must undergo the process called **digestion**. Digestion includes mechanical action, such as chewing and contractions of the intestinal tract; chemical action, such as the secretion of hydrochloric acid (HCl) in the stomach and bile in the small intestine; and action of enzymes such as maltase, lactase, and sucrase (which act on disaccharides), lipase (which acts on lipids), and peptidases (which act on proteins). Enzymes are produced either by the various parts of the digestive tract or by microorganisms. The effect of digestion is to reduce the size of molecules so that they may be absorbed into the blood.

CARNIVOROUS, OMNIVOROUS, AND HERBIVOROUS ANIMALS

Animals are classed as **carnivores, omnivores,** or **herbivores** according to the types of feed that they normally eat. Carnivores, such as dogs and cats, normally consume animal tissues as their source of nutrients; omnivores, such as humans and pigs, consume both plant and animal tissues; herbivores, such as cattle, sheep, horses, and rabbits, consume plant tissues. The carnivores and omnivores are monogastric animals, meaning that the stomach is simple and has only one compartment. Some herbivores, such as guinea pigs, horses, and rabbits, are also monogastric. Other herbivores, such as cattle, sheep, and goats, are ruminant animals, meaning that the stomach is complex and contains four compartments. The classification of animals as carnivores, omnivores, and herbivores does not mean that they cannot use certain feeds that they do not normally consume. For example, animal products can be fed to herbivorous ruminants, and certain cereal products can be fed to carnivores.

The digestive tracts of pigs and people are similar in **anatomy** and **physiology**; therefore, much, though not all, of the information gained from studies on pig nutrition and digestive physiology can be applied to the human. Both the pig and the human are omnivores and both are monogastric animals. Neither of them can synthesize the B-complex vitamins or amino acids to a significant extent. Both pigs and humans tend to eat large quantities and become obese. Humans can control obesity by exercising as a means of using, rather than storing, excess energy. Also, humans can control their food intake as a means of preventing obesity, whereas pigs eat freely as long as feed is available. Obesity in swine is controlled by limiting the amount of feed available to them or through genetic selection of leaner animals.

DIGESTIVE TRACT OF THE PIG, A MONOGASTRIC ANIMAL

The anatomy of the digestive tract varies greatly from one species of animal to another. The basic parts of the digestive tract are mouth, stomach, small intestine and large intestine, or colon. The primary function of the parts preceding the intestine is to reduce the size of feed particles. The small intestine functions in the splitting of food molecules and absorption, and the large intestine absorbs water and forms indigestible wastes into solid form called feces. In a mammal having a simple stomach (such as the pig), the mouth has teeth and lips for grasping and holding feed that is **masticated** (chewed), and salivary glands that secrete saliva for moistening feed so it can be swallowed. Feed passes from the mouth to the stomach through the esophagus. A sphincter (valve) is at the junction of the stomach and esophagus. It can prevent feed from coming up the esophagus when stomach contractions occur. The stomach empties its contents into that portion of the small intestine known as the duodenum. The pyloric sphincter, located at the junction of the stomach and the duodenum, can be closed to prevent feed from moving into or out of the stomach. Feed goes from the duodenum to the jejunum portion and then to the ileum portion of the small intestine. It then passes from the small intestine to the large intestine, or colon. The ileocecal valve, located at the junction of the small intestine and the colon, prevents material in the large intestine from moving back into the small intestine. The small intestine actually empties into the side of the colon near, but not at, the **anterior** end of the colon. The blind anterior end of the **colon** is the **cecum**, or, in some animals, the vermiform appendix. The large intestine empties into the rectum. The anus has a sphincter, which is under voluntary control so that defecation can be prevented by the animal until it actively engages in defecation. The structures of the digestive system of the pig are shown in Figure 12-1.

Animals such as pigs, horses, rabbits, and poultry are classed as **monogastric** animals, but they differ markedly in certain ways. For example, the rabbit (Figure 12-2) and the horse have large cecal structures in which much fermentation occurs. Because the cecum is posterior to the area where most feed is absorbed, horses and rabbits do not obtain all of the nutrients made by microorganisms in the cecum unless they eat their own feces. Rabbits are known to eat their feces regularly, and horses eat their feces when they are on diets of low-quality roughage and need some of the nutrients in the feces.

The digestive tracts of most species of farm poultry differ from that of the pig in several respects. Because they have no teeth, poultry birds break their feed into a size that can be swallowed by pecking with their beaks or by scratching with their feet. Feed goes from the mouth through the esophagus to the crop, which is an enlargement of the esophagus in which feed can be stored. Some fermentation may occur in the crop, but it does not act as a

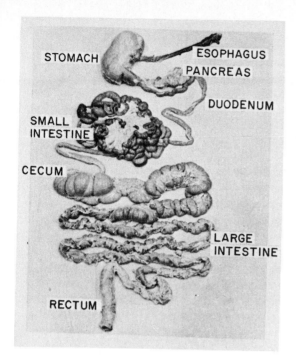

Figure 12-1. Digestive tract of the pig as an example of the digestive tract of a monogastric animal.

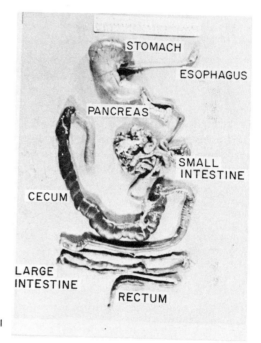

Figure 12-2. Digestive tract of the rabbit, a monogastric animal with highly developed ceca where fermentation occurs.

fermentation vat. Feed passes from the crop to the proventriculus, which is a glandular stomach in birds that secretes gastric juices and HCl but does not grind feed. Feed then goes to the gizzard, where it is ground into finer particles by strong muscular contractions. The gizzard apparently has no function other than to reduce the size of feed particles, because birds from which it is removed digest a finely ground ration. Feed moves from the gizzard into the small intestine. Material from the small intestine empties into the large intestine. At the junction of the small and large intestines are two ceca which contribute little to digestion. Material passes from the large intestine into the cloaca, into which urine also empties. Material from the cloaca is voided through the vent (Figure 12-3).

Figure 12-3. Digestive tract of the chicken showing crop, proventriculus, and gizzard, all of which are characteristic of poultry.

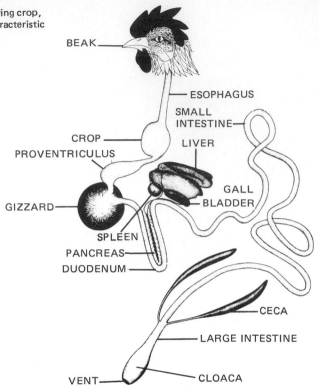

STOMACH COMPARTMENTS OF RUMINANT ANIMALS

In contrast to the single stomach of monogastric animals, the stomachs of cattle, sheep, and goats have four compartments—**rumen, reticulum, omasum**, and **abomasum**. The rumen is a large fermentation vat where bacteria and protozoa thrive and break down roughages to obtain nutrients for their use. It is lined with numerous papillae, which give it the appearance of being covered with a thick coat of short projections. The papillae increase the surface area of the rumen lining. The microorganisms in the rumen can digest cellulose and can synthesize amino acids from nonprotein nitrogen and can synthesize the Vitamin B-complex vitamins. Later, these microorganisms are digested in the small intestines to provide these nutrients for the ruminant animal's use. The reticulum has a lining with small compartments similar to a honeycomb, so it is often called the honeycomb portion. The omasum has many folds, so it is often called the manyply portion. The abomasum, or true stomach, corresponds to the stomach of monogastric animals.

Animals that have the four-compartment stomach (Figure 12-4) eat forage rapidly and later **regurgitate** it and chew it again leisurely. The regurgitation and chewing of undigested feed is known as rumination. Animals that ruminate are known as ruminants. As feeds are fermented by microorganisms in the rumen, large amounts of gases (chiefly methane and carbon dioxide) are produced. The animal normally can eliminate the gases

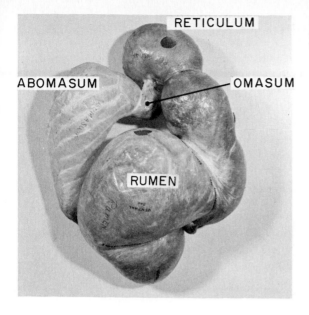

RETICULUM

ABOMASUM — OMASUM

RUMEN

Figure 12-4. The four compartments of the stomach of the sheep, an animal in which pregastric fermentation occurs.

by belching, also called **eructation**. If gases are not eliminated by eructation, bloating occurs; the rumen and reticulum expand drastically when bloated.

DIGESTION IN MONOGASTRIC ANIMALS

Feed that is ingested (taken into the mouth) stimulates the secretion of saliva. Chewing reduces the size of the ingested particles and saliva moistens the feed. The enzyme amylase, which is present in saliva, acts on starch and dextrins in monogastric animals. However, very little actual breakdown of starch into simpler compounds occurs in the mouth, primarily because feed is there for a short time.

An **enzyme** is an organic catalyst that speeds a chemical reaction without being altered by the reaction. Enzymes are rather specific; that is, each type of enzyme acts on only one or a few types of substances. Therefore, it is customary to name enzymes by giving the name of the substance on which it acts and adding the suffix **-ase**, which, by convention, indicates that the molecules so named are enzymes (Table 12-1). For example, lipase is an enzyme that acts upon **lipids** (fats); maltase is an enzyme that acts upon maltose to convert it into two molecules of glucose; lactase is an enzyme that acts upon lactose to convert it into one molecule of glucose and one molecule of galactose; and sucrase is an enzyme that acts upon sucrose to convert it into one molecule of glucose and one molecule of fructose. Some lipase is present in saliva but little hydrolysis of lipids into fatty acids and glycerides occurs in the mouth.

As soon as it is moistened by saliva and chewed, feed is swallowed and passes through the esophagus to the stomach. The stomach secretes HCl, mucus, and the digestive enzymes pepsin and gastrin. The strongly acidic environment in the stomach favors the action of pepsin. Pepsin breaks proteins down into polypeptides. The HCl also assists in coagulation, or curdling, of milk. Little hydrolysis of proteins into amino acids occurs in the stomach. Mucous secretions help to protect the stomach lining from the action of strong acids.

In the stomach, feed is mixed well and some digestion occurs; the mixture that results is called **chyme**. The chyme passes next into the duodenum, where it is mixed in with secretions from the pancreas, bile, and enzymes from the intestine.

Secretion from the pancreas and discharge of bile from the gall bladder are stimulated by secretin, pancreozymin, and cholecystokinin, three hormones that are released from the duodenal cells. The enzymes from the pancreas are lipase, which hydrolyzes fats into fatty acids and glycerides; trypsin, which acts on proteins and polypeptides to reduce them to small peptides; **chymotrypsin**, which acts on peptides to produce amino acids; and amylase, which breaks starch down to disaccharides, after which the disaccharides are broken down to monosaccharides. The liver produces bile that helps emulsify fats; the bile is strongly alkaline and so helps to neutralize the acidic chyme coming from the stomach. Some minerals that are important in digestion also occur in bile.

By the time they reach the small intestine, amino acids, fatty acids, and monosaccharides are all available for absorption. Thus, the small intestine is the most important area for both digestion and absorption of feed. **Absorption** of feed molecules may be either passive or active. Passive passage results from diffusion, which is the movement of molecules from a region of high concentration of those molecules to a region of low concentration. Active transport of molecules across the intestinal wall may be accomplished through a process in which cells of the intestinal lining engulf the molecules and then actively transport these molecules to either the bloodstream or to the lymph. Energy is expended in accomplishing the active transport of molecules across the gut wall.

When molecules of digested feed enter the capillaries of the blood system, they are carried directly to the liver. Molecules may go to the lymphatic system and pass directly to the heart, after which they go to various parts of the body including the liver. The liver is an extremely important organ both for metabolizing useful substances and for detoxifying harmful ones.

DIGESTION IN RUMINANT ANIMALS

In ruminant animals (cattle, sheep, and goats), **predigestive fermentation** of feed occurs in the rumen and the reticulum. The bacteria and protozoa in the rumen use the roughages consumed by the animal as feed for their growth and multiplication; consequently, billions of these microorganisms develop. The environment of the rumen is ideal

Table 12-1. Important enzymes in the digestion of feed

Enzyme	Substrate	Substances resulting from enzyme action
Amylase	starch, dextrin	disaccharides
Chymotrypsin	peptides	amino acids
Lactase	lactose	glucose and galactose
Lipase	lipids	fatty acids and glycerides
Maltase	maltose	glucose and glucose
Pepsin	protein	proteoses and peptones
Peptidases	peptides	amino acids
Sucrase	sucrose	glucose and fructose
Trypsin	protein	polypeptides

for the microorganisms because moisture, a warm temperature, and a constant supply of feed are present. Excesses of microorganisms are continuously removed along with the feed that passes into the abomasum. When the feed passes into the abomasum, strong acids destroy the bacteria and protozoa. The ruminant animal then digests the microorganisms and uses them as a source of nutrients. Thus, ruminant animals and microorganisms mutually benefit from one another. All digestive processes in ruminants are the same as those in monogastric animals after the feed reaches the abomasum, which corresponds to the stomach of monogastric animals.

A young ruminant does not consume roughages. Consequently, at this stage in life it acts as a monogastric animal. Milk is directed immediately into the abomasum in young ruminants. When roughage is consumed, it is directed into the rumen where bacteria and protozoa break it down into simple form for their use. The rumen starts to develop functionally as soon as roughage enters it, but some time is required before it is completely functional. Complete development of the rumen, reticulum, and abomasum requires about 2 months in sheep and about 8 months in cattle. One can influence the development of the rumen by the type of feed one gives to the animal. If only milk and concentrated feeds are given, the rumen shows little development. If very young ruminants are forced to live on forage, the rumen develops much more rapidly. "**Roughages**" and "concentrated feeds" are defined and discussed in Chapter 13.

PREGASTRIC AND POSTGASTRIC FERMENTATION

In some animals, such as the rabbit and the horse, **postgastric** (cecal) **fermentation** of roughages occurs. In these animals, all the feed that can be digested by a monogastric animal is digested and absorbed before the remainder reaches the cecum. These animals are perhaps more efficient than ruminants in their use of feeds such as concentrates. In the ruminant animals, the **concentrates** given along with roughages are used by the bacteria and protozoa. Because the microorganisms in ruminants use starches and sugars, little glucose is available to ruminants for absorption. The microorganisms do provide volatile fatty acids, which are absorbed by the ruminant and used in place of glucose as an energy source. The **postgastric fermentation** that occurs in horses and rabbits breaks down roughages, but this takes place posterior to the areas where nutrients are actively absorbed; consequently, all of the nutrients in the feed are not obtained by the animal in postgastric fermentation. Postgastric fermentation ferments roughages less efficiently than pregastric fermentation.

STUDY QUESTIONS

1. You should learn the parts of the digestive tract of a monogastric animal such as the pig. How does the anatomy of the digestive tract of the pig differ from that of the rabbit, chicken, and horse?
2. a. What are the four compartments of a ruminant stomach?
   b. In which compartments of the ruminant stomach does fermentation occur?
   c. Which part of the ruminant stomach is comparable to the stomach of the monogastric animal?
   d. What is meant by pregastric and postgastric fermentation?
3. a. What enzymes are produced by the stomach and upon what substances do they act?
   b. What function does hydrochloric acid produced by the stomach perform in digestion?

    c. What is digestion?

    d. What are the functions of bile?

    e. What enzymes are produced by the pancreas and upon what substances do they act?

4.   a. What enzymes are produced by the small intestine and upon what substances do they act?

    b. Where does most absorption of nutrients occur and into what systems do absorbed nutrients go?

    c. What is absorbed from the large intestine?

    d. By what methods may molecules pass across the intestinal wall?

    e. What is the function of the cecum?

SELECTED REFERENCES

Church, D. C. 1969. *Digestive Physiology and Nutrition of Ruminants.* Vol. 1. Corvallis, Oregon: Published by D. C. Church, distributed by O and B Books.

Church, D. C. and Pond, W. G. 1974. *Basic Animal Nutrition and Feeding.* Corvallis, Oregon: Published by D. C. Church, distributed by O and B Books.

Church, D. C. 1975. *Digestive Physiology and Nutrition of Ruminants.* Second Edition. Corvallis, Oregon: O and B Books.

Ensminger, M. E. and Oletine, C. G., Jr. 1978. *Feeds and Nutrition-Complete.* Clovis, California: The Ensminger Publishing Co.

13

The Functions of Nutrients

Animal feeds are classified as concentrates and roughages. Concentrates include cereal grains (corn, wheat, barley, oats, and milo), oil meals (soybean meal, linseed meal, and cotton seed meal), molasses, and dried milk products. They are high in energy, low in fiber, and highly (80% to 90%) digestible. Roughages include legume hays, grass hays, straws from the production of grass seed and grain, **silage**, stovers (dried corn, cane, or milo stalks and leaves with the grain portion removed), and **soilage** (cut green feeds). Roughages are less digestible than concentrates. Roughages are typically 50% to 65% digestible, but the digestibility of some straws is drastically lower. Young animals lack the capacity to consume enough low-quality roughage to achieve normal growth. Dairy cows are also unable to consume low-quality roughage in sufficient quantity to supply the nutrients required for high-level milk production.

NUTRIENTS AND THEIR FUNCTIONS

A **nutrient** is a chemical substance in either a simple or compound form that is absorbed from the digestive tract and travels through the bloodstream to all parts of the body for use in metabolism. Thus, proteins as such are not nutrients, but their components, the **amino acids** are. Nutrients that can be supplied only through the diet are called essential nutrients.

Ruminant animals require fewer essential nutrients than monogastric animals: micro-organisms (bacteria and protozoa) in the rumen of ruminants can make many nutrients, including all of the B-complex vitamins, all of the amino acids if elemental or compound forms of nitrogen, sulphur, and carbon are present, and protein. By contrast, monogastric animals, which lack a rumen, must obtain all of these nutrients (with the exception of some amino acids) from their feed.

Amino Acids

Amino acids are nitrogen-containing compounds that bind chemically in many combinations in chains to form proteins which are the building blocks of the body. The chemical, or peptide bonding, of amino acids is illustrated using alanine and serine which results in the formation of a dipeptide:

$$
\begin{array}{ccc}
\underset{\text{alanine}}{\overset{\displaystyle NH_2 \quad O}{CH_3-\overset{|}{\underset{|}{C}}-\overset{\diagup\!\diagup}{\underset{\diagdown}{C}}}} & + & \underset{\text{serine}}{\overset{\displaystyle NH_2 \quad O}{HO-CH_2-\overset{|}{\underset{|}{C}}-\overset{\diagup\!\diagup}{\underset{\diagdown}{C}}}} \longrightarrow \underset{\text{dipeptide}}{\overset{\displaystyle CH_3 \quad O \quad H \quad HO-CH_2 \quad O}{NH_2-\overset{|}{\underset{|}{C}}-C\text{-}O\text{-}N-\overset{|}{\underset{|}{C}}-\overset{\diagup\!\diagup}{\underset{\diagdown}{C}}}}
\end{array}
$$

It can be seen that amino acids have both a basic portion NH_2 and an acid portion, $C\!\!\overset{\diagup\!\diagup O}{\diagdown OH}$, and it is because of these that they can combine into long chains to make proteins. When digestion occurs, the action is at the peptide linkage to free amino acids from one another.

Some amino acids are provided in abundance in many proteins but others are quite limited. Thus, monogastric animals may have difficulty obtaining scarce amino acids that they cannot synthesize. Some amino acids (alanine, aspartic acid, glutamic acid, hydroxyproline, proline, and serine) can be synthesized by monogastric animals if carbon, hydrogen, oxygen, and nitrogen are available. These amino acids are called nonessential amino acids. Other amino acids, however, either cannot be synthesized at all (arginine, histidine, isoleucine, leucine, lysine, methionine, phenylalanine, threonine, tryptophan, and valine) or are produced so slowly that they are called semiessential and must be provided in the feed (arginine, cystine, glycine, and tyrosine). The student is referred to Table 13-1 which shows all the nutrient materials that are known to be used by animals. Some other minerals may be useful to animals but their functions have not yet been determined (Ba, Br, Cr, F, Sr, and V).

All amino acids are needed by all animals; the terms *essential* and *nonessential* merely refer to whether or not they must be supplied through the diet. The shortage of any particular amino acid can prevent an animal from using other amino acids for needed functions; a protein deficiency results.

Proteins

Proteins are complex molecules that form body tissues and act as enzymes. The building blocks for growth (including growth of muscle, bone, and connective tissue), milk production, and cellular and tissue repair are amino acids that come from proteins in feed. The interstitial (between cells) fluid, blood, and lymph require amino acids to regulate body water and to transport oxygen and carbon dioxide. All enzymes are proteins, so amino acids are also required for enzyme production.

Proteins are, on the average, 16% nitrogen. Knowing this allows us to calculate the amount or percent of protein in a feed if we know how much nitrogen is present in a given quantity of the feed. Because 16% into 100% = 6.25, we can multiply the amount of nitrogen by 6.25 to obtain the amount of protein present. If, for example, a feed is 3% nitrogen, 100 g

Table 13-1. Nutrients in feeds

INORGANIC

Macro	Micro
Ca, Mg, K, Na, Cl, P, S	Co, Cu, Cr, Mn, Mo, Zn, Fe, I, Se, Si, F, V, Ni, As

ORGANIC

Carbohydrates	Nitrogenous substances	Lipids
N-Free (Soluble)	Amino Acids	Simple
Monosaccharides	Essential	Fatty acids
Fructose	Arginine, histidine, isoleucine,	Butyric, caprilic, linoleic,
Glucose	leucine, lysine, methionine,	linolenic, oteic, palmitic,
Galactose	phenylalanine, threonine,	stearic
Disaccharides	tryptophan, valine	Sterols
Lactose	Essential (Under special conditions)	Cholesterol, egosterol,
Maltose	Glycine, cystine, tyrosin	steroid hormones
Sucrose	Not essential (For domestic animals)	Compound
Polysaccharides	Alanine, aspartic, glutamic acid,	Neutral fats
Starch	hydroxy proline, proline, serine	Esters of glycerol and
Crude Fiber (Insoluble)		fatty acids
Polysaccharides	Other N-compounds	Phospho lipids
Cellulose	Urea, amines	Colamine-cephalins,
Lignin		lecithin, sphingomyelins
Hemicellulose		Waxes
		Esters of fatty acids with
		long chain aliphatic
		alcohol or with cyclic
		alcohols

VITAMINS

Fat soluble	Water soluble
A, D, E, and K	Nicotinic acid, biotin, pantothenic acid
B-complex	Ascorbic acid (C)
Thiamine (B_1),	
riboflavin (B_2),	
pryidoxin (Bo), folic acid complex,	
cyanocubalamine (B_{12})	

of the feed contains 3 g nitrogen. Multiplying 6.25 × 3 gives 18.75, meaning that 100 g of this feed contains 18.75 g protein.

In analyses of feeds, the percent of nitrogen is determined chemically and the percent of protein in the feed is determined in the manner just described. The estimate thus obtained of the amount of protein in the feed may be too high because feeds may contain an appreciable amount of nonprotein nitrogen. The presence of a large amount of nonprotein nitrogen and a relatively small amount of protein in the feed is quite detrimental to the nutrition of monogastric animals because they cannot use some of the nonprotein nitrogen.

Rations in which the amount of protein present has been overestimated due to the presence of nonprotein nitrogen could likely be successfully developed for ruminants because ruminants can use nonprotein nitrogen quite efficiently. In fact, such sources of nitrogen as urea are often used to provide a portion of the protein needs of ruminant animals. This procedure is perfectly satisfactory so long as only a portion of the protein needs are supplied by urea—animals do poorly or may not survive when urea is the only source of protein. Although some cattle have been raised with urea as the only protein (nitrogen) source and cows raised on urea as a protein source have produced calves, it is generally advised not to use urea at a level in excess of 50% of the protein needs.

The digestibility of protein is determined by measuring the grams of nitrogen consumed and the grams of nitrogen in the feces. Protein digestibility, expressed as the percentage of protein consumed that is actually digested, is calculated as follows:

$$\frac{\text{g Nitrogen consumed} - \text{g nitrogen in feces}}{\text{g nitrogen consumed}} \times 100 = \%\text{ protein digestibility}$$

Carbohydrates and Fats

Carbohydrates and fats are composed of carbon, hydrogen, and oxygen. Complete oxidation (burning or taking on oxygen) of carbon gives much energy, and the primary function of both carbohydrates and fats is to provide energy for the body. Energy is the force, or power, that is used to drive a variety of systems. It can be used to power movement of the animal, but most of it is used as chemical energy to drive reactions necessary to convert feed into animal products and to keep the body warm. The unit of measurement of energy is the **calorie**, which is the amount of heat required to raise the temperature of 1 g water 1 °C from 14 to 15 °C. This is a very small amount of energy; therefore, most nutritional expressions are in **kilocalories**:

(1 kcal = 1,000 calories) or megacalories (1 mcal = 1,000,000 calories)

Carbohydrates are absorbed into the body from the small intestine as monosaccharides (simple sugars). Animals appear to derive energy from all of the common monosaccharides readily. Fats are absorbed into the body as fatty acids and glycerides. The energy that is available to the animal from the feed eaten depends on carbons that can be oxidized. As measured by the amount of heat liberated when feed substances are burned, proteins and carbohydrates yield approximately 4 kcal per g and fat yields approximately 9 kcal per g. Thus, fat has 2.25 times the energy value of protein or carbohydrates. (Although these figures are frequently cited as the physiological fuel value of protein, carbohydrates, and fats, much variation occurs depending on the particular feedstuff.) Although protein supplies as much energy as carbohydrates, carbohydrates and fats are used more often as sources of energy because they can be supplied more economically.

MEASUREMENTS OF ENERGY IN FEEDS

Energy in feeds may be measured in various ways depending on the particular interest of the individual desiring the information. It is important, however, that one understand the biological meaning of the information that is obtained by the different methods.

The functions of various forms of energy in the body are outlined in Table 13-2. Table 13-2 shows the fate of gross energy that is ingested as the energy passes through the digestive system and is metabolized to provide energy for productive functions.

Total Digestible Nutrients

Total digestible nutrients **(TDN)** is an overall estimate of the energy of a feed. If TDN is expressed as a percentage (% TDN), it is the sum of the percentages of digestible protein, crude fiber, nitrogen-free extract (carbohydrates), ether extract (fat or lipid) × 2.25 and minerals. All lipid materials are soluble in ether, thus, ether extract is what is measured but it is the lipid material. One gram of TDN is roughly equivalent to 4 kcal energy. Because much of the loss of the total energy of a feed occurs through digestion, a determination of TDN is one method of evaluating the energy of a feed.

TDN can be determined by analyzing on a dry-weight basis the feed given the animal and the feces produced by the animal for fats, carbohydrates (soluble sugars and starch and insoluble or fiber materials), protein (nitrogen × 6.25) and minerals. The digestible fat, carbohydrate, protein, and minerals must be summed to give total digestible nutrients. The digestibility of protein is obtained by determining the digestibility of nitrogen in a feed. Digestibility is expressed as a percentage for nitrogen, for example, as follows:

$$\frac{\text{Nitrogen in feed} - \text{Nitrogen in feces}}{\text{Nitrogen in feed}} \times 100 = \% \text{ digestibility}$$

As an example, if 100 g feed contains 3.2 g nitrogen and 100 g feces contains 0.8 g nitrogen, the percent digestibility of nitrogen is:

$$\frac{3.2 - 0.8}{3.2} \times 100 = 75\%$$

Note that the determination of 3.2 g nitrogen in 100 g feed enables one to estimate the percentage of protein in the feed as 20% (3.20 × 6.25 = 20).

By substituting digestible fat, carbohydrate, and minerals for nitrogen in the formula for percent digestibility, one can calculate the percent digestibility of each of these nutrients. After the digestibility of each of the nutrient materials has been determined, their quantities in the feed and their digestibilities are used to compute total digestibility for the feed. For example, 100 g of feed containing the following nutrient materials that have digestibility percentages of those given below would have a total digestible nutrient value of:

Nutrient		Digestibility	Digestible nutrient
Protein	20 g	.75	15.00
Carbohydrates			
Soluble (NFE)	55	.85	46.75
Insoluble (fiber)	10	.20	2.00
Fat 5 (× 2.25)	11.15	.85	9.50
Minerals	3.85	.35	1.35
			74.60

This feed has a TDN of 74.6%.

If the digestible protein of a feed is determined by use of a ruminant animal, it may be much higher than if it were determined by use of a pig, particularly if the feed contains considerable nonprotein nitrogen that is useful to ruminants but not to pigs.

The six most important factors that influence **digestibility** are as follows:

1. *Balance of the ration.* ("Balance" here refers to the amount of protein in relation to the amount of other nutrients in the feed). Rations that are quite low in protein may be difficult to digest.

2. *Combinations of feed ingredients.* Fiber in the ration tends to reduce digestibility of the concentrates. The combination of different sources of proteins and of grains included in a ration usually enhances digestibility.

3. *Rapidity of movement of substances through the digestive tract.* Magnesium sulphate, which is often found in alkaline water, or large amounts of molasses or lush pasture that are often provided for cattle cause feed to pass through the digestive tract so rapidly that digestion and absorption are reduced. Farmers often feed molasses to cattle on lush pastures to provide energy, even though digestibility is lowered.

4. *High-energy sources in feed.* High-energy sources reduce fiber digestibility in ruminants. When ruminants are given high-energy feeds along with roughages, the rumen microorganisms obtain their nutrients from the high-energy feeds instead of digesting the roughages. High-energy feeds for pigs enhance digestibility.

5. *Feed intake.* Digestibility is reduced when large amounts of feed are consumed.

6. *Preparation of feed.* Grinding, cracking, or rolling breaks the kernels of grains open, aiding the actions of digestive enzymes and helping prevent such feeds as barley and wheat from passing through the animal as whole kernels. Hay that is finely cut may approach some concentrates in level of digestibility.

Gross Energy

The amount of **gross energy** in a feed may be measured by combusting the feed and measuring the amount of heat that it liberates. This method of measuring energy provides information on the total amount of energy in a feed which is quite useful when feeds are not composed of large quantities of indigestible materials.

Digestibility of Energy

Digestibility of energy provides information on the percentage of gross energy that is available to the animal for metabolic functions. Digestibility of energy is determined from measurements of the kilocalories of energy ingested and the kilocalories voided through the feces. The kilocalories in the feed and feces are determined by measuring the heat produced when feed and feces are combusted. The digestibility of energy is calculated as follows:

$$\text{Digestibility of energy} = \frac{\text{kcal ingested} - \text{kcal in feces}}{\text{kcal ingested}} \times 100$$

Digestible Energy

Digestible energy is expressed in kilocalories per kilogram (kg) of feed. It is calculated by multiplying the total kilocalories in 1 kg feed by the digestibility of energy. This

Table 13-2. The use of energy

Food intake (Gross energy)				
	1. Apparent digestible energy	1. Metabolizable energy	1. Net energy	1. Production energy
				A. Energy storage
				a. Storage of energy in fetus and in body of pregnant female
				b. Semen in male
				c. Growth
		2. Gaseous products of digestion	2. Heat increment	d. Milk
			*A. Heat of fermentation	e. Eggs
			*B. Heat of nutrient metabolism	f. Fur
				g. Wool
				B. Work
	2. Fecal energy	3. Urinary energy		
	A. Food origin	A. Food origin		2. Maintenance energy
	B. Metabolic origin	B. Body origin		*A. Basal metabolism
				*B. Voluntary activity
				*C. Heat to keep body warm
				*D. Energy to keep body cool

*Expended as heat.

method of expressing digestible energy provides information on how much energy of the feed is digested, but it does not provide information on how much energy is available to the animal for body processes.

Metabolizable Energy

Metabolizable energy is digestible energy minus energy lost when energy-containing substances are passed in urine and when combustible gases are lost in eruction or in defecation. Energy losses through combustible gases may be sizable in ruminants because ruminants eruct considerable quantities of methane.

Net Energy

Net energy is the energy available for use by the animal. It is the metabolizable energy minus the loss from heat increment. **Heat increment** is the amount of heat lost as a result of the physical and chemical process that is associated with metabolism. Its components are the heat of nutrient metabolism and the heat of fermentation. Net energy is available to the animal for such body processes as maintenance, growth, reproduction, lactation, and work.

MEASUREMENTS OF ENERGY EXPENDITURE

Energy expenditure may be measured by determining the amount of oxygen (O_2) consumed and the amount of carbon dioxide (CO_2) released. The ratio of volume of CO_2 produced to the volume of O_2 consumed can be used to differentiate between oxidation of carbohydrates and fats. Energy expenditure is a measure of the **metabolism** that is occurring in an animal's body. This ratio, called respiratory quotient (RQ) is 1.0 for carbohydrates but lower for fats, with long-chain fats having an RQ of 0.7. **Basal metabolism** is the metabolism that is measured when the animal is quiet and is not digesting food; consequently, it is a measure of metabolism that avoids the complications that are introduced when energy expended for digestion and muscular activity must be measured. Basal metabolism can be determined for monogastric animals but not for ruminants. It is highly unlikely that a ruminant does not ever have feed passing from the rumen to the abomasum. Some feed is still present in the rumen even in cattle that have died of starvation.

The slaughter technique is sometimes used to determine what use growing animals make of the energy they consume. Three groups of similar animals are used. Animals of one group are slaughtered and their bodies are analyzed to obtain a measure of how much energy is present in the bodies of the members of all three groups. The amount of energy in the bodies of animals can be determined through combustion or through an analysis of the amounts of protein, fat, and ash, after which the energy can be calculated. The amount of carbohydrate in the animal body is generally low, but it should also be determined for a precise assessment. The second group is allowed only enough feed for each individual to maintain its body weight, so that the amount of energy required for body maintenance can be estimated. The third group is fed **ad libitum** (that is, given free access to all the feed desired). Its members are finally slaughtered and their bodies are analyzed in the same way as the members of group 1.

Through the slaughter technique, the energy required for gain can be calculated. The quantity of energy initially present in the bodies of animals in group 1 subtracted from the quantity of energy in the bodies of animals in group 3 at the time of slaughter gives the energy stored. Energy consumed by group 3 (fed ad libitum) minus the energy consumed by group 2 (maintenance) gives an estimate of how much energy was used by group 3 animals to achieve an increase in weight.

MINERALS

Minerals are inorganic substances that perform a variety of functions in an animal's body. Animals need some minerals (called **macrominerals**) in relatively large amounts and others (called **microminerals**) in small amounts. (Note that the terms *macromineral* and *micromineral* refer to the quantity of the mineral needed rather than to its atomic size.) Microminerals are also called "trace" minerals because only a trace, or minute quantity, is needed.

The macrominerals include calcium, chlorine, magnesium, phosphorus, potassium, sodium, and sulphur. Calcium and phosphorus are required in certain amounts and in a certain ratio to each other for bone growth and repair and for other body functions. The blood plasma contains sodium chloride; the red blood cells contain potassium chloride. The osmotic relations between the plasma and the red blood cells are maintained by proper concentrations of sodium chloride and potassium chloride. Sodium chloride may be depleted by excessive sweating that results from heavy physical work in hot weather. It is essential that salt and plenty of water be available under such conditions. The acid-base balance of the body is maintained at the proper level by minerals.

The microminerals include arsenic, chromium, cobalt, copper, fluorine, iodine, iron, manganese, molybdenum, nickel, selenium, silicon, vanadium, and zinc. Microminerals may become a part of the molecule of a vitamin (for example, cobalt is a part of vitamin B_{12}) and they may become a part of a hormone (for example, thyroxine, a hormone made by the thyroid gland, requires iodine for its synthesis).

Hemoglobin of the red blood cells carries oxygen to tissues and carbon dioxide from tissues. Iron is required for the production of hemoglobin because it is a part of the hemoglobin molecule. A small quantity of copper is also necessary for the production of hemoglobin (even though it does not normally become a part of the hemoglobin molecule in farm animals) because it apparently is necessary for normal iron absorption from the digestive tract and for release of iron to the blood plasma.

Certain important metabolic reactions in the body require the presence of minerals. Selenium and vitamin E both appear to work together to help prevent white muscle disease, which is a calcification of the striated muscles, the smooth muscles and cardiac muscles. Both vitamin E and selenium are more effective if the other is present. Excesses of certain minerals may be quite harmful. For example, excess amounts of fluorine, molybdenum, and selenium are highly toxic.

VITAMINS

A vitamin is a nutrient needed in a small amount that provides a specific function for the animal. Vitamins may be classed as either fat-soluble or water-soluble. The fat-soluble vitamins are vitamins A, D, E, and K. Vitamin A helps maintain proper repair of internal and external body linings. Because the eyes have linings, lack of vitamin A adversely affects the eyes. Vitamin A is also a part of the visual pigments of the eyes. Vitamin D is required for proper use of calcium and phosphorus in bone growth and repair. Recent studies have shown that a major function of vitamin D is to regulate the absorption of calcium and phosphorus from the intestine. Vitamin D is produced by the action of sunlight on steroids of the skin; therefore, animals that are exposed to sufficient sunlight make all the vitamin D they need. Vitamin K is important in blood clotting; hemorrhage may occur if the body is deficient in vitamin K. Aspirin, which is sometimes given to pet horses, dogs, and cats that are old and arthritic, may react with vitamin K to prevent its function; therefore, the taking of large amounts of aspirin may cause the body to require more vitamin K than normal. Vitamin E was once thought to be important in reproduction in animals because laboratory rats have usually failed to reproduce when fed a ration deficient in vitamin E. Vitamin E has not been shown to be important in reproduction in farm animals.

The water-soluble vitamins are ascorbic acid (vitamin C), biotin, choline, cyanacobalamin (vitamin B_{12}), folic acid, inositol, niacin, pantothenic acid, para-ami-nobenzoic acid, pyridoxine (vitamin B_6), riboflavin (vitamin B_2), and thiamine (vitamin B_1). More diseases caused by inadequate nutrition have been described in the human than in any other animal, and among the best known are those caused by a lack of certain vitamins: **beri-beri** (lack of thiamine); **pellagra** (lack of niacin); **pernicious anemia** (lack of vitamin B_{12}); **rickets** (lack of vitamin D); and scurvy (lack of vitamin C).

In ruminant animals, all of the water-soluble vitamins are made by microorganisms in the rumen. Water-soluble vitamins also appear to be readily available to horses; perhaps some are made by fermentation in the cecum. The water-soluble vitamins cannot be synthesized by monogastric animals and must therefore be in their feed. Most fat-soluble vitamins are not synthesized by either ruminants or monogastrics and must be supplied in the ration of both groups (an exception is vitamin K, which is synthesized by rumen bacteria in ruminants). Many vitamins are supplied through feeds normally given to animals.

Some feeds occasionally react with certain vitamins in such a way as to prevent those vitamins from being available to the animal. For example, thiaminase, which splits thiamine into simpler constituents that lack vitamin activity is present in the tissues of certain species of fish. If fish of these species are included in the diet of animals without the meat being cooked, it will cause a serious deficiency of thiamine. Cooking the meat will destroy thiaminase activity; therefore, cooked fish of these species create no problem.

Study Questions

1. Classify the following feeds as concentrates or roughages: clover hay, corn, tankage, straw, soybean meal, silage, molasses, and corn stover.
2. a. What is meant by essential amino acids?
   b. Do ruminant animals need to be fed essential amino acids?
   c. How is the protein content of a feed determined? Is it possible that the chemical determination of protein content of a feed may not indicate the actual protein content of the feed? Would an erroneous estimate of the amount of protein present in a feed be more important in feeding monogastric animals or in feeding ruminants?
   d. What are the functions of carbohydrates and fats?
3. List five factors that may influence digestibility of a ration.
4. Differentiate the following: gross energy, digestible energy, net energy, and metabolizable energy.

5. What is meant by macrominerals and microminerals? List five minerals that are classed as macrominerals and five that are classed as microminerals.
6. a. List three fat-soluble and three water-soluble vitamins. Indicate how each is important.
    b. What vitamins that can be made by ruminants must be supplied to monogastric animals?

Selected References

Brody, S. 1945. *Bioenergetics and Growth, with special reference to the efficiency complex in domestic animals.* New York: Reinhold Publishing Co. (Note — This is a classic publication and is used widely today even though it was published many years ago.)

Campbell, J. R. and Lasley, J. F. 1969. *The Science of Animals That Serve Mankind.* New York: McGraw-Hill Book Co.

Cullison, A. E. 1979. *Feeds and Feeding.* Second Edition. Reston, Virginia: Reston Publishing Co., Inc., A. Prentice-Hall Co.

Ensminger, M. E. and Olentine, C. G., Jr. 1978. *Feeds and Nutrition.* Complete. Clovis, California: The Ensminger Publishing Co.

Harris, L. 1962. *Glossary of Energy Terms.* Washington, D.C.: National Academy of Science, National Research Council Publication 1040.

Jurgens, M. H. 1978. *Animal Feeding and Nutrition.* Fourth Edition. Dubuque, Iowa: Kendall-Hunt Publishing Co.

Mertz, W. 1981. The Essential Trace Elements. *Science* 213: 1322-1338.

14

Providing Needed Substances For Body Functions

The feeding of animals is of fundamental importance to any farm production program, because animals must be healthy to function efficiently and yield maximal benefits to the producer. One of the basic tasks of the producer, then, is to supply animals with feed that will satisfy their needs for body maintenance, growth, reproduction, lactation, and work. The profits derived from any feeding program must be assessed against the costs, and knowledgeable producers can increase their profit ratio by feeding their animals nutritiously but nevertheless economically.

BODY MAINTENANCE

The **maintenance** of the body requires that tissues be repaired, that energy for body activities be available, and that molecules lost through metabolic processes be replaced. Water, proteins, energy (supplied by carbohydrates or fats), and minerals provide the basic nutrients that animals need to maintain their bodies; adult animals that are reproducing, lactating or working require these same nutrient materials but the proteins and energy are needed in much greater amounts.

Water. The need for water exists in all animals because the feed eaten and digested must be in a liquid solution for enzyme action of digestion and for absorption. The nutrients absorbed into the body are carried in solution to the body parts where they are used, and the waste materials of metabolism are carried in solution to the kidneys for excretion. Water is constantly lost through respiration and perspiration. In fact, water is vital in all body functions; it is most important however, when animals are eating dry feed and, particularly, when the weather is hot. Animals that are feeding on lush pasture in the rainy season may obtain all the water they need from the pasture.

Protein. As discussed in Chapter 13, proteins provide amino acids from which body structures and enzymes are built. Thus, protein is needed in body maintenance in the production of new blood cells to replace those that are lost, in repairing muscle tissue, and in replacing enzymes that are lost in metabolism.

To determine how much protein is required for body maintenance, protein can be omitted from its diet. Some nitrogen will be voided through the urine even when none is fed, and the amount of nitrogen excreted, when multiplied by 6.25 (See Chapter 13) indicates the minimum amount of protein needed for maintenance:

$$5 \text{ g} \times 6.25 = 31.25 \text{ g}$$

Minerals. The minerals needed by animals have been given in Chapter 13, however, certain of the minerals are more likely to be in short supply than others. Salt (sodium chloride) is important in maintenance because sodium and potassium maintain an equilibrium between the solution outside and inside the cells. Potassium is present in the cells and sodium is present in the fluids outside the cells. Sodium is lost through fluid excretions (urine and perspiration); therefore, there is a constant need for sodium chloride (salt).

Animals have the capacity to conserve salt when the supply is low and to eliminate salt when slight excesses are consumed but it is desirable to provide salt free-choice to all animals.

Calcium and phosphorous are used in bone repair and they must be in a proper ratio of 3 parts calcium to 1 part phosphorus. The need for either of these is dependent upon the feeds that are provided. If some grain is given along with hay, the calcium and phosphorus needs should be adequately supplied but if animals are maintained on hay alone, there may be a need for additional phosphorus.

Ruminant animals and horses can be maintained on pastures with no additional feed needed except salt. They can also be maintained on reasonably nutritious hay. In areas where hay may be deficient in certain nutrients, the appropriate minerals may be needed as a supplement. Although pigs will consume some pasture or legume hay, they cannot be maintained on pasture or hay alone. Because pigs, rabbits, and chickens are kept in enclosed areas, the feed given them must supply all of their needs for maintenance.

BODY SIZE AND MAINTENANCE

Maintenance needs are related to body size. A large animal obviously needs more feed than a small one, but maintenance requirements are not linearly related to body weight. Small animals require more feed *per unit of body weight* for maintenance than large ones. Studies have shown (Brody, 1945) that the approximate maintenance requirement in relation to weight is expressable as $Wt^{0.75}$ rather than $Wt^{1.00}$. Thus, if a 500-lb animal requires 15 lbs of feed per day for maintenance, a 1,000-lb animal of the same type does not require twice as much feed even though the latter animal weighs twice as much as the first. The quantity of a $1,000^{0.75}$ can be determined and applied to show that the 1,000-lb animal requires approximately 1.7 times as much feed for maintenance as the 500-lb animal.[1] The 1,000-lb animal requires, therefore, approximately 25.5 lbs daily (15 lbs $\times$ 1.7 = 25.5 lbs).

1. The calculations can be summarized as follows:
     Log 500 = 2.69897   Antilog 2.0242275 = 105.750
     Log 1,000 = 3.00000   Antilog 2.2500000 = 177.925
     2.69897 $\times$ 0.75 = 2.0242275   177.925 $\cong$ 1.70 $\times$ 105.75, showing that the
     3.00000 $\times$ 0.75 = 2.2500000   1,000-lb animal requires approximately 1.7 times
                                           as much feed as the 500-lb animal.

GROWTH

Growth has previously been defined as protein synthesis in excess of its breakdown (see Chapter 9). Growth is accomplished by an increase in cell numbers, an increase in cell size, and a combination of these. There are several important requirements for growth, including protein, minerals, vitamins, and energy. Growth at the cellular level is accomplished primarily through the building of muscle and connective tissue. These tissues are composed largely of protein; therefore, young, growing animals, which need feed to sustain growth in addition to maintenance, require much protein. One might think of a young, growing animal as a muscle-building factory and protein in the feed as the raw material for the manufacturing process. If provided with only a maintenance allowance for an extended period, a young animal may be permanently stunted.

Monogastric animals not only need a certain quantity of protein, but they must also have certain amino acids for proper growth. The protein needs of hogs, for example, are usually supplied by feeding them soybean meal in combination with some source of animal protein such as tankage, or by supplying soybean meal containing the animal protein factor, vitamin B_{12}. Young ruminant animals cannot consume enough roughages to make maximum growth. If young ruminant animals are being nursed by dams that produce adequate amounts of milk, the young will do well on good pastures, good quality hay, or both together.

The mineral needs of a young, growing animal include calcium and phosphorus for proper bone growth, salt, for a normal sodium level in the body, and any mineral that may be deficient in the area in which the animal lives. Calcium is usually plentiful in legume forages, and phosphorus is usually plentiful in grains, so a combination of hay and grain should provide all the calcium and phosphorus that young ruminant animals need. Animals fed on hay alone may need additional phosphorus, and those fed on concentrates alone may need additional calcium. Some producers feed a mixture of steamed bone meal or dicalcium phosphate and salt at all times to assure that their animals have the necessary calcium and phosphorus.

Two minerals, iodine and selenium, require special consideration. Some areas may be deficient in one or both of these elements. An insufficient amount of iodine in the ration of pregnant females may cause an iodine deficiency in the fetus, which prevents thyroxine from being produced and thus causes a goiter (see Figure 9-2) of the newborn. Young with goiters die shortly after they are born. Iodized salt can be easily provided to the pregnant females to avoid the iodine deficiency. A lack of selenium may cause the young to be born with white muscle disease; it can be prevented by giving the pregnant female an injection of selenium. The injectable selenium is distributed commercially and directions for proper dosages that are supplied by the distributor should be followed closely because an overdose can kill the animal.

Vitamins are needed by young, growing animals. Young ruminant animals are usually on pasture with their dams and are thus exposed to sunshine. The action of ultraviolet rays from the sun converts steroids in the skin into vitamin D, thus providing the animal with this vitamin. Vitamin D is needed for the proper use of calcium and phosphorus in bone growth. Because pigs, poultry, and rabbits are often raised inside where sunshine is limited or lacking, they need some dietary source of vitamin D. It can be supplied through irradiated materials such as yeast and sun-cured hays, or by cod liver oil. If cod liver oil is used to supply this vitamin for pigs, rabbits, or poultry, it must be removed from the feed prior to slaughtering for a length of time sufficient to eliminate the bad flavor it creates in the meat.

Most vitamins must be supplied to pigs and poultry through feeds. The only vitamin commonly fed to ruminant animals is vitamin A, and then only when they are on dry pasture, are fed hay that is quite mature, or are fed hay that has been moistened in processing and has, therefore, been dried in the sun for several days. The activity of vitamin A in silage is usually quite high because this vitamin is usually preserved by the acid fermentation that takes place when silage is made. Silage, however, is usually quite low in vitamin D because the plants used in making silage do not make this vitamin through the action of sunlight and they are not exposed to sunlight for long periods after they are cut.

ENERGY

Young animals need much energy to sustain their growth, high metabolic rate, and activities. They obtain some energy from their mother's milk, and more is supplied as carbohydrates (starch and sugar) and fats. Feed grains are high in carbohydrates and also contain some fats. Young ruminants on pasture typically obtain sufficient energy from pasture and from milk. Poor pastures yield forages of low quality, which in turn cause a decrease in milk productivity by dams. In these circumstances, the young may require supplemental feeding to sustain normal growth. The energy needs of young pigs and poultry are generally supplied by feeding them grains such as corn, barley, or wheat. Rabbits are usually fed a commercially prepared ration which contains sufficient grain to provide energy needs of young rabbits. Young horses can usually obtain their energy needs when they are running on pasture with their dams since most mares produce much milk; however, when foals are weaned, they need some grain in addition to good pasture or good quality hay.

The total digestible nutrients (TDN) needed for maintenance increases as animals become larger (Table 14-1). If the feed given were 60% digestible, the amount of feed required for maintenance would be approximately 1.7 times the TDN figures given in Table 14-1.

REPRODUCTION

The requirements for **reproduction** fall into two categories—the requirements for gamete production and the requirements for growing the fetus in the uterus of the mother. In general, healthy males and females are capable of producing gametes. The energy needs

Table 14-1. TDN needed for maintenance of cattle in the growing-finishing period

Weight of cattle (lbs)	TDN needed daily for maintenance (lbs)
400	5.7
600	7.7
800	9.7
1,000	11.4
1,100	12.3

for germ cell production are no greater than those needed to keep animals in a normal, healthy condition. For example, ruminant animals grazing on pastures of mixed grass and legumes are generally neither deficient in phosphorus nor lacking in fertility. A lack of phosphorus may cause irregular estrous cycles and impaired breeding in females.

Animals that are losing weight rapidly because of poor feed conditions and animals that are overly fat may be abnormally low in fertility. To attain maximum fertility from female animals, they should be in moderately low to moderate condition as breeding season approaches, but should, ideally, be increasing in condition for 2 to 3 weeks prior to and during the breeding season. The feeding of females so that they gain in condition and attain maximum rates of ovulation and conception is known as **flushing**. Flushing has proven of value in some breeding programs; however, when females that are already in a rather high state of fleshing (fatness) are fed to gain during the breeding season, the results are often disappointing because fertility is unimproved or is actually lowered. For animals on dry pasture to attain maximum fertility, it may be necessary to supplement their diet with vitamin A. Pigs and poultry are usually fed rations that are adequate for germ cell production.

The nutrients required by the **fetus** for growing are much greater in the last one-third of pregnancy than earlier. Because the fetus is growing, its requirements are the same as those for growth of a young animal after it is born. Healthy females can withdraw nutrients from their bodies to support the growing fetus temporarily while the amount or quality of their feed is low, but damage to both the female and the fetus may occur if nutrition is inadequate for a lengthy time of 2–3 months in cattle or 4–6 weeks in sheep. No figures are available on swine because they are usually well fed in confinement. For example, ewes that experience a severe shortage in feed after being bred are known to give birth to about 20% fewer lambs than normal. It is assumed in such cases that the embryo dies and is absorbed.

LACTATION

Among common farm animals, cows and dairy goats produce the most milk. Milk production requires considerable protein, minerals, vitamins, and energy. The need for protein is great because milk contains more than 3% protein. As an example, a cow that weighs 1,500 lbs needs at least 30 lbs of feed that is 15% protein per day, this gives her 4.5 lbs protein for her body and the milk she produces. If the protein she eats is 60% digestible, there is 2.7 lbs of digestible protein of which 1.5 lbs is present in her milk. If a cow of this weight is to produce 100 lbs of milk per day, she must now consume 50 lbs of feed containing 15% protein to compensate for the 3 lbs of protein in the 100 lbs of milk that she gives.

Calcium and phosphorus are the two most important minerals needed for lactation. Milk is rich in these minerals, and their absence or imbalance may result in decreased lactation or even death in cattle. The lactating cow can withdraw calcium and phosphorus from her bones, but if her ration is inadequate she reduces her level of milk production or injures herself and may develop milk fever shortly after giving birth. Cows afflicted by milk fever may become comatose and die if not treated. An intravenous injection of calcium gluconate helps achieve complete recovery in less than a day. Recovery after treatment with calcium gluconate is phenomenal; cows are often up and normal in appearance within

2 hours. Milk fever rarely occurs in species or animals that produce relatively small quantities of milk.

Dairy cows produce milk that contains considerable quantities of vitamins A and D and most of the B-complex vitamins. Because cows are ruminants, it is unnecessary to feed them B-complex vitamins, and they require vitamin D supplementation only if confined indoors.

Exceptions occur in situations in which beef cows give large amounts of milk. If they do not have adequate feed while they are nursing their calves, they may not conceive for the next calf crop. In those parts of the world where sheep and dairy goats are the principal dairy animals, the energy needs of these animals are quite similar to those of dairy cows that produce much milk.

The requirements for milk production in sows are usually provided by increasing the percentage of protein in the ration, by increasing the amount of feed allowed, and by providing a mineral mix (a combination of minerals that usually contains calcium, phosphorus, salt, and some trace minerals) free-choice (that is, having feed available at all times so animals can choose how much to eat).

Energy is perhaps the most vital requirement for the production of much milk. A lactating cow needs energy for body maintenance while she also produces milk and provides the energy stored in it. She cannot eat enough hay to obtain the quantity of energy needed so she must receive high-energy feeds such as concentrates; even then her production may be limited by the amount of feed that she can eat. A cow that is a high-level milk producer needs double the energy of a nonlactating cow of the same size. Even when fed large amounts of concentrates, a cow that is producing much milk often loses weight and fleshing because she cannot consume enough feed to produce at her maximum level; therefore, she draws on her own body reserves to supply part of her energy needs.

WORK

Animals that are used for work, either for pulling heavy loads or for being ridden, require large amounts of energy in addition to the needs for maintenance. Horses are the primary work animals in the United States, but elsewhere, donkeys, cattle, and water buffaloes are used.

Horses, mules, and donkeys rely partly on perspiration to remove nitrogenous wastes. If a horse is used for hard work for 5 days of the week and is not allowed to exercise the next 2 days, its kidneys are put to a strain and illness may result.

COMPOSITION OF FEEDS

Table 14-2 compares the composition of some important concentrates and roughages. Among concentrates, for example, grains are relatively high in phosphorus and low in calcium, whereas, among the roughages, **legume** hays (alfalfa and clover) are relatively high in calcium and low in phosphorus. Cattle and sheep that are fed only legume hay will likely need a phosphorus supplement. Pigs that are fed grain liberally may need a calcium supplement.

As Table 14-2 shows, wheat straw provides very little protein but, because of its high crude fiber content, does provide a considerable amount of energy (much of which is used as energy to digest it). Certainly, cows that are wintered on wheat straw need some source

Table 14-2. Composition of certain concentrates and roughages used for feeding farm animals

	Digestible protein	Crude fiber	Ether extract	Nitrogen free extract	Ash	Calcium	Phosphorus	Total digestible nutrients			Energy cal/kg metabolizable energy	
				Cattle & Swine				Cattle	Sheep	Swine	Cattle	Sheep
	%	%	%	%	%	%	%	%	%	%		
Concentrate feeds												
Barley	8.5	5.3	1.7	67	3.0	0.07	0.40	65	72	77	2.65	2.91
Corn	6.5	2.0	3.8	70	1.1	0.03	0.27	71	75	79	2.57	2.73
Oats	9	10.6	4.5	58	3.4	0.09	0.33	86	72	67	3.12	2.62
Wheat	10	2.5	1.9	71	1.9	0.08	0.34	78	78	78	2.83	2.83
Wheat bran	12	8.6	4.4	54	5.7	0.16	1.31	62	58	56	2.23	2.11
Cottonseed meal	35	10.9	5.6	28	6.1	0.19	1.09	74	69	—	2.36	2.50
Linseed meal	32	8.9	5.1	35	5.6	0.39	0.87	74	73	70	2.66	2.65
Tankage (digester)	50	1.8	9.0	2.4	21.2	11.47	5.25	63	62	73	2.37	2.37
Roughage feeds												
Alfalfa hay	10[b]	28	1.7	37	9.0	1.41	0.24	48	51	—	1.74	1.83
Red clover hay	9.0[a]	29	2.3	38	7.1	1.30	0.20	54	53	—	1.95	1.91
Corn silage	1.4[a]	7.6	1.2	15	1.9	0.08	0.06	16	16	—	0.49	0.59
Timothy hay	1.9	30	2.3	45	4.5	0.60	0.24	45	48	—	1.98	2.12
Tall fescue hay	9.8[a]	28	2.2	40	8.0	0.12	0.08	50	50	—	1.77	1.81
Wheat straw	0.4[a]	38	1.4	38	6.3	0.14	0.16	44	36	—	1.55	1.25

[a]Not reported for swine.
[b]Swine 2.3%.
—not reported.
Percentage figures are on the as-fed basis.
Modified from Atlas of Nutritional Data on U.S. and Canadian Feeds, 1972, with the permission of the National Academy of Science, Washington, D.C.

of protein. Cottonseed, linseed, and soybean meals are all very high in digestible protein and are excellent protein supplements. In general, soybean meal and tankage are the most common used protein supplements for swine whereas soybean, cottonseed, and linseed meals are all satisfactory as protein supplements for cattle, sheep, and goats. Although cattle, sheep, and goats eat tankage, they generally prefer other supplements.

Ruminants readily eat all silages, which, because of their high moisture (and consequent low nutrient) percentages, fail to supply the nutritional needs of animals that are growing rapidly or producing large quantities of milk. Such animals need some concentrates along with a limited amount of silage.

Concentrates contain considerably more energy and more digestible nutrients than roughages do. Some concentrates, such as grains, are high in energy but relatively low in protein. Others, such as tankage, soybean, cottonseed, and linseed meals are all high in protein as well as energy. Roughages differ in protein content, with legume hays being relatively high in protein and grasses and cereal grain straws relatively low.

STUDY QUESTIONS

1. How would you recommend that a farmer provide what is needed for body maintenance in cattle, sheep, and horses?
2. What are the important nutrients needed by rapidly growing animals? Suggest how you would provide for the nutrient needs of rapidly growing calves and pigs following weaning.
3. How would you feed female animals prior to and during the breeding season in order to obtain maximum fertility?
4. What are the primary nutrients needed for a high level of milk production? What will happen to a dairy cow which produces milk in great quantity if she does not have an adequate supply of calcium and phosphorus?
5. How would you provide proper nutrition for cows and sheep that are nursing their young?
6. What is the greatest nutritional requirement of horses that are doing hard physical work?
7. Under what conditions and with which species of farm animals would you depend on roughages as the sole feed source?
8. Compare the important concentrates and roughages in terms of their digestible protein, total digestible nutrients, energy content and percentage digestibility.
9. Why do pigs often need calcium supplementation and dairy cows often need phosphorus supplementation?

SELECTED REFERENCES

Brody, S. 1945. *Bioenergetics and Growth*. New York: Reinhold Publishing Corp. Note—This is a classic publication.

Cullison, A. E. 1979. *Feeds and feeding*. 2nd edition. Reston, Virginia: Reston Publishing Co., Inc. A Prentice-Hall Co.

Ensminger, M. E. and Olentine, C. G., Jr. 1978. *Feeds and Nutrition*. Complete. Clovis, California: The Ensminger Publishing Co.

Jurgens, M. H. 1978. *Animal Feeding and Nutrition*. 4th edition. Dubuque, Iowa: Kendall/Hunt Publishing Co.

McDonald, P., Edwards, R. A., and Greenhalgh, J. F. D., 1981. *Animal Nutrition*. 3rd edition. New York: Longman College and Professional Book Division.

National Research Council. 1979. *Atlas of Nutritional Data on United States and Canadian Feeds*. Washington, D.C.: National Academy of Sciences.

15

Genetics

Body tissues of animals and plants are composed of cells, which can be seen with the microscope. These cells, with certain exceptions, have an outer membrane, internal cytoplasm, and a nucleus. This nucleus contains rodlike bodies called **chromosomes**. Body cells contain these chromosomes in pairs. When the cells divide to produce more body cells, the chromosomes replicate, line up in the center of the cell, and separate in such a way that each of the two cells produced (daughter cells) contains the same chromosome composition as the mother cells. This process is called **mitosis** (Figure 15-1).

THE PRODUCTION OF GAMETES

The testicles of the male and the ovaries of the female produce cells that become gametes (sex cells) by a process called **gametogenesis**. The gametes produced by the testicles are called sperm; the gametes produced by the ovaries are called eggs, or ova. Specifically, the production of sex cells that will become sperm is called **spermatogenesis**; the production of ova, **oogenesis**. Fundamental to gametogenesis is **meiosis**, a special type of nuclear division in which the germ cells, each of which contains one member of each of the pairs of chromosomes present in the body cells, are formed.

The chromosomes are normally in pairs in an individual. Let us examine gametogenesis in a theoretical species whose members have two pairs of chromosomes in each cell. Meiosis occurs in the primordial germ cells (cells capable of undergoing meiosis) located near the outer wall of the **seminiferous tubules** of each testicle and near the surface of each ovary. The chromosomes **replicate** themselves so that each chromosome is doubled. Then each pair of chromosomes comes together in extremely accurate pairing called synapsis and they synapse (come together). The cell after replication and synapsis is called a primary spermatocyte in the male and a primary oocyte in the female. Because subsequent differences exist between spermatogenesis and oogenesis, the two will be described separately.

Figure 15-1. Mitosis.

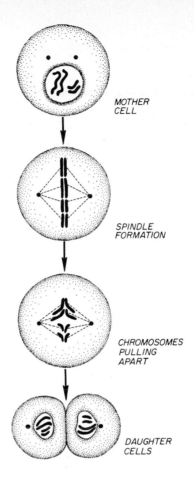

*MOTHER
CELL*

*SPINDLE
FORMATION*

*CHROMOSOMES
PULLING
APART*

*DAUGHTER
CELLS*

SPERMATOGENESIS

The process of spermatogenesis in our theoretical species is shown in Figure 15-2. The primary spermatocyte contains two pairs of chromosomes in synapsis. Each chromosome is doubled following replication. Thus, the primary spermatocyte contains two bodies or structures formed by four parts. Each body is called a **tetrad** (*tetra*- means "four"; -*ad* means "body"). By two rapid divisions, in which no further replication of the chromosomes occurs, four cells, each of which contains two chromosomes, are produced. The first of these two divisions is called the first maturation division. From it are produced two secondary spermatocytes, each of which contains a chromosomal body of two parts, called a dyad (*di* means "two"; -*ad* means "body"). The second maturation division, in which each of the two secondary spermatocytes divides, results in the production of four **spermatids**, each of which contains two chromosomes.

After meiosis and spermatogenesis, the cells each lose much of their cytoplasm and develop a tail. This process, **spermiogenesis**, results in formation of sperm. Four sperm are

Figure 15-2. Spermatogenesis and spermiogenesis involving two pairs of chromosomes. Note that four sperm cells result from one primordial germ cell.

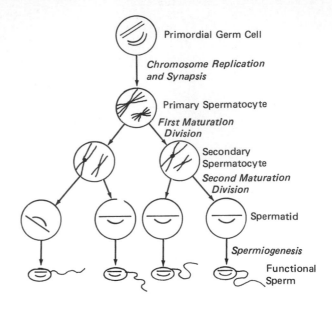

produced from each primary spermatocyte. Whereas four chromosomes in two pairs are present in the primordial germ cell, only two chromosomes and no pairs are present in each sperm. Thus, the number of chromosomes in the sperm has been reduced to half the number in the primordial germ cell.

OOGENESIS

The process of **oogenesis** in our theoretical species is shown in Figure 15-3. Like the primary spermatocyte of the male, the primary oocyte of the female contains a tetrad. The first maturation division produces one relatively large nutrient-containing cell, the secondary oocyte, and a smaller cell, the first polar body. Each of these two cells contains a dyad. The unequal distribution of nutrients that results from the first maturation division in the female serves to maximize the quantity of nutrients in one of the cells (the secondary oocyte) at the expense of the other (the first polar body). The second maturation division produces the egg (ovum), and the second polar body, each of which contains two chromosomes. The first polar body may also divide, but all polar bodies soon die and are reabsorbed. Note that the egg, like the sperm, contains only one chromosome of each pair that was present in the primordial germ cell. The gametes each contain one member of each pair of chromosomes that existed in each primordial germ cell.

FERTILIZATION

When a sperm and an egg of our theoretical species unite to start a new life, each contributes one chromosome to each pair of chromosomes of the fertilized egg, now called a zygote. Fertilization is defined as the union of the sperm and the egg along with the establishment of the paired condition of the chromosomes. The zygote is termed **diploid** (diplo means "double") because it has chromosomes in pairs, one member of each pair

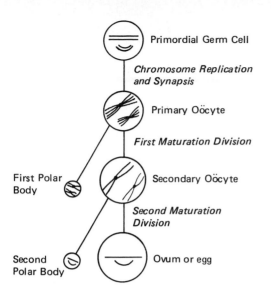

Primordial Germ Cell

Chromosome Replication and Synapsis

Primary Oöcyte

First Maturation Division

First Polar Body

Secondary Oöcyte

Second Maturation Division

Second Polar Body

Ovum or egg

Figure 15-3. Oogenesis in which two pairs of chromosomes are illustrated. Note that only one egg results from a primordial germ cell and that it contains considerable amounts of nutrients.

having come from the sire and one member having come from the dam. Gametes have only one member of each pair of chromosomes; therefore, gametes are termed **haploid** (haplomeans "half"). Gametogenesis thus reduces the number of chromosomes in a cell to half the diploid number. Fertilization reestablishes the diploid state.

DNA, GENES AND CHROMOSOMES

Two members of each typical pair of chromosomes in a cell synapse with each other in meiosis, are alike in size and shape, and carry **genes** that affect the same hereditary characteristics. Such chromosomes are said to be **homologous**. The genes are points of activity found in each of the chromosomes that govern the way in which traits develop. The genes form the coding system that directs enzyme and protein production. Thus, they control the development of traits.

Chromosomes in advanced organisms, such as poultry and livestock, are composed basically of a protein sheath surrounding **deoxyribonucleic acid** (DNA for short). Segments of DNA are the genes.

DNA is itself composed of three parts: deoxyribose sugar, phosphate, and four nitrogenous bases. The combination of deoxyribose, phosphate, and one of the four bases is called a **nucleotide**; when many nucleotides are chemically bonded to one another, they form a strand that composes one-half of the DNA molecule. (A molecule formed by many repeating sections is called a polymer.) Two of these strands wind around each other in a double helix to form the DNA molecule.

The bases of DNA are the parts that hold the key to inheritance. The four bases are adenine (A), thymine (T), guanine (G), and cytosine (C). In the two strands of DNA, A is always complementary to (pairs with) T, and G is always complementary to C (Figure 15-4). During meiosis and mitosis, the chromosomes are replicated by the unwinding and pulling apart of the DNA strands and a new strand is formed alongside the old. The old strand

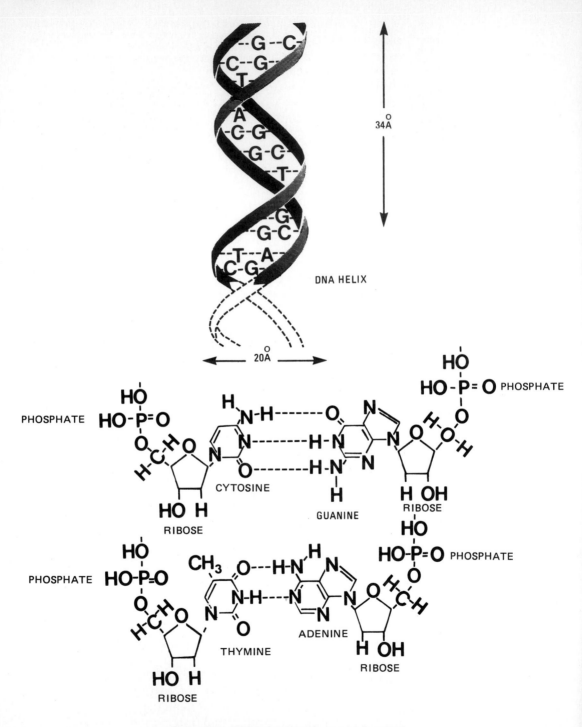

Figure 15-4. DNA helix and structure of nucleotides.

served as a template, so that wherever an A occurs on the old strand, a T will be directly opposite it on the new; and wherever a C occurs on the old strand, a G will be placed on the new. Complementary bases pair with each other until two entire double-stranded molecules are formed where originally there was one.

Nearly all genes code for proteins, which are also polymers. It is important to realize that DNA and protein are both polymers. For each of the 26 amino acids of which proteins are made, there is at least one "triplet" sequence of three nucleotides. For example, two DNA triplets, TTC and TTT, code for the amino acid called lysine; four triplets, CGT, CGA, CGG, and CGC, all code for the amino acid called alanine. If we think of a protein molecule as a word, and amino acids as the letters of the word, each triplet sequence of DNA can be said to code for a letter of the word and the entire encoded message, the series of base triplets, is the gene.

The processes by which the code is "read" and protein synthesized are called **transcription** and translation. To understand these processes, another group of molecules, the **ribonucleic acids** (RNA's), must be introduced. There are three types of RNA: transfer RNA (tRNA), which identifies both an amino acid and a base triplet in mRNA; **messenger RNA (mRNA)**, which carries the information codes for a particular protein; and ribosomal RNA (rRNA) which is essential for ribosome structure and function. All three RNA's are coded by the DNA template.

The first step in protein synthesis is that of transcription. Just as the DNA molecule serves as a template for self-replication using the pairing of specific bases, it can also serve as a template for the mRNA molecule. Messenger RNA is similar to DNA, but is single-stranded rather than double-stranded and has the base uracil (U) in place of the base thymine found in DNA. It is shorter, coding for only one or a few proteins. The triplet sequence that codes for one amino acid in mRNA is called a codon. Through transcription, the encoded message held by the DNA molecule becomes transcribed onto the mRNA molecule. The mRNA then leaves the nucleus and travels to an organelle called a ribosome, where protein synthesis will actually take place. The ribosome is composed of rRNA and protein.

The second step in protein synthesis is the union of amino acids to their respective tRNA molecules. The tRNA molecules are coded by the DNA. They have a structure and contain an anti-codon that is complimentary to an mRNA codon. Each RNA unites with one amino acid. This union is very specific such that, as an example, lysine never links with the tRNA for alanine but only to the tRNA for lysine.

The mRNA attaches to a ribosome for translation of its message into protein. Each triplet codon on the mRNA (which is complementary to one on DNA) associates with a specific tRNA bearing its amino acid, using base-pairing mechanism similar to that found in DNA replication and mRNA transcription. This matching of each tRNA with its specific mRNA triplet begins at one end of the mRNA and continues down its length until all the codons for the protein-forming amino acids are aligned in the proper order. The amino acids are each chemically bonded to each other by so-called peptide bonding as the mRNA moves through the ribosome, and the fully formed protein disassociates from the tRNA-mRNA complex and is ready to fulfill its role as a part of a cell or as an enzyme to direct metabolic processes. Figure 15-5 summarizes the steps of protein synthesis. It depicts the amino acids arginine (arg), leucine (leu), threonine (thr), and valine (val) being moved into

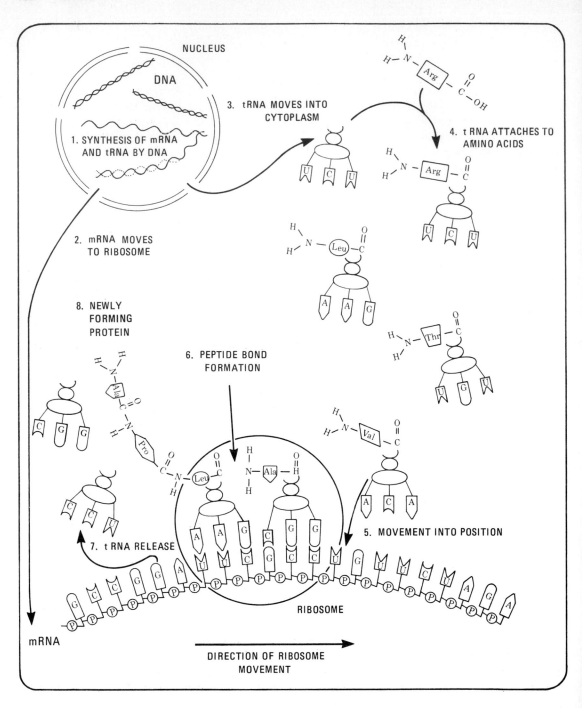

Figure 15-5. Protein synthesis in the cell.

position to join a chain that already includes the amino acids alanine (ala), proline (pro), and leucine (leu).

GENES AND CHROMOSOMES

Because the chromosomes are in pairs, the genes are also in pairs. The location of a gene in a chromosome is called a **locus** (plural, loci). For each locus in one of the members of a pair of homologous chromosomes, a corresponding locus occurs in the other member of that chromosome pair. (A special pair of chromosomes, the so-called "sex chromosomes," exist as a pair in which one of the chromosomes does not correspond entirely to the other in terms of what gene loci are present. For a discussion of sex chromosomes, the reader should consult an introductory genetics text.) The transmission of genes from parents to offspring depends entirely on the transmission of chromosomes from parents to offspring.

The genes located at corresponding loci in **homologous chromosomes** may correspond to each other in the way that they control a trait, or they may contrast. If they correspond, the individual is said to be **homozygous** at that locus (*homo-* means "alike"; *-zygous* refers to the individual); if they differ, the individual is said to be **heterozygous** (*hetero-* means "different"; *-zygous* refers to the individual). Those genes that occupy corresponding loci in homologous chromosomes but affect the same character in different ways are called **alleles**. Genes that are alike and, thus, affect the character developing in the same way are called "identical genes". Some authorities speak of alike genes in homologous chromosomes as identical alleles.

The geneticist usually illustrates the chromosomes as lines and indicates the genes by alphabetical letters. When the genes at corresponding loci on homologous chromosomes differ, one of the genes often overpowers, or dominates, the expression of the other. This allele is called **dominant**. The allele whose expression is prevented is called **recessive**. The dominant allele is symbolized by a capital letter. The recessive allele is symbolized by a lower case letter. For example, in cattle, black hair color is dominant to red hair color, so we let *B* = black and *b* = red. Three combinations of genes are possible:

$$\dashv B \dashv B \qquad \dashv B \dashv b \qquad \dashv b \dashv b$$

Both **BB** animals and **bb** animals are homozygous for the genes that determine hair color, but one is homozygous dominant (**BB**) and the other is homozygous recessive (**bb**). The animal that is **Bb** is heterozygous; it has allelic genes.

THE SIX FUNDAMENTAL TYPES OF MATING

With three kinds of individuals (homozygous dominant, heterozygous, and homozygous recessive) and one pair of genes considered, six types of matings are possible. Using the genes designated as *B* = black and *b* = red in cattle, the six mating possibilities are: *BB* × *BB*; *BB* × *Bb*; *BB* × *bb*; *Bb* × *Bb*; *Bb* × *bb*; and *bb* × *bb*. Keep in mind that the discussion based on these genes is applicable to any other pair of genes in any species.

Homozygous dominant × *homozygous dominant* (*BB* × *BB*). When two homozygous dominant individuals are mated, each of them can produce only one kind of gamete, namely, the gamete carrying the dominant gene. In the particular example that we are using, this gamete carries gene *B*. The union of gametes from two homozygous dominant parents

Figure 15-6. Mating of homozygous dominant × homozygous dominant.

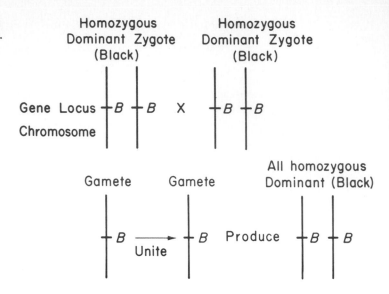

results in a zygote that is homozygous dominant (**B** × **B** = **BB**). Thus, homozygous dominant parents produce only homozygous dominant offspring (Figure 15-6).

Homozygous dominant × *heterozygous* (**BB** × **Bb**). The mating of a homozygous dominant with a heterozygous individual results in an expected ratio of 1 homozygous dominant: 1 heterozygous. The homozygous dominant parent produces only one kind of **gamete**, the one carrying the dominant gene (**B**). The heterozygous parent produces two kinds of gametes, one carrying the dominant gene (**B**) and one carrying the recessive gene (**b**). The latter two kinds of gametes are produced in approximately equal numbers. The chances that a gamete from the parent producing only the one kind of gamete (the one having the dominant gene) will unite with each of the two kinds of gametes produced by the heterozygous parent are equal; therefore, the number of homozygous dominant offspring (**B** × **B** = **BB**) and heterozygous offspring (**B** × **b** = **Bb**) produced should be approximately equal (Figure 15-7).

Homozygous dominant × *homozygous recessive* (**BB** × **bb**). The homozygous dominant individual can produce only gametes carrying the dominant gene (**B**). The recessive individual must be homozygous recessive to show the recessive trait; therefore, it produces only gametes carrying the recessive gene (**b**). When the two kinds of gametes unite, all offspring produced receive both the dominant and the recessive gene (**B** × **b** = **Bb**) and are thus all heterozygous (Figure 15-8).

Heterozygous × *heterozygous* (**Bb** × **Bb**). Each of the two heterozygous parents produces two kinds of gametes in approximately equal ratios: one kind of gamete carries the dominant gene (**B**); the other kind carries the recessive gene (**b**). The two kinds of gametes produced by one parent each have an equal chance of uniting with each of the two kinds of gametes produced by the other parent. Thus, four equal-chance unions of gametes are possible. If the gamete carrying the **B** gene from one parent unites with the gamete carrying the **B** gene from the other parent, the offspring produced are homozygous dominant (**B** × **B** = **BB**). If the gamete carrying the **B** gene from one parent unites with the

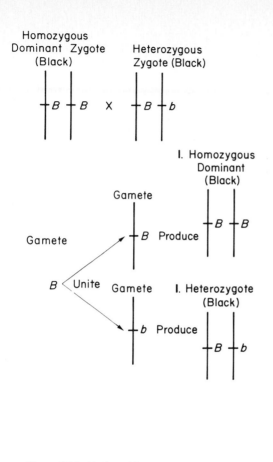

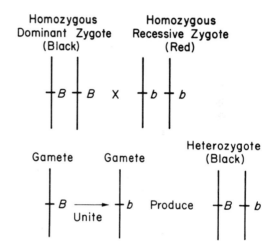

Figure 15-7. Mating of homozygous dominant X heterozygote.

Figure 15-8. Mating of homozygous dominant X homozygous recessive.

gamete carrying the **b** gene from the other parent, the offspring are heterozygous (**B** X **b** = **Bb**). The latter combination of genes can occur in two ways: the **B** gene can come from the male parent and the **b** gene from the female parent, or the **B** gene can come from the female parent and the **b** gene from the male parent. When the gametes carrying the **b** gene from both parents unite, the offspring produced are recessive (**b** X **b** = **bb**).

The total expected ratio among the offspring of two heterozygous parents is 1**BB**:2**Bb**:1**bb** (Figure 15-9). This 1:2:1 ratio is the genetic, or *genotypic*, ratio. The appearance of the offspring is in the ratio of 3 dominant: 1 recessive because the **BB** and **Bb** animals are all black (dominant), and cannot be genetically distinguished from one another by looking at them. This 3:1 ratio, based on external appearance, is called the *phenotypic* ratio.

Heterozygous X homozygous recessive (**Bb** X **bb**). The heterozygous individual produces two kinds of gametes, one carrying the dominant gene (**B**) and the other carrying the recessive gene (**b**), in approximately equal numbers. The recessive individual produces only the gametes carrying the recessive gene (**b**). There is an equal chance that the two kinds of gametes produced by the heterozygous parent will unite with the one kind of

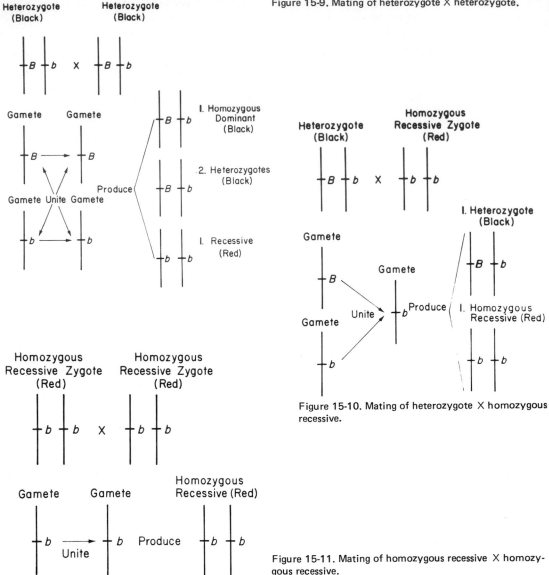

Figure 15-9. Mating of heterozygote × heterozygote.

Figure 15-10. Mating of heterozygote × homozygous recessive.

Figure 15-11. Mating of homozygous recessive × homozygous recessive.

gamete produced by the recessive parent (**B × b = Bb**; **b × b = bb**). The offspring produced when these gametes unite thus occur in the expected ratio of 1 heterozygous: 1 homozygous recessive (Figure 15-10).

Homozygous recessive × homozygous recessive (**bb × bb**). The recessive individuals are homozygous; therefore, they can produce gametes carrying only the recessive gene (**b**). When these gametes unite (**b × b = bb**), all offspring produced will be recessive (Figure 15-11). This example illustrates the principle that recessives, when mated together, breed true.

The knowledge of what results from each of the six fundamental types of matings provides a background for understanding complex crosses. When more than one pair of genes are considered in a mating, one can understand the expected results because one can combine each of the combinations of one pair of genes with each of the other combinations of one pair of genes to obtain the expected ratios.

PARENTAL AND FILIAL GENERATIONS

In genetics research, the dominant is often mated with the recessive, after which the offspring are intermated. If we use the same genes as before (*B* = black and *b* = red), the mating sequence is:

First parental (P$_1$) generation (or P$_1$ zygotes)	*BB* × *bb*
P$_1$ gametes	*B* *b*
First filial (F$_1$) generation (or F$_1$ zygotes)	*Bb*
F$_1$ zygotes	*Bb* × *Bb*
F$_1$ gametes	*B, b, B, b*
F$_2$ zygotes	*BB, Bb, Bb, bb*
	3 black 1 red

MORE THAN ONE PAIR OF GENES

Suppose that there are two pairs of genes, each pair independently affecting a particular trait, to be considered. Let us consider two pairs of genes, one pair of which determines coat color in cattle, and the second of which determines whether the animal is hornless (polled) or horned. The genes are designated as follows: *B* = black (dominant), *b* = red (recessive), *P* = polled (dominant), and *p* = horned (recessive).

If a bull that is heterozygous for both traits (*BbPp*) is mated to cows that are also heterozygous for both traits (*BbPp*), one can determine the expected phenotypic and genotypic ratios. We have already shown that a cross of *Bb* × *Bb* gives a ratio of 3 black:1 red in the offspring, or 3/4 black and 1/4 red. Similarly, a cross of *Pp* × *Pp* gives a ratio of 3 polled:1 horned. Combined, the *BbPp* × *BbPp* produces the following phenotypes:

BbPp × *BbPp*

3 black
- 3 polled   =   9 black polled
- 1 horned   =   3 black horned

1 red
- 3 polled   =   3 red polled
- 1 horned   =   1 red horned

Of the 3/4 which are black, 3/4 will be polled and 1/4 will be horned. Therefore, 3/4 of 3/4 or 9/16 will be black and polled; and 1/4 of 3/4, or 3/16, will be black and horned. Similarly, of the 1/4 which are red, 3/4 will be polled and 1/4 will be horned, and 3/4 of

1/4, or 3/16, will be red and polled and 1/4 of 1/4, or 1/16, will be red and horned. This distribution of phenotypes is more often expressed as a 9:3:3:1 ratio instead of a set of fractions:

$$1 \; BB \quad \begin{cases} 1 \; PP & = & 1 \; BBPP \\ 2 \; Pp & = & 2 \; BBPp \\ 1 \; pp & = & 1 \; BBpp \end{cases}$$

$$2 \; Bb \quad \begin{cases} 1 \; PP & = & 2 \; BbPP \\ 2 \; Pp & = & 4 \; BbPp \\ 1 \; pp & = & 2 \; Bbpp \end{cases}$$

$$1 \; bb \quad \begin{cases} 1 \; PP & = & 1 \; bbPP \\ 2 \; Pp & = & 2 \; bbPp \\ 1 \; pp & = & 1 \; bbpp \end{cases}$$

GENE INTERACTIONS

A gene may interact with other genes in the same chromosome (linear interaction), with its corresponding gene in a homologous chromosome (allelic interaction), or with genes in nonhomologous chromosomes (epistatic interaction). In addition, genes interact with the cytoplasm and with the environment. The environmental factors with which genes interact are internal, such as hormones, and external, such as nutrition, temperature, and amount of light.

Linear Interactions

The order in which genes occur in the chromosome is significant, because a reversal or alteration of that order may produce an entirely different phenotype. It has been demonstrated in the common fruit fly (*Drosophila*) that different phenotypes can result after the arrangement of genes on chromosomes has been altered; this has not been demonstrated in farm animals, perhaps because little is known about the location of genes in their chromosomes.

Allelic Interactions

When contrasting genes occupy corresponding loci in homologous chromosomes, each gene exerts its influence on the trait, but the effects of each of the genes depend on the relationship of dominance and recessiveness. Allelic interactions may also be termed dominance interactions. When unlike genes occupy corresponding loci, one might show complete dominance over the other. In this situation, only the effect of the dominant gene is expressed. A good example is provided by cattle, in which the hornless (polled) condition is dominant to the horned condition. The heterozygous animal is polled and is phenotypically indistinguishable from the homozygous polled animal.

There may be lack of dominance, in which heterozygous animals show a phenotype which is different from either homozygous phenotype and is usually intermediate between them. A good example is observed in sheep, in which two alleles for ear length lack dominance to each other. Sheep which are **LL** have long ears, **Ll** have short ears and **ll** are earless. Thus the heterozygous **Ll** sheep are different from both homozygotes and are intermediate in phenotype between them.

Sometimes heterozygous individuals are superior to either of the homozygotes. This condition, in which the heterozygotes are said to show overdominance, is an example of a selective advantage for the heterozygous condition. "Overdominance" means that heterozygotes possess greater vigor or are more desirable in other ways (such as producing more milk or being more fertile) than either of the two homozygotes that produced the heterozygote. Because the heterozygotes of breed crosses are more vigorous than the straightbred parents, they are said to possess **heterosis**. This greater vigor or productivity of crossbreds is also said to be an expression of hybrid vigor.

The three types of dominance can be illustrated by lines to show the possible expression of a trait:

┼ *AA, Aa*	┼ *AA*	┼ *Aa*
	┼ *Aa*	┼ *AA*
┼ *aa*	┼ *aa*	┼ *aa*
Complete dominance	Lack of dominance	Overdominance

These illustrations show that when complete dominance exists, the homozygous and heterozygous dominant individuals express the same phenotype: a phenotype that is, in fact, quite different from the one expressed by the homozygous recessive individual. With lack of dominance, the phenotypes of the two homozygotes are quite different from each other and the phenotype of the heterozygote is intermediate between the phenotypes of the two homozygotes. With overdominance, the phenotype of the overdominant heterozygote is superior to either of the two homozygotes, and one of the homozygotes is superior to the other.

The effects of dominance on the expression of traits that are important in livestock production ("production traits," such as fertility, milk production, growth rate, feed conversion efficiency, carcass merit, and freedom from inherited defects) can be illustrated by bar graphs (Figure 15-12). Figure 15-12A, which illustrates complete dominance, shows

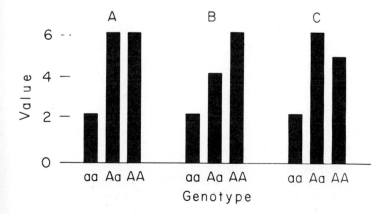

Figure 15-12. Bar graphs illustrating: **(A)** Complete dominance; **(B)** Lack of dominance; **(C)** Overdominance.

equal performance of the homozygous dominant and heterozygous individuals, both of which are quite superior to the performance of the recessive individual.

With lack of dominance, one homozygote is superior, the heterozygote is intermediate, and the other homozygote is inferior (Figure 15-12B). When traits are controlled by genes that show the effects of overdominance, the heterozygote is superior to either of the homozygous types (Figure 15-12C).

Allelic interactions play a role in determing heritability in situations in which many pairs of genes interact. We can best explan heritability by describing realized heritability, which is the portion of what one attempts to obtain through selection that is actually obtained. To illustrate the concept of **realized heritability**, let us suppose that a farmer has a herd of beef cattle that gain, on the average, 2.0 lbs per head per day. If the farmer selects from this original population a breeding herd whose members gain, on the average, 2.5 lbs per head per day, the farmer is selecting for an increase in daily gain of 0.5 lb per head. Suppose that the calves produced by members of the selected herd gain an average of 2.3 lbs per head per day. An increase of 0.3 lb over the amount gained per head per day by members of the original herd has thus been obtained. To find the portion obtained of what was reached for in the selection, one divides 0.3 by 0.5, which equals 0.6. This figure is the realized heritability. If one multiplies 0.6 by 100%, the result is 60%, which is the percentage obtained of what was selected for in this generation.

The practice of **selection** (differentially reproducing what one desires) is vital in modern livestock operations. Through selection, certain animals are allowed to produce more offspring than others, thereby influencing the characteristics of future generations. For example, if two animals are kept, one of which leaves 10 offspring in the herd and the other which leaves only 5, the former animal has been given double the selection pressure that was applied to the latter animal. If that animal leaves no offspring, it has been culled, that is, eliminated from contributing inheritance to the population.

The type of dominance that determines the expression of a trait and the frequency of the two alleles for that trait in the herd affect heritability. When there is a lack of dominance, the three genotypes **AA**, **Aa**, and **aa** can all be distinguished; therefore, selection is highly effective because the heritability is high. Both complete dominance and overdominance result in a high heritability when the frequency of **A** is low, because **AA** and **Aa** can be distinguished from **aa**. As the frequency of **A** approaches 0.5, selection for traits controlled by complete dominance diminishes greatly and selection for traits controlled by overdominance effects becomes zero. Selection for traits controlled by genes having overdominance effects is for the heterozygous condition; therefore, selection becomes totally ineffective when the frequency of **A** is 0.5. Selection is modestly effective for traits controlled by genes having complete dominance because the animals that are recessive can be prevented from reproducing (culled).

Epistatic Interactions

A gene or a pair of genes in one pair of chromosomes may alter or mask the expression of genes in other chromosomes. A gene that interacts with other genes that are not allelic to it is said to be **epistatic** to them. In some cases epistatic effects result when one gene or a pair of genes influences the expression of one gene or a pair of genes in another pair of chromosomes. In other cases a gene or one pair of genes in one pair of chromosomes may influence many other genes in many pairs of chromosomes. An example in which a pair of recessive genes alters the expression of many other genes is dwarfism in cattle. A dwarf animal may possess genes for rapid and economical gains and desirable carcass characteristics, but, because the animal is homozygous for dwarfism, none of these genes fully express themselves. We noted in previous discussions that when an animal that is heterozygous for two traits is mated to another animal that is heterozygous for two traits, offspring are produced in the phenotypic ratio of 9:3:3:1. The derivation of this ratio is illustrated in the following example, which uses the genes B = black, b = red, P = polled, and p = horned. When **BbPp** animals are intermated the results are as follows:

$$BbPp \times BbPp$$

3 black
- 3 polled   =   9 black polled
- 1 horned   =   3 black horned

1 red
- 3 polled   =   3 red polled
- 1 horned   =   1 red horned

Let us explore some examples in which the expression of certain genes is altered by **epistatic** genes.

One dominant epistatic gene. In cattle, B = black, b = red, Bs = blackish, and bs = nonblackish. Blackish (often called mahogany) is the dark color that is observed in Jersey and Ayrshire cattle. The gene for black, acting as a dominant epistatic gene, prevents the **Bs** and **bs** genes that are present from expressing themselves:

$$BbBsbs \times BbBsbs$$

3 black
- 3 blackish
- 1 nonblackish   =   12 black

1 red
- 3 blackish   =   3 blackish
- 1 nonblackish   =   1 red

One of the groups of 3 has been placed along with the group of 9 to give 12 black because the black epistatic gene prevents the 9 black individuals that carry the **Bs** gene from being phenotypically different from the 3 nonblackish individuals that have **bsbs**.

A pair of recessive epistatic genes. In rabbits, B = black, b = chocolate, C = color, and c = albino (white). An albino is white because there is no pigment in the skin or hair. It has pink eyes because there is no pigment in the eyes and the blood vessels in the eyes therefore give them a pink appearance:

$$CcBb \times CcBb$$

3 colored
- 3 black   =   9 black
- 1 chocolate   =   3 chocolate

1 albino
- 3 black
- 1 chocolate   =   4 albino

The effect of the homozygous combination of *cc* genes is to remove the pigment from both those individuals that would otherwise be black and those that would otherwise be chocolate. Thus the *cc* genes act as a pair of recessive epistatic genes. One group of 3 and 1 individual have the same (albino) **phenotype**. Therefore, they are combined into a group of 4 albinos out of the 16.

Dominant and recessive epistatic genes. In chickens, *I* = inhibition of color or white that is found in Leghorns, *i* = noninhibited (colored), *W* = colored, and *w* = white as is present in White Rock chickens. *I* is epistatic to *W* and *ww* is epistatic to *ii*:

$$IiWw \times IiWw$$ (Both of these heterozygous genotypes result in a white phenotype.)

3 inhibited $\left\{ \begin{array}{l} 3 \text{ colored} \\ \\ 1 \text{ white} \end{array} \right.$ 12 $\left. \begin{array}{l} \\ \\ \end{array} \right\}$ = 13 white

1 noninhibited $\left\{ \begin{array}{l} 1 \text{ white} \\ \\ 3 \text{ colored} \end{array} \right.$ = 3 colored

The group of 9, one group of 3, and 1 individual are white. Therefore, they are combined to give 13 white. The resulting ratio is 13 white:3 colored.

Two pairs of recessive epistatic genes. In White Rock chickens, *C* = color, *c* = albino, *W* = color, and *w* = white. Homozygous *cc* prevents *W* from showing and *ww* prevents *C* from showing:

$$CcWw \times CcWw$$ (Both of these heterozygous genotypes result in a colored phenotype.)

3 colored $\left\{ \begin{array}{ll} 3 \text{ colored} & 9 \\ \\ 1 \text{ white} & 3 \end{array} \right.$ = 9 colored

1 white $\left\{ \begin{array}{ll} 3 \text{ colored} & 3 \\ \\ 1 \text{ white} & 1 \end{array} \right.$ = 7 white

Some *WW* and *Ww* chickens are albinos, and they and the albino White Rock *ww* have no pigment in the eyes. The *CC* or *Cc* chickens that are *ww* are true White Rocks, having pigment in the eyes.

Interactions Between Genes and Environment

Genes interact with both the external and the internal environments. The external environment includes temperature, light, altitude, humidity, and feed supply. Some breeds of cattle (the Brahmans) can withstand high temperatures better than others. Some breeds (Scottish Highland) can withstand the rigors of extreme cold better than others.

Perhaps the most important external environmental factor is feed supply. Some breeds of cattle can survive when feed is in short supply for considerable periods of time, and they may consume almost anything that can be eaten. Other breeds of cattle select only the relatively good feeds and are severely affected when good feed is not available. Generally, when plenty of good feed is available, the animals that select the good forages gain more rapidly than the animals that eat any type of forage and survive when good forages are not available.

An important contributor to the status of the internal environment of animals is their hormonal condition. When the gene for blackish described previously is in the heterozygous condition (***Bsbs***), its expression depends on the amount of androgens present. Thus, heterozygous bulls are dark mahogany color whereas heterozygous females are light mahogany. It has been shown that castration of heterozygous bulls will prevent them from being dark and that injections of testosterone into heterozygous females will cause them to become dark mahogany. Homozygous (***BsBs***) blackish animals are dark regardless of androgen levels; therefore, both males and females that are ***BsBs*** are dark. Both males and females that are homozygous recessive (***bsbs***) lack the gene ***Bs***; therefore, both are red in color.

Allelic, epistatic, and environmental interactions all influence heritability and, as a consequence, all influence the progress that can be made through selection. When the external environment has a large effect on production traits, **heritability** is quite low. For example, if animals in a population are maintained at different nutritional levels, those that are best fed obviously grow faster. In such a case, much of the difference in growth may be due to the nutritional status of the animals rather than to differences in their genetic composition. The producer has two alternatives; either standardize the environment so that it causes less variation among the animals, or maximize the expression of the production trait by improving the environment. The first approach is geared at increasing heritability so that selection is more effective. Improvements made by this approach are permanent. The latter approach does nothing to improve the genetic qualities of the animals, but it may do a great deal to bring about immediate increases in productivity.

STUDY QUESTIONS

1. Diagram spermatogenesis and spermiogenesis using one pair of chromosomes and label the following: primary spermatocyte, secondary spermatocyte, spermatid, sperm, first maturation division, and second maturation division.
2. How does oogenesis differ from spermatogenesis?
3. In cattle ***B*** = black and ***b*** = red. A heterozygous (***Bb***) bull is mated to a herd of heterozygous (***Bb***) cows. What genotypic ratio would you expect among the offspring? What phenotypic ratio would you expect?
4. In cattle, ***B*** = black, ***b*** = red, ***P*** = polled, and ***p*** = horned. A black, polled bull is bred to a black, polled cow and produces a red, horned calf. What is the genotype of the bull, the cow, and the calf?
5. You should learn six fundamental types of matings that can be made when one pair of genes is involved.
6. a. Explain how DNA codes for the production of protein.
   b. What other function besides comprising part of the animal body do proteins serve?
7. Diagram the types of dominance.
8. What is meant by "epistatic"?
9. If a trait is controlled largely by genes that lack dominance, how effective will selection be? Why?

10. Suppose a trait is controlled largely by genes that show overdominance and the frequency of A and a is 0.5 for each. How effective will selection be?
11. What is meant by heritability?
12. Give an example in cattle in which one gene in the homozygous recessive condition acts epistatically to most of the important genes that control production traits.
13. Are all genes equally affected by the environment?
14. How would you breed animals to obtain maximum productivity when most of the production traits are governed by genes showing overdominance effects?

SELECTED REFERENCES

Gardner, E. J. and Snustead, D. P. 1981. *Principles of Genetics.* 6th edition. New York: John Wiley and Sons.

Hartl, D. L. 1980. *Principles of Population Genetics.* Sutherland, Massachusetts: Sinauer Associates, Inc. Publishers.

Lasley, J. F. 1978. *Genetics of Livestock Improvement.* 3rd edition. Englewood Cliffs, New Jersey: Prentice-Hall.

Strichberger, M. W. 1976. *Genetics.* 2nd edition. New York: Macmillan Publishing Co., Inc.

Suzuki, A., Griffiths, J. F. and Lewoltin, R. C. 1981. *Introduction to Genetic Analysis.* 2nd edition. San Francisco: W. H. Freeman.

16

Selection

Selection, the process of retaining animals that show superiority for the traits we want and differentially reproducing these superior animals, is an important part of modern livestock programs.

SELECTION USING ONLY ONE TRAIT

The single trait likely to give the greatest total genetic improvement as a result of selection varies with the species and the desire of the producer—examples are weight at 18 months for beef cattle; amount of milk produced for dairy cattle; 100-day weight for sheep; and 154-day total litter weight for swine. Using these single traits has risks, however, because of the possible correlation or lack of correlation with other characters. For example, selection for weight of beef cattle at 18 months should tend to increase the rate of gain prior to and following weaning, the efficiency with which feed is converted into body tissue, and the leanness of the carcass, but will not decrease the occurrence of unfavorable traits that are inherited independently, such as unsound feet and legs and defects of the reproductive tract, mouth or eyes.

Selecting dairy cattle only on the basis of amount of milk produced does not prevent a decline in milk quality or an increase in udder breakdown, nor does it prevent the occurrence of unsound feet and legs when cows are closely confined on hard floors (concrete). Selecting sheep on the basis of 100-day weight does not assure that they are sufficiently finished to make good carcasses and it does not improve wool quantity or quality. Selecting swine only on the basis of 154-day total litter weight does not assure that those produced will have a high percentage of lean in the carcass, and no selection is made against abnormalities that do not interfere with rapid growth.

SELECTION BASED ON CONSIDERATION OF ALL IMPORTANT TRAITS

The usual procedure in selecting farm animals is to consider all of the traits that are economically important. Any unimportant trait should not be included when selecting animals, because the inclusion of an unimportant trait reduces the amount of selection pressure that can be applied to the important traits. Traits that will be important in the future should be considered as well as those that are important at present.

In selecting beef cattle, sheep, and swine, the primary objective is to produce meat of high quality rapidly and economically, because lean meat will always be valuable as a nutritional source for humans. Thus, in these three species, fertility, mothering or milking ability, rate of growth, feed efficiency, carcass merit, and freedom from inherited defects are all economically important and will likely be important in the future. Perhaps the most important consideration in selecting all farm animals in the future will be the rate and efficiency of protein production.

When several traits must be considered, it is difficult to compare individual animals. Those animals that possess the highest over-all merit are desirable for selection. Three methods of selection are used when several traits are considered; these are tandem method, minimum culling method, and index method.

The Tandem Method

Suppose six traits are pertinent to a particular program of selection. Selection for these six traits can be done by first selecting for only one trait for a generation or two, then selecting for a second trait for a generation or two, and then subsequently selecting for a third, fourth, fifth, and sixth trait in successive generations. This method of selection is called the **tandem method**. It is rather ineffective, particularly if more than two traits are selected for, or if any of the traits selected for are associated with each other in such a way that the desirable aspect of one trait tends to be associated with the undesirable aspect of another. Perhaps this method can be considered so ineffective in most conditions that it should not be used. If, however, traits are genetically associated such that the desirable aspect of one is associated with the desirable aspect of another, selection for either trait improves both.

The Minimum Culling Method

The basis of the **minimum culling method** is to set a minimum level for each trait considered that an animal must meet if that animal is to be kept for breeding. It is most useful when the number of traits being considered is small and when only a small percentage of the offspring is needed to replace the parents. For example, with poultry, 100 or more chicks might be produced from each hen and 1,500 or more chicks from each rooster. Here only 1 female out of 50 or 1 male out of 750 is needed to replace the parents. It is simple using this method to select from the upper 10% of the females and the upper 1% of the males.

Errors in selection for animals such as beef cattle can be made when using the minimum culling method, because a considerable proportion of the offspring is needed for replacement. Let us take an example, with three bulls, of the type of error that can occur when the minimum culling method is used. Suppose that the minimum levels of performance to be met are:

Suckling gain (gain made by young during   = 2.0 lbs/per day
    nursing period)

Feed test gain (gain made by animal during feed test) = 3.0 lbs/per day

Feed per unit of gain (amount of feed consumed per unit of body weight gained) =6.5 lbs/per lb gain

Score for carcass (15 = perfect, 10 = average, and 1 = cull) = 12

The score for carcass is based on an examination of living animals and is derived from the viewer's subjective opinion of how good a carcass the animal would make. A score of 15 means that the viewer thinks that the animal's carcass would be perfect; 10 indicates average; and 1 indicates a carcass that no one would want. A carcass rated at 12 is desirable but not perfect.

When the performances of the three bulls in our example are compared with the minimum levels sought, the following comparisons are seen:

Trait	Minimum	Bull A	Bull B	Bull C
Suckling gain	2.0	1.9	2.0	2.8
Feed test gain	3.0	3.9	3.1	3.6
Feed per unit gain	6.5	5.0	6.4	5.3
Score for carcass	12.0	14.5	12.1	11.9

Using the minimum culling method, bulls A and C would be culled because A is below minimum in suckling gain and C is below minimum in score for carcass merit. However, bull A is superior to bull B in feed test gain, feed per unit of gain, and score for carcass, and bull C is superior to bull B in feed test gain, suckling gain, and feed per unit of gain.

The minimum culling method does have a management advantage in that one can eliminate by castration or sale a certain portion of the animals for a particular trait as soon as that trait has been measured. Because steers sell at higher prices than bulls, producers typically castrate most young bulls at weaning time. This procedure is subject to a hazard, however, because suckling gain does not tell whether the calves will gain rapidly or efficiently thereafter or whether they will make desirable carcasses; heavy culling by castration at weaning prevents assured selection of the best overall animals. If 80% of all male calves are castrated, only 20% are left to select from in terms of feed-test gain, feed per unit of gain, and score for carcass value. Thus, selection pressure applied to suckling gain is far greater than that which can now be applied to the other three traits.

The Index Method

In the **index method** animals are rated numerically in comparison to each other for each of the traits that are under consideration. For each trait, the animal that performs best is assigned the number 1, the animal that performs second best is assigned the number 2 and so on. The animals in the sample are numbered in this way so the highest number is assigned to the animal that performs the worst. If five animals are compared with each other for a given trait, (as in the following example) the animal that performs worst is assigned a 5. The animals are each rated for all traits, and the total score achieved by each animal is then calculated. The animal that attains the lowest total score is the best. The animal that attains the highest total score is the worst. Let us examine the records of five bull calves to illustrate how the index method is used.

Animal	Suckling gain	Test gain	Feed per unit gain	Score for carcass	Total (based on numerical ranking for each trait)
A	2.5	3.6	5.0	14.0	
Rating	1	2	2	3	8
B	1.9	3.4	5.6	14.1	
Rating	5	3	4	2	14
C	2.3	3.2	5.5	12.0	
Rating	3	4	3	5	15
D	2.4	3.7	4.8	14.2	
Rating	2	1	1	1	5
E	2.0	3.0	6.0	13.6	
Rating	4	5	5	4	18

Bull D has the lowest total; therefore, he is the best of the group. Bull E has the highest total; therefore, he is the worst of the group. Errors in comparing animals with each other through the index method can occur in situations in which large differences between individuals exist for one trait but only minor differences exist between them for other traits. This rating method does not allow one to discriminate between large and small differences between animals for any one trait.

In the example just given in which five bulls were compared through the index method, each of the four traits considered was given the same emphasis. If there is evidence from heritabilities of traits that more genetic progress can be made if a particular trait is emphasized to a greater extent than others, this emphasis can be achieved by multiplying the rating of that trait by the proper amount. For example, if one trait has double the heritability of each of the other three, one can multiply the ratings for that trait by 2. Also, evidence may suggest that one trait may be much more important economically. Again, that trait can be emphasized by multiplying ratings for that trait by the appropriate amount. If, for example, an improvement in suckling gain will result in twice the income that a similar improvement in each of the other traits will yield, the value of suckling gain for each animal can be multiplied by 2.

The best index method, called the overall merit index, is computed with the traits combined into a listing of overall merit based on the product of heritability and the economic importance of each of the traits. It is complicated, and is calculated by electronic computers. Most breed associations are developing services for breeders so that data can be sent to the associations and animals can be indexed.

SELECTING FOR COMBINING ABILITY

General combining ability is determined by the way in which a particular line or herd combines with a large array of other herds, lines or populations. In general, breeding a line or herd with other herds, lines, or populations is a method of determining, by progeny testing, which line is most homozygous for incompletely dominant genes. One can apply selection to a particular herd by mating males from that herd to females from several other herds and using those males for breeding whose offspring show the highest performance. This method of selection, called **recurrent selection**, is effective in developing a line that will combine well with most other herds.

Specific combining ability is determined by how offspring from specific lines perform when lines are combined with one another. The objective in combining two lines with each other as a selection method is to obtain two herds, lines, or populations that will, when combined, produce offspring that give outstanding performance. The combining of two lines with each other has been used effectively as a method of selection for poultry.

Suppose that there are two populations, A and B, and that the objective is to develop these two populations through selection so that when they are combined they will produce outstanding offspring. Males of population A are mated to females of population B and males of population B are mated to females of population A. Males of population A that sired the best offspring when mated to females of population B and females of population A that produced the best offspring when mated to males of population B are intermated to perpetuate population A. Similarly, males of population B that sired the best offspring when mated to females of population A and females of population B that produced the best offspring when mated to males of population A are intermated to perpetuate population B. None of the A × B offspring are used for breeding. After several generations, there are two populations, A and B, which will produce outstanding offspring when crossed because they have been selected for their favorable crossing ability. The two populations may themselves be maintained through mating within each population. This method of selection, called **reciprocal recurrent selection,** is one in which one generation is used for testing and another generation is then used to perpetuate the populations. It is a method of progeny testing because the animals needed to perpetuate the lines are selected on the basis of performance of their offspring.

Details of selection within each species are presented in subsequent chapters in which breeding, feeding, and management are discussed.

SOME USEFUL TOOLS IN SELECTION

The person who is breeding farm animals can use the following tools, or aids, in selection: appearance, pedigree, production records, progeny tests, and family records. Each of these tools has a place and each has certain strengths and weaknesses. Let us evaluate these tools to show where they have value and where they might be of no value.

Appearance

The appearance of an animal being considered for selection is important in (1) telling whether or not the animal has any inherited anatomical abnormalities, and in (2) providing an estimate of the merit of the carcass that could be made from the animal. Any animal that has such inherited abnormalities as crooked legs, undershot or overshot jaws, **cryptorchidism** (one testicle retained in the abdomen), or **hernia** (rupture in either the naval region or the scrotal region) should not be kept for breeding. Environmentally caused abnormalities (such as loss of tail and bad wire cuts) that do not interfere with the animal's breeding should not be a cause for culling the animal. The use of appearance of the live animal for estimating the merit of the carcass the animal would make has been discussed in Chapter 6.

Pedigree

A **pedigree** is the record of the ancestry of an animal. It is most useful as a guide to selection when the traits considered are low to moderate in their degree of heritability. Selection on the basis of production records is most effective for highly heritable traits. Unless production records of the ancestors of the animal are included with the pedigree, the pedigree tells very little about its genetic merit. The pedigree provides information on the relationship of an animal's parents, from which its extent of inbreeding can be determined. If two animals are of equal merit and one is inbred whereas the other is outbred, the inbred animal is more likely to transmit its phenotype than the other.

Some producers emphasize the occurrence in a pedigree of an ancestor that has placed high in livestock shows. The occurrence of such an ancestor in a pedigree may or may not be an indication of an animal's genetic value for show standards. Some great show animals are produced by fortunate "**nicking**" in the mating that was made (meaning that the mating resulted in an outstanding offspring). Such show animals are unlikely to transmit their desirable conformation to their offspring because nicking usually results from heterozygosity that is produced in obtaining an outstanding animal.

When production records concerning the ancestors are available, the pedigree becomes more valuable. However, certain considerations must be taken into account if accurate evaluations based on such a pedigree are to be made. Some pedigrees have records of some of the ancestors but not all of them. Unless it is known why some records are missing, the logical conclusion is that the performance of certain animals was so poor that their records were not entered. Also, outstanding records on only a few ancestors that lived six or eight generations ago are poor evidence that the animal in question will inherit outstanding traits. If, however, an individual ancestor has a good record and this individual occurs several times in the pedigree, the chances are good that the animal being considered has inherited some genetic material from the outstanding ancestor. In general, the most attention should be given to records of parents and grandparents. Another important consideration is that low records are as likely to be transmitted to the animal under consideration as are high records.

The value of records in a pedigree depends on the accuracy with which they were recorded and on the integrity of the person who kept the records. Most breeders of purebred stock are highly honest. Some honest mistakes will be made, but not many. Unfortunately, there are a few people in the purebred business who think of the pedigree only as a marketing gimmick, and they make the pedigree as appealing as possible.

Production Records

Production records are useful for selection when traits are moderate to high in their degree of heritability, and little progress will be made from selection based on production records of traits that are low in their degree of heritability. Production records can be influenced to make them appear better than they actually are. For example, if a calf is weaned and put on a roughage ration for a period of 2 to 3 months, it will make a rather high rate of gain and will be quite efficient in feed conversion when it is placed on a feed-testing program because it is compensating for the austerity it previously experienced. Attention should be given to weight-per-day of age (the weight that an animal has for each day that it has lived: weight-per-day of age = $\dfrac{\text{weight}}{\text{age in days}}$ as well as to rate of gain during the feed-testing program. Even though an animal has a good rate of gain during the feed-testing program, it is not highly desirable if its weight-per-day of age is low. The accuracy with which records are established and the honesty of those who obtain records are important considerations. For traits that are moderately to highly heritable, selection based on production records will give more favorable results than selection on any other basis. The breed associations for most farm animals have record of performance programs and will make the computations that are needed for the breeder to select the most productive animals.

The evaluation of an animal on the basis of its own record is called **performance testing.** The evaluation of an animal on the basis of the performance of its offspring is called **progeny testing**. It is important to differentiate clearly between these two methods of evaluation.

Progeny testing is most useful in species in which the trait being tested is expressed in only one sex, such as milk production by dairy cows and goats or egg production by poultry. Progeny testing is also useful in evaluating carcass merit. It is the most accurate method of evaluating an animal's breeding value. One must keep in mind that even though progeny testing is highly accurate, it may not be the most useful method; when, for example, traits are highly heritable, more progress can be made from selection based on the individual's record than by progeny testing because of the long time required for progeny testing.

To progeny test, the animal must be grown to breeding age and size, bred to several animals, after which it must wait for these animals to have young, and obtain records on the young. Thus, two generations could be produced by selecting an individual on the basis of its own record in the same time that is required to produce one generation through progeny testing. Progeny testing also requires the production of several offspring to evaluate accurately the breeding value of an individual, and several animals need to be progeny tested to assure finding one that is good. Progeny testing is expensive, requiring a large investment if it is to be done properly. Quite often, purebred breeders who have males to be progeny tested contract with commercial producers who need males for breeding to their females. Records are kept on the offspring of such matings, at the expense of the purebred breeders.

Progeny testing to evaluate the breeding value of an animal for carcass merit requires fewer offspring for accuracy than progeny testing to evaluate milk-producing ability. Eight steers by a sire should provide sufficient information for evaluating that sire's breeding value for carcass merit, whereas 40 to 50 offspring are needed to properly evaluate a bull's transmitting ability for milk production.

With most farm animals, progeny testing can be done on only the males because female cattle, horses, and sheep produce too few offspring to be progeny tested and still be available later for breeding in the herd. Poultry and sows can be progeny tested because they produce relatively large numbers of offspring.

Another use of progeny testing is detection of undesirable recessive genes present in herds and flocks. It may be important to know that a male being used for breeding is free from undesirable traits caused by recessive genes. If the undesirable trait does not impair reproduction or cause death, its presence can be tested for in a male by breeding the male to females who show the trait. Only eight offspring, all not showing the undesirable trait are needed to give reasonable assurance that the male being tested does not carry a particular undesirable gene. Keep in mind that one never proves that a male does not carry an undesirable gene. Even if the odds against a particular gene being present are 999:1, the gene could still be present. If an undesirable recessive gene impairs fertility or causes death, the undesirable recessive animal cannot be used for breeding. A male can be tested for all possible undesirable and lethal genes by mating him to his daughters. To obtain reasonable assurance that the male carries no undesirable genes, 30 to 35 offspring resulting from mating a sire to his daughters are required.

Family Selection

In some herds where several unrelated sires have been used to produce the females of the herd, several small families exist in the herd. Observations on performance in each of the families may reveal that performance is high in most or all animals of some families but that the performance of many animals in other families is quite low. An effective way to improve such a herd is to cull all families having low performance and keep replacements only from families having good performance (**family selection**). Of course, this type of program is done only once, after which animals must be selected on some other basis.

STUDY QUESTIONS

1. a. If you were using only one trait on which to base your selection of beef cattle, which trait would you choose and why would you choose it?
   b. What types of errors are likely to result from the use of the minimum culling method of selection?
   c. In what situations is the minimum culling method of selection reasonably effective?
2. a. What potential error exists in developing an overall merit rating for animals based on their individual traits?
   b. How can proper emphasis be given each trait that is considered when using an index method to develop an overall merit rating?
   c. What does general combining ability mean? Is the use of general combining ability a specialized type of progeny testing?
3. a. What is meant by specific combining ability?
   b. What is the objective of reciprocal recurrent selection?
   c. Does selection for specific combining ability have more usefulness for beef cattle or for poultry? Why?
4. a. What information of value in selecting animals can be obtained examining the appearance of individual animals?
   b. Of what value is a pedigree for selection purposes if no production records of ancestors are included in the pedigree?
   c. If only one animal four generations back of the one being evaluated has a really good production record, how much can this record be relied on to assure good production by the animal under consideration?
   d. What must be assumed about the ancestors of an animal if no production records concerning them are available in the pedigree?

5. a. For traits of high heritability, how important are production records of the animal under consideration?
    b. What safeguard should be considered in evaluating production records?
    c. What is the most accurate method of evaluating the genetic worth of an animal?
    d. For what types of traits is progeny testing most valuable?
6. a. To find an animal of high genetic value about whose performance one can be confident, what two things are needed?
    b. What is family selection and of what value is it?
    c. Can cows and ewes be progeny tested? Why?

SELECTED REFERENCES

Hazel, L. N., and Lush, J. L. 1942. The efficiency of Three Methods of Selection. *Journal of Heredity.* 33:393-99.

Lasley, J. F. 1978. *Genetics of Livestock Improvement.* 3rd edition. Englewood Cliffs, New Jersey: Prentice-Hall, Inc.

Warwick, E. J. and J. E. Legates. 1979. *Breeding and Improvement of Farm Animals.* 7th Edition. New York: McGraw-Hill.

17

Systems of Breeding

The functions and objectives of breeders of **purebred** livestock and breeders of commercial livestock (usually called, respectively, "purebred breeders" and "commercial breeders"), differ so greatly that it is well to state them as a background before discussing breeding systems. The function of the purebred breeder is to develop breeding stock processing the highest predictability for transmitting the most desirable inheritance possible. By contrast, the function of the commercial breeder is to make use of the available genetic material in a manner that will give the most efficient, rapid, and economical production of the quality of product most desired by the consumer.

Both purebred and commercial breeders use several systems of breeding. As examples, let us discuss two general systems that purebred breeders use (continuous **outbreeding** and closed-herd breeding) and two that commercial breeders use (**grading up** and **outcrossing**).

PUREBRED BREEDERS

Breeders of purebred livestock may continuously bring in and use males that are unrelated to the females in a herd (continuous outbreeding) or they may select breeding males and females from within the herd (closed-herd breeding).

Continuous outbreeding. The continuous use of unrelated males in a purebred herd maintains the greatest possible amount of genetic impurity, or heterozygosity. As a result, the performance in the herd may be at a high level. However, because of the lack of genetic purity, breeding animals from such a purebred herd may not transmit their desirable performance with a high degree of certainty.

Closed-herd breeding. When a herd is closed to outside breeding stock and all breeding animals are selected from within the herd, a certain closeness of inbreeding (the mating of related animals) occurs. The amount of inbreeding depends on the number of sires used within the herd. The fewer the males, the closer the inbreeding.

Inbreeding tends to produce genetic purity. It does not change the genes or determiners in any other way. The genetic purity, or homozygosity, produced by inbreeding, may result in inbreeding depression (loss of vigor and the appearance of some individuals of poor quality if the herd carries genes for undesirable recessive traits) because such traits show only when the genes determining them are in the homozygous, or pure, condition.

Breeding animals have been marketed largely on the basis of their appearance and production records. This method of choosing breeding stock is probably sound if the breeding animals are selected from outbred animals. However, breeding-stock animals from closed herds may not appear as desirable as outbred animals even though they are likely to be superior to the outbreds in terms of the offspring they produce. When a breeder who is operating a closed-herd breeding program sells breeding males to a commercial producer, the animals should be advertised on the basis of how well their offspring will perform rather than on the basis of the appearance and performance of the males being sold.

In general, the traits affected most adversely by inbreeding are those having to do with the fitness of the animal to survive in its environment, such as fertility, milking ability, and vigor of young in early life.

Some breeders may find it necessary to introduce breeding stock into a closed herd to overcome certain weaknesses. Breeding stock that is introduced into a closed herd should come from another superior closed herd. The introduction of unrelated, outbred breeding stock into a closed herd may result in the loss of improvements already made, rather than in the achievement of improvements. Any introduction of breeding stock into a closed herd should be done with caution. Some breeders mate a few of their best females to an unrelated male from a superior closed herd, producing young males to go back into the herd.

A herd developed by closed-herd breeding transmits hereditary traits with a great deal of certainty. Males from closed herds in which rigorous selection has been practiced, when bred to unrelated females, generally sire offspring that show superior performance.

Factors that influence success of inbreeding. Four factors generally determine the degree of success of inbreeding:

1. *Genetic merit of foundation stock.* If the herd is relatively free from deleterious and undesirable recessive genes, inbreeding should not cause detrimental effects to show. However, a herd possessing deleterious and undesirable genes in high frequency shows immediate detrimental effects from inbreeding. If a herd is closed and selection of all breeding stock is from within the herd, the extent to which undesirable inheritance is present can be determined in a few years. If undesirable or abnormal offspring occur at a high frequency, unrelated males from a herd in which closed-herd breeding has been successful can be introduced as a means of reducing the frequency of undesirable genes. The herd can be closed again, without detrimental results, following the introduction of genes from males from outside the herd.

2. *Effectiveness of selection.* In a closed herd it is essential to select for traits that contribute directly to fitness, such as fertility, suckling ability, and vigor in early life, as well as to select for other traits deemed important to production. Effective selection aimed at reducing the genetic material that interferes with successful inbreeding greatly aids such a program. The achievement of effective selection may entail a

careful study of related animals, some progeny testing, and the culling of whole segments of a herd that carry a high frequency of undesirable genes.

3. ***Rate of inbreeding.*** Intense inbreeding fixes the genes (creates homozygosity) very rapidly. As homozygosity is increased, selection becomes less effective because genetic variation is reduced (progress from selection depends on genetic variability). If intense inbreeding reduces the amount of effective selection that can be done, inbreeding at a rapid rate is not desirable. The use of several males in a closed herd results in a very slow increase in inbreeding. A herd large enough to justify the use of 6 to 10 males (150 to 300 females) from an economic standpoint is of sufficient size to be closed without fear of a great increase in inbreeding. In fact, so little inbreeding should result in such a closed herd that little or no manifestation of inbreeding depression should occur. At the same time, the use of closed-herd breeding in a herd of this size, along with selection, should result in fixing of genetic material so that breeding stock should transmit hereditary traits with a high degree of predictability. It is doubtful that a breeder of a small herd can afford to use a sufficiently large number of males to close the herd and yet keep the rate of inbreeding at a level low enough that selection can be practiced sufficiently to make improvements. Breeders of small herds (1 or 2 sires and 20 to 40 females) are better advised to use males continuously from a closed herd in which the size of the breeding operation permits the program to be carried out without markedly increasing inbreeding.

Perhaps the closed herd in which only 4 males and 100 females are used is as small a herd as one can safely use in a closed-herd breeding program because, when fewer males are used, the rate of inbreeding increases so rapidly that selection becomes markedly less effective. A herd in which 4 males are used would normally have approximately 100 breeding females because keeping 4 males to breed fewer than 100 cows would be expensive. However, it might be practical to breed a smaller number of females than 100 by using the 4 best young males in the herd each year and then selling them for breeding the following year. The use of young males is a sound policy even in a larger closed herd because of the acceleration in rate of improvement. It is wise, however, to retain a male for more service in the herd if he has sired outstanding offspring. Keep in mind that use of an outstanding male at the expense of other males is sound from the standpoint of selection, but may result in inbreeding at a rate that will interfere with future selection. Thus, judgment must be used in determining how extensively a superior male should be used in a closed herd.

4. ***Care and management.*** Proper care and management are exceedingly important in a closed herd because of the tendency for inbreeding to reduce fertility and vigor in early life. The selection that can be done depends on the number of suitable animals raised. Thus, it follows that every effort should be made through proper care and management to obtain and raise a large number of offspring.

The purebred breeder is the key to success in livestock production. The purebred breeder can estimate the breeding value of animals selected to go into his or her herd and of those selected for sale to commercial producers. The equation for calculating **breeding value** for any measured trait is $B = H(I - C)$, where B = breeding value, H = heritability of that trait, I = the individual's record, and C = average of the contemporaries for that trait (Cundiff and Gregory, 1977). Heritability for most of the production traits range from 30%

to 45% but is lower for fertility (10%) and higher for final weight (60%) and ribeye area (70%). If the purebred breeder develops breeding stock of high genetic merit that will transmit outstanding performance in a predictable manner, such breeding stock can be used in "grading up" and in cross breeding in commercial operations. Commerical producers cannot produce animals of high performance unless the males they use have the genetic traits that make high performance possible. Thus, the purebred breeder has grave responsibilities to the livestock industry.

COMMERCIAL BREEDERS

Commercial breeders must obtain maximum productivity of a desirable product at minimum costs. Normally, therefore, inbreeding would not be used in a commercial production program. Inbred males, not related to the females of the herd, however, might be used in the breeding program.

Grading up. The continuous use of purebred sires of the same breed in a grade herd or flock is called **grading up**. Keeping replacement females from within the herd causes the inheritance of the purebred animals used to be substituted for that of the herd on which such purebred males are used. The percentage of inheritance of the desirable purebred is 50%, 75%, 84.5%, and 94% for four generations when grading up is practiced. The fourth generation resembles the purebred sires so closely in genetic composition that it approximates the purebred level. The grading up system is useful in the breeding of cattle and horses, but it has little or no value in the breeding of sheep, swine, or poultry. High-producing purebred sheep, swine, and poultry breeding stock are available at economic prices; therefore, the breeder can buy them for less than he can produce them by grading up.

The use of production-tested males that are above average in performance in a commercial herd can grade the herd up not only to a general purebred level but to a high level of production.

The commercial breeder should not use males from the same herd continuously, because this tends to lower the performance level of certain traits. Males from two good production-tested herds should be used alternately. In this way the general inheritance level of the female herd can be increased for production traits, and the offspring produced show greater hybrid vigor. If males can be obtained from production-tested herds in which closed-herd breeding is practiced, even more hybrid vigor can be expected.

Outcrossing. The crossing of strains or lines within a breed, the crossing of breeds, and the crossing of two species, such as Brahman x Hereford cattle, may be classed as different kinds of outcrossing. Several types of crosses can be made. In general, the genetic diversity is greater in wide crosses (such as crosses of breeds or species) than it is when strains or lines within a breed are crossed.

The continuous use of males that are unrelated to the females of the herd is also called outcrossing or outbreeding. If herds have been closed with all replacements selected from within the herd for three to five generations, such herds have become more genetically pure than they were before being closed, and are often called inbred lines. The crossing of inbred lines is called **line-crossing** which is one way of practicing outcrossing.

It is impractical to maintain lines of inbred beef cattle and converge them to produce hybrids in the same fashion as hybrid corn or chickens are produced. However, bulls from closed herds that are bred to groups of outbred cows transmit hereditary traits with a high degree of predictability.

Rotational use of males from closed herds. A commercial producer can use males from two closed herds by rotating the female offspring to males of a closed herd that differs from the herd that produced males used to produce these females. In this manner, the commerical producer can make use of the high transmitting ability of males from the closed herds. The producer can also obtain hybrid vigor by breeding females that are sired by males of closed herd A to males from closed herd B. Thus, the breeding of females to males from the closed herd that did not provide the sires of the females maintains in the females and in their offspring as much hybrid vigor as can be expected within a breed. This breeding system is schematically outlined as follows for a crossbreeding program using beef cattle.

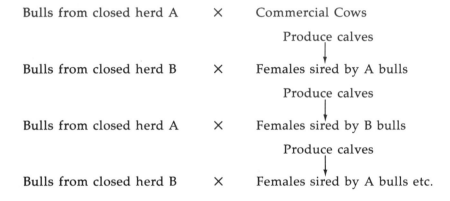

This system provides a means of obtaining a certain amount of increased vigor in the producing females and in their offspring without crossing breeds. Producers who prefer not to cross breeds may find this system advantageous.

Crossing breeds. Crossbreeding is utilized by many commercial producers for two primary reasons: (1) to combine the superiority of two or more breeds and (2) to obtain heterosis or hybrid vigor.

Since one breed is not superior to all other breeds in the economically important traits, crossing the breeds allows the producer to combine the superior performance of several breeds.

Heterosis measures the superiority of the offspring over the average of the breeds which go into the cross. For example, if the average weaning weights for breed A and breed B were 400 and 500 pounds respectively, the weaning weights of the A × B crossbred progeny would be approximately 475 pounds. This is 25 pounds above the average of the two parental breeds which is approximately 5% heterosis (25/450 = 5.6%). A logical question is, why produce 475 pound crossbreds when breed B would produce 500 pound calves? Remember that a breed is not superior for all traits. Another important trait is percent calf crop. If breed A, breed B, and the A × B crossbreds had average calf crop percentages of 90, 80, and 88 respectively, then combining weaning weight and percent calf crop would give the crossbreds a definite superiority over either breed A or breed B. The crossbreds would wean more pounds of calf per cow bred.

There are several types of crossbreeding systems available to the commercial producer. These systems are described in detail in the chapters involving the species where crossbreeding is used most extensively: beef - Chapter 18, pp. 190-209, swine - Chapter 22, pp. 243-256, sheep - Chapter 24, pp. 269-280.

STUDY QUESTIONS

1. a. What are or should be the functions of (1) the purebred breeder and (2) the commercial breeder?
    b. For what objectives are breeds of livestock crossed?
    c. If one crosses two breeds to capitalize on the strong points of the breeds, how is the cross made?
2. a. What is rotational crossbreeding?
    b. Draw a diagram to show how rotational crossbreeding is done. You select the species and breeds that you will use.
    c. What is "grading up"? In which species do you feel it should have the most use?
3. a. What is outbreeding? How predictable is the breeding value of outbred animals?
    b. The reason for crossbreeds is to obtain hybrid vigor. Why is rotational crossbreeding superior to crossing of straightbreds?
    c. Suppose you are considering rotational crossbreeding of beef cattle. What factors will you consider in deciding what breeds to use?

SELECTED REFERENCES

Agricultural Research Service, USDA. 1976. *Evaluating Germ Plasm for Beef Production.* Cycle 1. Clay Center, Nebraska: U.S. Meat Animal Research Center Progress Report No. 3. ARS-NC-41.

Agricultural Research Service, USDA. 1976. *Germ Plasm Evaluation Program.* Clay Center, Nebraska: U.S. Meat Animal Research Center Progress Report No. 4. ARS-NC-48.

Bogart, R., Landers, J. H., Jr., and Frischknecht, W. D. 1974. *Improving Beef Cattle Through Breeding.* Corvallis: Oregon State University Bulletin 802.

Cundiff, L. V., and Gregory, K. E. 1977. *Beef Cattle Breeding.* Agriculture Information Bulletin 286. Washington, D. C.: United States Department of Agriculture.

Dalton, D. C. 1980. *An Introduction to Practical Animal Breeding.* New York: Granada Publishers.

Hartl, D. L. 1980. *Principles of Population Genetics.* Sutherland, Massachusetts: Sinauer Associates, Inc. Publishers.

18

Beef Cattle Breeds and Breeding

A breed of cattle is defined as a race or variety, the members of which are related by descent and similar in certain distinguishable characteristics. More than 250 breeds of cattle are recognized throughout the world, and several hundred other varieties and types have not been identified with a breed name.

Some of the oldest recognized breeds in the United States were officially recognized as breeds during the middle to late 1800s. Most of these breeds originated from the crossing and combining of existing strains of cattle. When a breeder or group of breeders decided to establish a breed, distinguishing that breed from other breeds was of paramount importance; thus, major emphasis was placed on readily distinguishable visual characteristics, such as color, color pattern, polled or horned condition, and rather extreme differences in form and shape.

New cattle breeds, such as Brangus, Santa Gertrudis, and Beefmaster, have come into existence in the United States in recent years. These breeds have been developed by attempting to combine the desirable characteristics of several existing breeds. In most cases, however, the same identifying characteristics as already mentioned have been selected for in the breeding programs to give the new breeds visual identity. After some of these first breeds were developed, it was not long until the word **purebred** was attached to them. Herd books and registry associations were established to assure the "purity" of each breed and to promote and improve each breed. Purebred refers to purity of ancestry, established by the pedigree, which shows that only animals recorded in that particular breed have been mated to produce the animal in question. Purebreds, therefore, are the cattle within the various breeds that have pedigrees recorded in their respective breed registry associations.

When viewing a herd of purebred Angus, Herefords, or other breed, one notes uniformity, particularly uniformity of color or color pattern. Because of this uniformity of one or two characteristics, the word purebred has come to imply genetic uniformity (homozygosity) of all characteristics. Cattle within the same breed are not highly homozygous because high levels of homozygosity occur only after many generations of close inbreeding. This close inbreeding has not occurred in the cattle breeds. If breeds were uniform genetically, they could not be improved or changed even if changes were desired.

WHAT IS A BREED?

The genetic basis of cattle breeds and their comparison is not well understood by most livestock people. Often the statement is made, "There is more variation within a breed than there is between breeds." The validity of this statement needs to be carefully examined. Considerable variation does exist within a breed for most of the economically important traits. This variation is depicted in Figure 18-1, which shows the number of calves, of a particular breed, that fall within certain weight-range categories at 205 days of age. A bell-shaped curve is formed by connecting the high points of each bar. The breed average is represented by the solid line that separates the bell curve into equal halves. Most of the calves are near the breed average; however, at the outer edge of the bell curve are high- and low-weaning weight calves. Note that they are fewer in number at these extremes.

Figure 18-2 allows us to hypothetically compare three breeds of cattle in terms of weaning weight. The breed averages are different; however, the variation within each breed is comparable among all three. The statement, "There is more variation within a

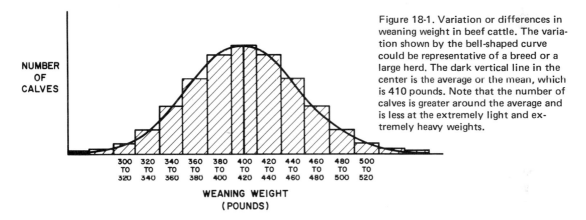

Figure 18-1. Variation or differences in weaning weight in beef cattle. The variation shown by the bell-shaped curve could be representative of a breed or a large herd. The dark vertical line in the center is the average or the mean, which is 410 pounds. Note that the number of calves is greater around the average and is less at the extremely light and extremely heavy weights.

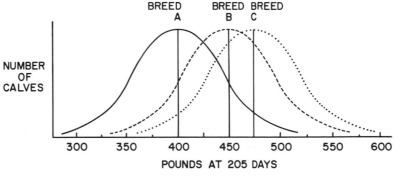

Figure 18-2. Comparison of the breed averages and the variation within each breed for weaning weight in beef cattle. The vertical lines are the breed averages. Note that some individual animals in Breed B and Breed C can be lower in weaning weight than the average of Breed A.

Figure 18-3. Breed differences in percentage of fat in the milk of Holstein and Jersey cows.

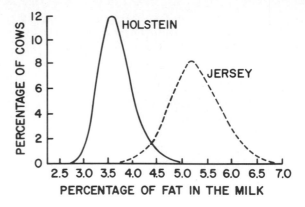

breed than between breed averages," is more correct than the statement, "There is more variation within a breed than there is between breeds."

Figures 18-1 and 18-2 are hypothetical examples; however, they are based on realistic samples of data obtained from the various breeds of cattle. Figure 18-3 shows percent milk fat differences in samples of cattle from the Holstein and Jersey breeds. It is obvious that by selecting certain Holsteins with high percent milk fat and low percent milk fat Jerseys, you could infer that Holsteins have a higher percent milk fat average than Jerseys. However, random sample selections of cows from each breed would show that Jerseys, on the average, do have a higher percent of fat in their milk than Holsteins. Occasionally a bull may seem to have a high level of productivity in a certain trait for which the information comes from selected data and not comparisons based on random samples. Only breed comparisons based on random samples are valid.

MAJOR BEEF BREEDS IN THE UNITED STATES

Until the 1960s and 1970s, the number of cattle breeds in the United States had remained relatively stable at around 15 to 20. Today, more than 60 breeds of cattle are available to the United States beef producers. Why the large importation of the different breeds from several different countries? Following are several possible reasons:

1. Feeding more cattle larger amounts of grain, a practice started in the 1940s, resulted in many overfat cattle—cattle that had been previously selected to fatten on forage rations. Therefore, a need was established for grain-fed cattle that could produce a higher percentage of lean to fat at the desired slaughter weight.
2. Economic pressures to produce more weight in a shorter period of time demonstrated a need for cattle with more milk and more growth.
3. An opportunity was available for some promoters to capitalize on merchandising a certain breed as being the ultimate in all production traits. This opportunity could be realized easily because there was little comparative information on breeds.

Figures 18-4, 18-5, and 18-6 identify the most numerous breeds of beef cattle in the United States. Table 18-1 gives some distinguishing characteristics and brief background information for each of these breeds.

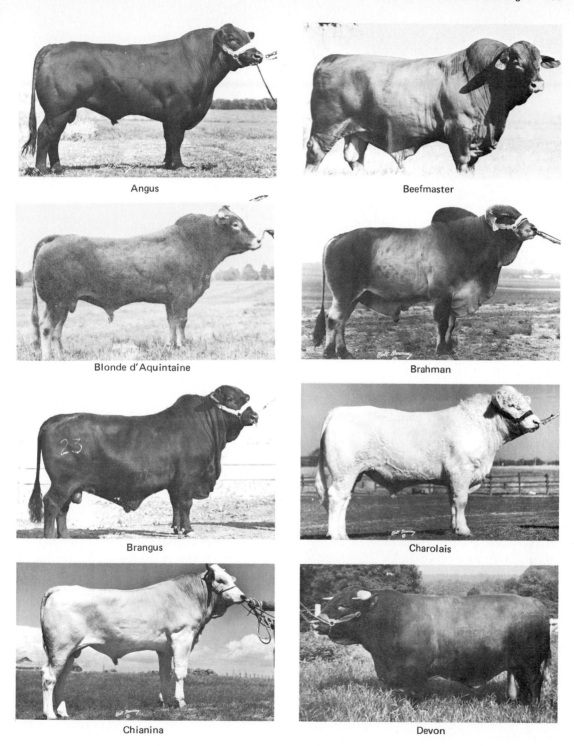

Figure 18-4. Some of the breeds of cattle used for beef production in the United States.

Galloway

Gelbvieh

Hereford

Limousin

Longhorn

Maine-Anjou

Murray Grey

Pinzgauer

Figure 18-5. Some of the breeds of cattle used for beef production in the United States.

Figure 18-6. Some of the breeds of cattle used for beef production in the United States.

Table 18-1. Background and distinguishing characteristics of major beef breeds in the United States

Breed	Distinguishing characteristics	Brief background
Angus	Black color and polled.	Originated in Aberdeenshire and Angushire of Scotland. Imported into the U. S. in 1873 to cross with Longhorn.
Beefmaster	Has various colors.	Developed in the U.S. from Brahman, Hereford, and Shorthorn breeds. Selected for its ability to reproduce, produce milk, and grow under range conditions.
Blonde d'Aquitaine	Fawn colored, sometimes with a reddish tinge, heavily muscled.	Originated in France and live cattle imported in the U.S. in 1973. Semen imported prior to that time.
Brahman	Various colors, with gray predominant. They are one of the Zebu breeds which have the hump over the top of the shoulder. Most Zebu breeds also have large, drooping ears and excess skin in the throat and dewlap.	Major importations to the United States from India and Brazil. Largest introductions in early 1900s. These cattle are heat tolerant and well adapted to the harsh conditions of the Gulf Coast region.
Brangus	Black and polled predominate, although there are Red Brangus.	Breed developed around 1912 in the U.S. from three-eighths Brahman and five-eighths Angus.
Charolais	White color with heavy muscling, horned or polled.	One of the oldest breeds in France. Brought into the U.S. soon after World War I, but its most rapid expansion occurred in the 1960s.
Chianina	White color with black eyes and nose. Extremely tall cattle.	An old breed originating in Italy. Acknowledged to be the largest breed, with mature bulls weighing more than 3,000 pounds. First used for breeding in U.S. in 1971. Early use was for draft.
Devon	Dark red (North Devon) to light red or brown (South Devon).	North Devon is an old breed originating in England. South Devon is more of a dual-purpose breed. The Devon first came to America with the Pilgrims in 1623.
Galloway	Polled with majority solid black in color, some dun-colored, and others white with black noses, ears, and feet. The Belted Galloway is black with distinctive white belt.	An old breed from the Scottish province of Galloway. Long, burly hair has made it adaptable to the harsh climates of the north. The Belted Galloway has a separate breed association. It has the same origin as the Galloway but had an infusion of the belted cattle in the seventeenth or eighteenth century.
Gelbvieh	Golden colored.	Originated in Austria and West Germany. Dual-purpose breed used for draft, milk, and meat.

Breed	Description	History
Hereford	Red body with white face and horned.	Introduced in the U.S. in 1817 by Henry Clay. Followed the Longhorn in becoming the traditionally known range cattle.
Limousin	Golden color with marked expression of muscling.	Introduced into the U.S. in 1969, primarily from France.
Longhorn	Multi-colored with characteristically long horns.	Came to West Indies with Columbus. Brought to the U.S. through Mexico by the Spanish explorers. Longhorns were the noted trail-drive cattle from Texas into the Plains States.
Maine-Anjou	Red and white spotted.	A large breed developed in France and introduced into Canada in late 1968.
Murray Grey	Silver gray and polled.	Developed in Australia from Shorthorns crossed on Angus.
Pinzgauer	Reddish chestnut color with white markings on rump, back, and belly.	A hardy breed developed in the Pinza Valley of Austria and areas of Germany and Italy. Introduced into North America in 1972.
Polled Hereford	Red body with white face and polled.	Bred in 1901 in Iowa by Warren Gammon who accumulated several naturally polled cattle from horned Hereford breeders.
Red Angus	Red color and polled.	Founded as a performance breed in 1954 by sorting the genetic recessives from Black Angus herd.
Red Polled	Red color and polled.	Introduced into the U.S. in the late 1800s. Originally a dual-purpose breed but now considered a beef breed.
Santa Gertrudis	Red color and horned.	First United States breed of cattle developed on the King Ranch in Texas. Combination of five-eighths Shorthorn and three-eighths Brahman.
Scotch Highland	Golden color with long, shaggy hair.	Bred in the highlands of Scotland. Imported into U.S. in 1894.
Shorthorn	Red, white, or roan in color. Both horned and polled.	Introduced into the U.S. in 1783 under the name "Durham." Most prominent in the U.S. around 1920.
Simmental	Yellow to red and white color pattern. Both polled and horned.	A prominent breed in parts of Switzerland and France. First bull arrived in Canada in 1967. Originally selected as a dual-purpose milk and meat breed.
Tarentaise	Sold wheat-colored hair ranging from a light cherry to dark blond.	Mountain cattle derived from an ancient Alpine strain in France. Originally a dairy breed in which maternal traits have been emphasized.

Table 18-2. Beef breeds with registration numbers of more than 10,000 per year

Breed	Registration numbers (1980)
Angus	257,578
Beefmaster	30,000
Brangus	24,500
Brahman	36,450
Charolais	26,907
Hereford	201,495
Polled Hereford	151,658
Limousin	13,793[a]
Santa Gertrudis	31,211
Shorthorn	19,426
Simmental	28,853[a]

Source: National Society of Livestock Record Association, personal correspondence, 1981.

[a]Reported by breed associations as purebred only. Registration of percentage cattle are not included in these numbers. Percentage refers to cattle that have other breeds represented in their breeding and that are not recognized as purebred by their respective breed associations. Total registrations, including percentage cattle, for the Limousin and Simmental breeds are 31,937 and 66,109, respectively.

The relative importance of the various breeds' contributions to the total beef industry is best estimated by the registration numbers of the breeds (Table 18-2).

Although registration numbers are for purebred animals, they reflect the commercial cow-calf producers' demand for the different breeds. The registration numbers show that Angus, Hereford, and Polled Hereford are the most important breeds of the beef cattle industry in the United States. The numbers of cattle belonging to various breeds in this country have changed over the past years as shown in Figure 18-7. It is reasonable to expect that in future years some breeds will become more numerous, while other breeds will decrease significantly in numbers. These changes will be influenced by economic conditions and by genetic changes being made in the economically important traits in the breeds.

IMPROVEMENT OF BEEF CATTLE THROUGH BREEDING METHODS

Genetic improvement in beef cattle can occur by selection and also by using a particular mating system. Significant improvement by selection results when the selected animals are superior to the herd average and the heritabilities of the traits are relatively high (40% and higher). It is important that the traits included in a breeding program be of economic importance and that the traits be measured as objectively as possible.

TRAITS AND THEIR MEASUREMENT

Most of the economically important traits of beef cattle can be classified under (1) reproductive performance, (2) weaning weight, (3) postweaning growth, (4) feed efficiency, (5) carcass merit, (6) longevity, (7) conformation, and (8) freedom from genetic defects.

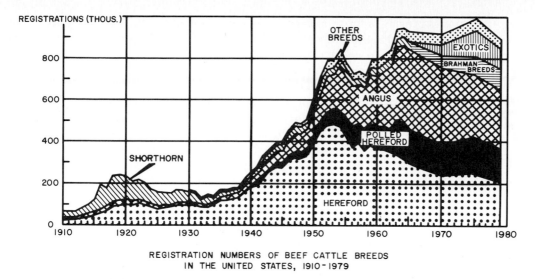

REGISTRATION NUMBERS OF BEEF CATTLE BREEDS
IN THE UNITED STATES, 1910-1979

Prior to 1963 — Some Polled Herefords were registered only in the American Hereford Association. Others were registered in the American Polled Hereford Association.

Prior to 1965 — Brahman breeds were included in other breeds. Very few exotics registered.

After 1965 — Shorthorn ends, included in other breeds.
Brahman breeds (Brahman, Brangus, Santa Gertrudis).
Exotics (Chianina, Limousin, Maine Anjou, Simmental).

Figure 18-7. Registration numbers of beef cattle in the United States, 1910 through 1979.

Reproductive performance has the highest economic importance when compared to the other traits. Most cow-calf producers have a goal for percent calf crop weaned (number of calves weaned compared to the number of cows in the breeding herd) of 90% or higher. Beef producers also desire each cow to calve every 365 days or less and have a calving season for the entire herd of less than 90 days. All of these are good objective measures of reproductive performance. The heritability of fertility in beef cattle is quite low (less than 20%), so little genetic progress can be made through selection. The most effective way to improve reproductive performance is to improve the environment through, for example, adequate nutrition and good herd health practices.

Reproductive performance can be improved through breeding methods by crossbreeding to obtain heterosis for percent calf crop weaned, using bulls with relatively light birth weights (heritability of birth weight is 40%) which decreases calving difficulty, and by selecting bulls that have a relatively large scrotal circumference. Scrotal circumference has a high heritability (40%), and bulls with a larger scrotal size (over 30 cm for yearling bulls) produce a larger volume of semen and have half-sister heifers that reach puberty at earlier ages than heifers related to bulls with a smaller scrotal size.

Weaning weight, as measured objectively by the scales, reflects the milking and mothering ability of the cow and the preweaning growth rate of the calf. Weaning weight is commonly expressed as the adjusted 205-day weight, where the weaning weight is adjusted for the age of the calf and age of the dam. This adjustment puts all weaning weight records on a comparable basis since older calves will weigh more than younger calves and the mature cows (5 to 9 years of age) will milk heavier than the young cows (2 to 4 years of age).

Weaning weights of calves are usually compared by dividing the calf's adjusted weight by the average weight of the other calves in the herd. This is expressed as a ratio. For example, a calf with a weaning weight of 440 lbs, where the herd average is 400 lbs, has a ratio of 110. This calf's weaning weight ratio is 10% above the herd average. Ratios should be used primarily for selecting cattle within the same herd. Comparing ratios between herds is misleading from a genetic standpoint because most of the differences between herds are caused by differences in the environment. Weaning weight will respond to selection because it has a heritability of 30%.

Postweaning growth measures the growth from weaning to a weight that approaches slaughter weight. Postweaning growth may take place on a pasture or in a feedlot. Usually animals with relatively high postweaning gains make efficient gains at a relatively low cost to the producer.

Postweaning gain in cattle is usually measured in pounds gained per day after a calf has been on a feed test for 140 to 160 days. Weaning weight and postweaning gain are usually combined into a single trait, namely, adjusted 365-day weight (yearling weight). It is computed as follows:

Adjusted 365-day weight = (160 x average daily gain) + adjusted 205-day weight

Average daily gain for 140 days and adjusted 365-day weight both have high heritabilities (40%), so genetic improvement can be quite rapid when selection is practiced on postweaning growth or yearling weight.

Feed efficiency is measured by the pounds of feed required per pound of liveweight gain. Specific records for feed efficiency can only be obtained by feeding each animal individually and keeping records on the amount of feed consumed. With the possible exception of some bull testing programs, determining feed efficiency on an individual animal basis is seldom economically feasible.

The interpretation of feed efficiency records can be rather confusing depending on the endpoint to which the animals are fed. Feeding endpoint can be a certain number of days on feed (e.g., 140 days) to a specified slaughter weight (e.g., 1,200 lbs), or to a carcass compositional endpoint (e.g., low Choice quality grade). Most of the differences shown by individual animals in feed efficiency are related to the pounds of body weight maintained through the feeding periods and the daily rate of gain or feed intake of each animal. Cattle fed from a similar initial feedlot weight (e.g., 600 lbs) to a similar slaughter weight (e.g., 1,100 lbs) will demonstrate a high relationship between rate of gain and efficiency of gain. In this situation, cattle that gain faster will require fewer pounds of feed per pound of gain. Thus, a breeder can select for rate of gain and thereby make genetic improvement in feed efficiency. However, when cattle are fed to the same compositional endpoint (approximately the same carcass fat), there is little, if any, difference in the amount of feed required per pound of gain. This is true for different sizes and shapes of cattle and even for various breeds that vary greatly in skeletal size and weight.

The heritability of feed efficiency is high (45%), so selection for more efficient cattle can be effective. It seems logical to use the genetic correlation between gain and efficiency where possible.

Carcass merit is presently measured by quality grade and yield grade, which were discussed in detail in Chapter 5. Many cattle breeding programs have goals to produce cattle that will grade Choice and have yield grades from 2.0 to 3.5. Objective measurements of backfat thickness probes on the live animal and measurement of the height of the skeleton at the hips can assist in predicting the yield grade at certain slaughter weights. Visual appraisal, which is more subjective, can be used to predict the amount of fat or the predisposition to fat and the skeletal size. These visual estimates can be relatively accurate in identifying actual yield grades if cattle differ by as much as two yield grades.

Because quality grade cannot be evaluated accurately in the live animal, it is necessary to evaluate the carcass. Steer and heifer progeny of different bulls are slaughtered to best identify the genetic superiority or inferiority of the bulls for both quality grade and yield grade. The heritabilities of most beef carcass traits are high (over 40%), so selection can result in marked genetic improvement for these traits.

Longevity measures the length of productive life and it is an important trait, particularly for cows. Bulls are usually kept in a herd for only a few years, or inbreeding may occur. Some highly productive cows remain in the herd at age 15 years or older while other highly productive cows have been culled from the herd prior to reaching 8 or 9 years of age. These cows may have been culled because of such problems as skeletal unsoundness, poor udders, unhealthy eyes, or unhealthy teeth. Little selection opportunity for longevity exists because few cows remain highly productive past the age of 10 years. Most producers need to improve their average herd performance as rapidly as possible rather than to improve longevity. In this situation, a relatively rapid turnover of cows is needed. Some selection for longevity occurs because producers have the opportunity to keep more replacement heifers born to cows that are highly productive in the herd for a longer period of time than heifers born to cows that stay in the herd for a short time. Some beef producers attempt to identify bulls that have highly productive, relatively old dams. Some conformation traits, such as skeletal soundness and udder soundness, may be evaluated to extend the longevity of production.

Conformation is the form, shape, and visual appearance of an animal. How much emphasis to put on conformation in a beef cattle selection program has been and continues to be controversial. Some producers feel that putting a productive animal into an attractive package contributes to additional economic returns. It is more logical, however, to place more selection emphasis on traits that will produce additional numbers and pounds of lean growth for a given number of cows. Placing some emphasis on conformation traits such as skeletal, udder, eye, and teeth soundness is justified. Conformation differences such as fat accumulation or predisposition to fat can be used effectively to make meaningful genetic improvement in carcass composition.

Most of the conformation traits are medium to high (30% to 60%) in heritability, so selection for these traits will result in genetic improvement.

Genetic defects, other than those previously identified under longevity and conformation, need to be considered in breeding productive beef cattle. Cattle have numerous known hereditary defects; most of them, however, occur very infrequently and are of minor concern. Some of the defects increase in their frequency and selection needs to be directed against them. Most of these defects are determined by a single pair of genes that are usually recessive. When one of these hereditary defects occurs, it is a logical practice to cull both the cow and the bull.

Some of the most common occurring genetic defects in cattle today are double muscling, syndactyly (mule foot), arthrogryposis (palate-pastern syndrome), osteopetrosis (marble bone disease), hydrocephalus, and dwarfism. Double muscling is evidenced by an enlargement of the muscles with large grooves between the muscle systems of the hind leg. Double-muscled cattle usually grow slowly and their fat deposition in and on the carcass is much less than that of the normal beef animal. Syndactly is a condition in which one or more of the hooves are solid in structure rather than cloven. Mortality rate is high in calves with syndactly. Arthrogryposis is a defect in which the pastern tendons are contracted, and the upper part of the mouth has not properly fused together. Osteopetrosis is characterized by the marrow cavity of the long bones being filled with bone tissue. All calves having osteopetrosis have short lower jaws, protruding tongue, and impacted molar teeth. A bulging forehead where fluid has accumulated in the brain area is typical of the defect of hydrocephalus. Calves with arthrogryposis, hydrocephalus, or osteopetrosis usually die shortly after birth. The most common type of dwarfism is snorter dwarfism, in which the skeleton is quite small and the forehead has a slight bulge. Some snorter dwarfs exhibit a heavy, labored breathing sound. This defect was most common in the 1950s, and it has decreased significantly since that time.

SELECTING REPLACEMENT HEIFERS

Heifers, as replacement breeding females, can be selected for several traits at different stages of their productive life. The objective is to identify heifers that will conceive early in the breeding season, calve easily, give a flow of milk consistent with the feed supply, wean a heavy calf, and make a desirable genetic contribution to the calf's postweaning growth and carcass merit.

Beef producers have found that it is a challenge to determine which of the young heifers will make the most productive cows. Table 18-3 shows the selection process that producers are using increasingly to select the most productive replacement heifers. This selection process assumes that more heifers will be selected at each stage of production than the actual number of cows to be replaced in the herd. The number of replacement heifers that producers keep is based primarily on how many they can afford to raise. More heifers than the number needed should be kept through pregnancy-check time. Heifers are selected on the basis that they become pregnant early in life primarily for economic reasons rather than expected genetic improvement from selection.

COW SELECTION

Cows should be culled from the herd based on the productivity of their calves and additional evidence that they can be productive the following year, such as soundness of udders, eyes, skeleton, and teeth. Cow productivity is measured by pregnancy test, weaning and yearling weights (ratios) of their calves, and the **breeding values** of the cow. A breeding value combines into one figure a measurement of genetic potential based on the

Table 18-3. Replacement heifer selection guidelines at the different productive stages

Stage of heifer's productive life	Emphasis on productive trait	
	Primary	Secondary
Weaning (7 to 10 months of age)	Cull only the heifers whose actual weight is too light to prevent them from showing estrus by 15 months of age.	Weaning weight ratio Maternal Breeding Value (MBV) Predisposition to fatness Adequate skeletal frame Skeletal soundess
Yearling (12 to 15 months of age)	Cull heifers that have not reached the desired target breeding weight (e.g., minimum of 650 to 700 lbs for small to medium-sized breeds or cross, minimum of 750 to 800 lbs for large-sized breeds and crosses).	Yearling weight ratio Maternal Breeding Value (MBV) Yearling Breeding Value (YBV) Predisposition to fatness Adequate skeletal frame Skeletal soundness
After breeding (19 to 21 months of age)	Cull heifers that are not pregnant and those which will calve in the latter one-third of the calving season	Yearling weight ratio Maternal Breeding Value (MBV) Yearling Breeding Value (YBV) Predisposition to fatness Adequate skeletal frame Skeletal soundness
After weaning first calf (31 to 34 months of age)	Cull to the number of first-calf heifers actually needed in the cow herd based on the weaning weight performance of the first calf. Preferably all the calves from these heifers have been sired by the same bull. Heifers should also be pregnant.	

individual's performance and the productivity of related animals such as the sire, dam, and other relatives. The common breeding values reported for cows, heifers, or bulls are **Maternal Breeding Value** (MBV), **Weaning Breeding Value** (WBV), and **Yearling Breeding Value** (YBV). MBV measures the milking potential, WBV measures primarily preweaning growth, and YBV measures postweaning growth and weight at a year of age. Breeding values over 100 are above average and reflect high productivity and calf performance. Table 18-4 shows the breeding values of the highest- and lowest-producing cows in a herd. Cow 1 is the highest-producing cow in the herd and should be kept. Cow 2 is the lowest-producing cow in the herd. She should be culled and replaced with a heifer of higher breeding potential.

BULL SELECTION

Bull selection must receive the greatest emphasis if the genetic improvement of a herd is to be maximized. Bull selection accounts for 80 to 90% of the herd improvement over a period of several years. This does not diminish the importance of good beef females, because genetically superior bulls have superior dams. However, most of the genetic superiority or inferiority of the cows will depend on the bulls previously used in the herd.

**Table 18-4. Breeding values and calf performance
of a high- and low-producing cow in the same herd.**

Cow number	Breeding Values			Performance of Calves		
	MBV	WBV	YBV	Number of calves	Weaning weight ratio	Yearling weight ratio
1	109	106	107	4	109	108
2	95	97	95	3	93	91

Most commercial beef herds should put emphasis on adjusted yearling weights and their respective ratios in their genetic improvement programs. This emphasis will allow the producers to increase the pounds of calf sold at weaning and also provide the feeder with cattle that will have high feedlot gains. Caution needs to be exercised because large birth weights are associated with high yearling weights. Therefore, selection for yearling weight needs to be moderated by putting emphasis on birth weights that are consistent with the calving ease desired in the herd. Pounds of calf per cow bred is an excellent trait for the cow-calf producer because it helps balance the traits of a live calf and weight of the calf at weaning.

Other useful records in bull selection, particularly yearling bulls, are scrotal circumference (and a total breeding soundness examination), MBV, WBV, YBV, backfat probe, and skeletal hip height.

All commercial and purebred producers should identify breeders who have honest, comparative records on their cattle. Most performance-minded seedstock producers will record birth weights, weaning weights, and yearling weights of their bulls. Increasingly, breeders are obtaining feedlot and carcass data on their own bulls, or they are using bulls from sires for whom these test data are known.

Purebred breeders should provide accurate performance data on their bulls, and commercial producers should request the information. Excellent performance records can be obtained and made available on the farm or ranch. The trait ratios are useful and comparative if the bulls have been fed and managed in similar environments.

BREEDING PROGRAMS FOR COMMERCIAL PRODUCERS

Most commercial beef producers use a crossbreeding program because they can take advantage of the heterosis in addition to the genetic improvement from selection. A crossbreeding system should be determined according to which breeds are available and adapted to the commercial producers' feed supply, market demands, and other environmental conditions. A good example of adaptability is the Brahman breed, which is more heat and insect resistant than most other breeds. Because of this higher resistance, the level of productivity (in the southern and Gulf regions of the United States) is much higher for the Brahman and the other breeds that include Brahman.

Most commercial producers travel 150 or fewer miles to purchase their bulls. Therefore, a producer should assess the breeders with excellent breeding programs in a 150-mile radius, as well as which breeds are available. This assessment, in most cases, should have a greater priority than determining which breeds to use and in which combination to use them.

Breeds should be chosen for a crossbreeding system based on how well the breeds complement each other. Table 18-5 gives some comparative ranking of the major beef breeds for the productive characteristics. Although the information in Table 18-5 is useful, it should not be considered the final answer for decisions on breeds to use. First, a producer needs to recognize that this information reflects breed averages; therefore, there are individual animals and herds of the same breed that are much higher or lower than the ranking given. Producers need to use some of the previously described methods to identify superior animals within the breed. Second, it should be recognized that these average breed rankings can change with time, depending on the improvement programs used by the leading breeders within the same breed. Obviously, those traits that have high heritabilities would be expected to change most rapidly, assuming the same selection pressure for each trait. A careful analysis of the information in Table 18-5 shows that no one breed is superior in all of the important productive characteristics. This gives an advantage to commercial producers using a crossbreeding program if they select breeds whose superior traits complement each other. An excellent example of breed complementation is shown by the Angus and Charolais breeds which complement each other for both quality grade and yield grade.

Most of the heterosis achieved in cattle as a result of crossbreeding is expressed by weaning time. The cumulative effect of heterosis on pounds of calf weaned per cow bred is shown in Figure 18-8, in which one sees that maximum heterosis is obtained when crossbred calves are obtained from crossbred cows. The traits that express an approximate heterosis of a 20% increase in pounds of calf weaned per cow bred are early puberty of crossbred heifers, high conception rates in the crossbred female, high survival rate of calves, increased milk production of crossbred cows and a higher preweaning growth rate of crossbred calves.

Consistent high levels of heterosis can be maintained generation after generation if crossbreeding systems such as those shown in Figures 18-9, 18-10, and 18-11 are used. The crossbreeding system shown in Figure 18-11 combines a two-breed rotation with a terminal cross. In this system, the two-breed rotation is used primarily to produce the replacement females for the entire cow herd. In most cow herds approximately 50% of the cows are bred to sires to produce replacement females with the remaining 50% being bred to terminal cross sires. All terminal cross calves are sold. This crossbreeding system maintains heterosis as high as the three-breed rotation system.

In the rotational crossing, breeds with maternal trait superiority (high conception, calving ease, and milking ability) would be selected. The terminal cross sire could come from a larger breed where growth rate and carcass cutability are emphasized. A primary advantage of the rotational-terminal cross system is that smaller or medium-sized breeds can be used in the rotational crossing and a larger breed could be used in the terminal crossing.

Table 18-6 shows the advantage a commercial producer has over a purebred breeder in being able to use more of the breeding methods for genetic improvement. Commercial producers can use crossbreeding while purebred breeders cannot use crossbreeding if they maintain breed purity.

Table 18-5. Breed evaluation for the productive characteristics[a]

Breed	Age at puberty	Weight at puberty	Birth weight	Wean weight	Pounds of calf weaned per cow exposed	Post-wean gain	Feed Efficiency Equal age	Feed Efficiency Equal weight	Feed Efficiency Equal fat	Marbling Equal age	Yield grade Equal age	Yield grade Equal ribeye fat	Palatability Grain fed and equal age
Angus	(1)b	(2)	(2)	(4)	(3)	(4)	(3)	(3)	(2)	(1)	(4)	(4)	(1)
Beefmaster	3	3	4	2	2	2	—	2	—	3	3	3	2
Blonde d'Aquitaine	4	4	4	3	2	2	—	2	—	4	2	3	1
Brahman	(5)	(5)	(4)	(1)	(1)	(3)	(3)	(3)	(3)	(5)	(3)	(3)	(3)
Brangus	(3)	(4)	(3)	(3)	(3)	(3)	—	(3)	—	(4)	(3)	—	2
Charolais	(4)	(5)	(5)	(1)	(4)	(1)	(1)	(1)	(3)	(4)	(1)	(1)	(1)
Chianina	(4)	(5)	(5)	(1)	(4)	(1)	(1)	(1)	(4)	(5)	(1)	(1)	(1)
Devon	(2)	(3)	(3)	(4)	4	(2)	(2)	(3)	—	(3)	(4)	(3)	(1)
Galloway	3	3	3	4	(2)	3	—	3	—	—	3	3	1
Gelbvieh	(1)	(2)	(4)	(1)	(3)	(2)	(2)	(2)	(2)	(4)	(3)	(3)	(1)
Hereford	(3)	(3)	(3)	(4)	(4)	(3)	(1)	(2)	(3)	(4)	(4)	(4)	(1)
Limousin	(4)	(4)	(4)	(3)	5	(3)	(2)	(2)	—	(5)	(1)	(1)	(1)
Longhorn	2	2	1	5	5	5	5	5	—	—	4	3	1
Maine-Anjou	(3)	(4)	(5)	(1)	(3)	(1)	(2)	(1)	—	(4)	(2)	(2)	(1)
Murray Grey	2	2	3	3	3	4	—	4	—	3	4	4	1
Pinzgauer	(1)	(2)	(4)	(2)	(2)	(2)	(2)	(2)	—	(4)	(3)	(3)	(1)
Polled Hereford	(3)	(3)	(3)	(4)	(3)	(3)	(2)	(3)	(3)	(4)	(4)	(4)	(1)
Red Angus	1	2	2	(2)	3	3	2	3	3	—	4	4	1
Red Poll	(1)	(1)	(2)	(2)	(3)	(3)	(4)	(3)	(3)	(4)	(4)	(3)	(1)
Santa Gertrudis	(3)	(5)	(4)	(4)	(4)	(4)	—	—	—	(4)	(4)	3	(1)
Scotch Highland	3	3	2	4	4	4	—	3	—	3	3	3	1
Shorthorn	3	3	3	4	3	3	3	3	2	4	5	5	1
Simmental	(2)	(4)	(5)	(1)	(3)	(1)	(2)	(2)	(3)	(4)	(2)	(1)	(1)
Tarantaise	(1)	(2)	(3)	(2)	(2)	(3)	(3)	(4)	(3)	(4)	(3)	(3)	(2)

[a]Ranking based on 1 (most desirable) through 5 (least desirable).

[b]Circled numbers are based primarily on extensive breed comparison data from the Meat Animal Research Center (MARC), Clay Center, Nebraska. Most of the MARC data are based on the various breeds of bulls being bred to Angus and Hereford cows. Rankings are not made where there are insufficient comparative data. Uncircled numbers are judgments based on less extensive, comparative data.

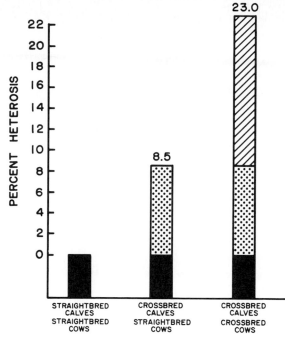

Figure 18-8. Heterosis, resulting from crossbreeding, for pounds of calf weaned per cow exposed to breeding.

Figure 18-9. Two-breed rotation cross. Females sired by Breed A are mated to Breed B bulls, and heifers sired by Breed B are mated to Breed A bulls. This will increase the pounds of calf weaned per cow bred by approximately 15%.

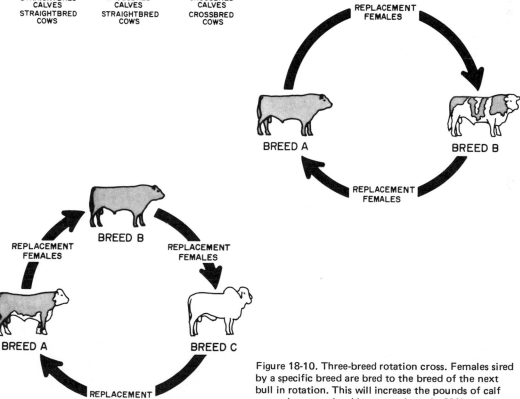

Figure 18-10. Three-breed rotation cross. Females sired by a specific breed are bred to the breed of the next bull in rotation. This will increase the pounds of calf weaned per cow bred by approximately 20%.

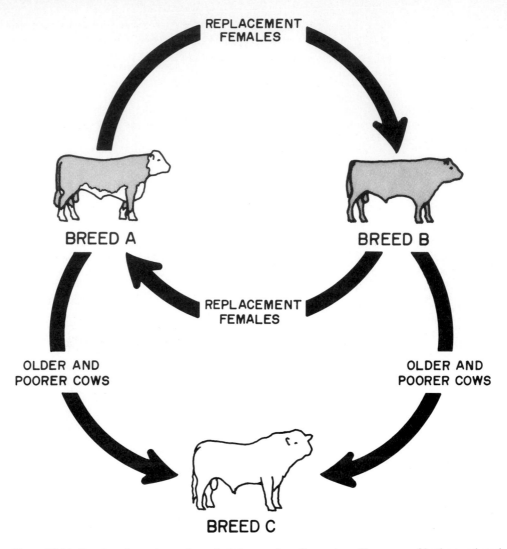

Figure 18-11. Two-breed rotation and terminal sire crossbreeding system. Sires are used in the two-breed rotation primarily to produce replacement heifers. Terminal cross sires are mated to the less productive females. This system will increase the pounds of calf weaned per cow bred by more than 20%.

Table 18-6. Heritability and heterosis for the major beef cattle traits

Traits	Heritability	Heterosis
Reproduction	low	high
Growth	medium	medium
Carcass	high	low

Traits with a low heritability respond very little to genetic selection, but they show a marked improvement in a sound crossbreeding program. The commercial producer needs to select sires carefully to improve the traits with a high heritability.

STUDY QUESTIONS

1. Which beef cattle breeds are most numerous in the United States? Why?
2. Which traits are important to consider in selection to improve beef cattle productivity? Which of these traits respond slowly and which rapidly to selection?
3. In considering a sound crossbreeding program, why must a producer consider how adaptable breeds are to a certain environment?
4. Explain what the term breed means.
5. Work out a sound crossbreeding program for a producer in your area.
6. Why is bull selection such an important part of breeding productive beef cattle?
7. Name and describe two of the most common genetic defects occurring in beef cattle.
8. Which breed is superior in all the important productive characteristics of beef cattle? Tell why.
9. Describe the procedure to select highly productive replacement beef females.
10. How would commercial producers in your area proceed to identify the most genetically superior bulls that could be used in their herd?

SELECTED REFERENCES

Briggs, H. M. and Briggs, D. M. 1980. *Modern Breeds of Livestock*. New York: The Macmillan Co.

Cundiff, L. V. and Gregory, K. E. 1977. *Beef Cattle Breeding*. Agriculture Information Bulletin No. 286, Washington, D. C.: USDA.

Gregory, K. E. and Cundiff, L. V. 1980. Crossbreeding in beef cattle: evaluation of systems. *Journal of Animal Science* 51:1224.

Lasley, J. F. 1978. *Genetics of Livestock Improvement*. Englewood Cliffs, New Jersey: Prentice-Hall, Inc.

Long, C. R. 1980. Crossbreeding for beef production: experimental results. *Journal of Animal Science* 51:1197.

Warwick, E. J. and Legates, J. E. 1979. *Breeding and Improvement of Farm Animals*. New York: McGraw-Hill.

19

Feeding and Managing Beef Cattle

Commercial beef cattle production typically occurs in three phases, or operations: cow-calf, stocker-yearling, and feedlot. The cow-calf operator raises the young calf from birth to 7 to 9 months of age (400 to 500 lbs), the stocker-yearling operator grows the calf to 600 to 800 lbs primarily on roughage, and the feedlot operator typically uses high-energy rations to finish the cattle to a desirable slaughter weight of approximately 1,000 to 1,300 lbs. Most of the steers and slaughter heifers are between 15 to 26 months of age when slaughtered. There are, however, alternatives to these three phases where several marketing transactions can occur. In an integrated operation, cattle may have only one owner from cow-calf through the feedlot, or ownership may change several times before the cattle are ready for slaughter. Alternative production and marketing pathways are shown in Figure 19-1.

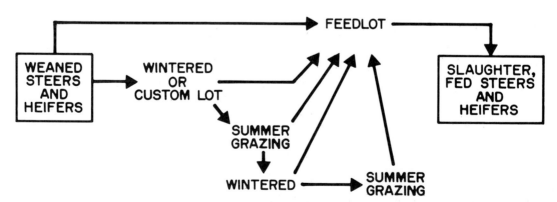

Figure 19-1. Alternative feeding and marketing pathways for weaned calves. Some steers and heifers go to slaughter after the summer grazing period; however, the numbers are relatively small.

COW-CALF PRODUCTION

Cow-calf production centers around 37 million head of beef cows spread across the United States. Most of the cows are concentrated in areas where forage is abundant. Figure 19-2 shows that the 11 states having over 1 million head of cows (58% of the United States total) are located primarily in the Plains and Corn Belt areas. Significant increases and decreases have occurred in numbers of cows in several states during the past 20 years (Figure 19-3). These changes in numbers are influenced by weather as it changes the forage supply, market price of cattle, and the relative values of different crops that can be grown only on land that can also produce hay and pasture.

There are two basic kinds of cow-calf producers: (1) commercial cow-calf producers, who raise most of the potential slaughter steers and heifers, and (2) purebred breeders, specialized cow-calf producers who produce primarily breeding cattle and semen.

Cow-calf producers are interested in managing their operations as economical units. The profitability of a commercial cow-calf operation can be assessed easily by analyzing the following three criteria: (1) calf crop percent weaned (number of calves produced per 100 cows in the breeding herd), (2) average weight of the calves at weaning (7 to 9 months of age), and (3) annual cow cost (the number of dollars to keep a cow each year).

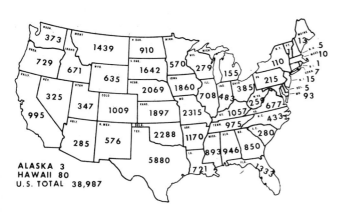

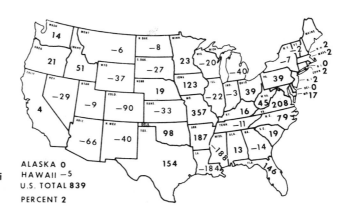

Figure 19-2. Beef cows that have calved during 1980 (1,000 head). For example, Texas is the leading state in beef cow numbers with 5,880,000 head.

Figure 19-3. Changes in beef calf crop, 1970 to 1980 (1,000 head). For example, Missouri produced 357,000 more calves in 1980 than 1970.

An example of the economic assessment of a commercial cow-calf producer who has an 80% calf crop weaned, 420 lb weaning weights, and a $300 annual cow cost would be as follows:

Calf crop percent (0.80) × weaning weight (420 lbs) = 336 lbs of calf weaned per cow in the breeding herd.

Annual cow cost (300) ÷ pounds of calf weaned (336) = $89.28 per hundredweight, which is the break-even price.

A break-even price of $89.28 per hundredweight means that the producer would have to receive over 89¢ per pound, or $89.28 per 100 pounds of calf sold, to cover the yearly cost of each cow in the breeding herd. Table 19-1 shows the break-even price for several levels of calf crop percent, weaning weight, and annual cow cost. This information reflects different management levels in which the break-even price ranges from more than $1.00 per pound to less than 50¢ per pound. Profitability of the commercial cow-calf operation is determined by comparing the market price of the calves at the time they are sold with the break-even price.

Cow-calf operations are managed best by operators who know the factors that affect calf crop percent, weaning weight, and annual cow cost. The primary management objective should be to improve pounds of calf weaned per cow and reduce or control the annual cow costs.

MANAGEMENT FOR HIGH CALF CROP PERCENTAGES

The eleven primary management factors affecting calf crop percent are as follows:
1. Heifers need to be fed adequate levels of a balanced ration to reach puberty at 15 months of age if they are to calve at the desired age of 2 years of age. Heifers of the English breeds and crosses (e.g., Angus and Hereford) should weigh 650 to 700 lbs at 15 months of age. Cattle of the larger exotic breeds or crosses should weigh 100 to 150 lbs more to assure a high percentage of heifers cycling at breeding.

Table 19-1. Break-even price (per hundred pounds) of commercial cow-calf operations with varying calf crop percentages, annual cow costs, and weaning weights

Calf crop percent weaned	Annual cow cost	Average calf weight at weaning (Pounds)		
		350	450	550
95	$350	$105.26	$ 81.87	$66.99
	300	90.23	70.18	57.42
	250	75.19	58.48	47.85
85	350	117.65	91.50	74.87
	300	100.84	78.43	64.17
	250	84.03	65.36	53.48
75	350	133.33	103.70	84.85
	300	114.28	88.89	72.73
	250	95.24	74.07	60.61

2. Heifers should be bred to calve early in the calving season. Heifers calving early are more likely to be pregnant as three-year-olds, while heifers calving late will likely not conceive during the next breeding season. Some producers save more heifers as potential replacements at weaning and as yearlings so selection for early pregnancy can be made at pregnancy test time (Figure 19-4).

3. Heifers typically have a longer postpartum interval than cows. This interval becomes shorter if first-calf heifers (heifers with their first calves) are separated from mature cows 60 days prior to calving and also after calving. This separation allows the heifers to obtain their share of the feed essential for a rapid return to estrus.

4. Feeding programs are designed to allow the cows and heifers to be in a moderate body condition (visually estimated by the fat over the back and ribs) at the time of calving. Thin cows at calving usually have a longer postpartum interval. Cows that are too fat reflect a higher feed cost than what is necessary for high and efficient production.

5. Cows, particularly first-calf heifers, should be observed every few hours at calving time. Some of the females will have difficulty calving (**dystocia**) and will need assistance in the delivery of the calf. Calving difficulties should be kept to a minimum to prevent potential death of calves and cows. Calving difficulty is also undesirable because cows given assistance will usually have longer postpartum intervals.

6. Calving difficulty should be minimized but not necessarily eliminated. A balance should be maintained between number of calves born alive and the weight of the calves at weaning. Heavier calves at birth usually have heavier weaning weights. However, when calves are too heavy at birth, the death loss of the calves increases. Birth weight is the primary cause of calving difficulty; therefore, management decisions should be made to keep birth weights of the calves only moderately heavy. Bulls of the larger breeds should not be bred to heifers, and large, extremely growthy bulls even in the breeds known for calving ease should not be bred to heifers. Birth weight within a herd is influenced by genetics. The genetic differences are more important than certain environmental differences such as amount of feed during gestation. Bulls, to be used artificially, should have extensive progeny test records for birth weight and calving ease in addition to an individual birth weight record.

7. The bulls' role in affecting pregnancy rate has a marked influence on calf crop percentage. Prior to breeding, bulls should be evaluated for breeding soundness by addressing physical conformation and skeletal soundness, palpating the genital organs, measuring scrotal circumference, and testing the semen for motility and morphology. **Libido** (sex drive) and mating capacity are additional important factors in how the bull affects pregnancy rate. These traits are not easily measured in individual bulls prior to breeding, or even after breeding in multiple-sire herds. The typical cow-to-bull ratio is quoted by most cattle producers as 30 to 1. However, some bulls can settle more than 50 cows in a 60-day breeding season. In some large pastures with rough terrain, the cow-to-bull ratio may have to be only 10 to 1 to assure a high calf crop percent.

8. Crossbreeding affects percent calf crop in several ways. Crossbred heifers usually cycle earlier and have higher conception rates than their straightbred counterparts. Crossbred calves are more vigorous and have a higher survival rate. An effective crossbreeding program can increase the calf crop by 8% to 12%.

Figure 19-4. Percent calf crop, as measured by a live calf born and raised per cow, is the most economically important trait for the cow-calf producer. The bull, cow, calf, and producer each make a meaningful contribution to the level of productivity for this trait.

9. The primary nutritional factor influencing calf crop percent is energy expressed in terms of total pounds of feed. The amount of feed (TDN) is important in helping initiate puberty, maintaining the proper body condition at calving, and keeping the postpartum interval relatively short. Other nutrients of major importance are protein, salt (sodium chloride), and phosphorous. Additional vitamins and minerals are only important in areas where the soil or feed are deficient in them.

10. Calf losses during gestation are usually low (2% to 3%) unless certain diseases are present in the herd. Serious reproductive diseases such as brucellosis, leptospirosis, vibriosis, and infectious bovine rhinotracheitis (IBR) can cause abortions which can markedly reduce the calf crop percent. These diseases can be managed by blood testing animals entering the herd or vaccinating for the diseases. Herd health programs are different for different operations depending on the incidence of the diseases in the area. The details of these programs should be worked out with the local veterinarian.

11. Calf losses after 1 to 2 days following birth are usually small (2% to 3%) in most cow-calf operations. Severe weather problems such as spring blizzards, can cause high calf losses where protection from the weather is limited. In certain areas and in certain years, health problems can also cause high death losses. Infectious calf scours and secondary pneumonia can occasionally reduce the potential calf crop 10% to 30%.

MANAGEMENT FOR HEAVY WEANING WEIGHTS

The seven primary management factors affecting calf weaning weights are as follows:
1. Calves born early in the calving season are heavier at weaning primarily because they are older. Calves are typically born over a several week period but are weaned

together on one specified day. Every time the cow cycles during the breeding season and fails to become pregnant, the weaning weight of her calf is reduced by 30 to 40 lbs at weaning. Most commercial producers have a breeding season of 90 days or less so that the calves are heavier at weaning and can be managed in uniform groups (Figure 19-5).

2. The amount of forage available to the cow and the calf has a marked influence on the weaning weights. The cow needs feed to produce milk for the calf. The calf, after about three months of age, will consume forage directly in addition to the milk it receives.

3. Growth stimulants, commonly given to nursing calves, will increase the weaning weight by an average of 8%. Zeranol is one of the most widely used products, and it is implanted as pellets under the skin near the base of the ear. The pellets dissolve over a period of several weeks and supply the growth-stimulating substance that is absorbed into the bloodstream. This implant, however, should not be used on bulls and heifers to be used for breeding purposes. The implant has been shown to sometimes interfere with the proper development and functioning of the reproductive organs (Figure 19-6).

4. Providing supplemental feed to the calves where the cows are unable to have access to it will increase the weaning weight of the calves. This practice of **creep feeding** needs to be used with caution because it is not always economical. It can be used on calves that have the ability to grow and not fatten. It helps calves make the transition of the weaning process and is a feasible practice under drought or marginal feed supply conditions. The creep feeding of breeding heifer calves can impair the development of their milk secretory tissue and subsequently reduce their milk production. This

Figure 19-5. This calf is approximately 7 months old and soon will be weaned or separated from its dam. Commercial cow-calf producers manage their cows so that the calves will be heavy at weaning time.

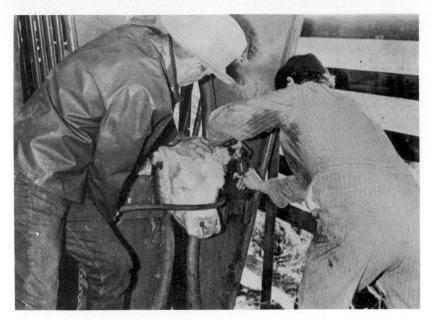

Figure 19-6. The squeeze chute is an essential piece of equipment for restraining cattle. The head can be restrained for inserting ear tags, treating eyes, implanting growth stimulants, or administering medication orally. Note the numbered ear tag, which is the most common form of individual animal identification.

impairment is believed to be caused by fat accumulating in the udder and crowding the secretory tissue.

5. Any diseases that affect the milk supply of the cow or the growth of the calf will cause a reduction in the weaning weight of the calf.

6. Genetic selection for milk production and calf growth rate will increase calf weaning weight. Adjusted 205-day weight is the trait most commonly used to select for milk and growth. Effective bull selection will account for 80% to 90% of the improvement in weaning weight, although weaning weight ratios and maternal breeding values can also be used in culling cows and selecting replacement heifers. It has been well demonstrated that effective selection can result in a 4-lb to 6-lb per year increase in weaning weight on a per calf basis.

7. Crossbreeding for the average cow-calf producer can result in a 20% increase in pounds of calf weaned per cow bred. Most of the increase occurs from improved reproductive performance; however, one-fourth to one-third of the 20% increase is due to the effect of heterosis on the growth rate of the calf and the increased milk production of the crossbred cow.

MANAGEMENT FOR LOW ANNUAL COW COSTS

In times of inflationary economic conditions, it may be difficult for a producer to lower annual cow costs. However, with careful attention, the increases can be moderated or, in some cases, kept at a similar level over several years. Adequate expense records must be maintained so that cost areas can be carefully analyzed (Figure 19-7).

Figure 19-7. Cattle producers must keep accurate income and expense records to assess annual cow costs and the profitability of their operations.

The greatest consideration should be given feed costs since they compose the largest part of annual cow costs, usually 60% to 75%. The period from the weaning of a calf until the last one-third of gestation in the next pregnancy is the time when cows can be maintained on comparatively small amounts of relatively cheap, low-quality feeds. Cow-calf operations having available crop aftermath feeds (e.g., corn stalks, grain stubble, and straw) will usually have the greatest opportunity to keep feed costs lower than other operations (Figure 19-8). Nutrient costs of the available feeds should be evaluated continually.

Figure 19-8. Corn stalk fields are available to cattle after the grain has been harvested. There are millions of acres of stalk fields and other crop aftermath feeds that can be grazed by cattle and other livestock.

Labor costs will usually compose 15% to 20% of the annual cow costs. Labor costs are usually lower on a per cow basis as herd size increases and in areas where moderate weather conditions prevail. It typically takes 15 to 20 hours of labor per cow unit per year. Operations that use labor inefficiently require twice as much.

Interest charges on operating capital account for another 10% to 15% of the annual cow cost. Producers can keep interest charges reduced by carefully analyzing the costs of different credit sources.

Cows and heifers should be palpated at approximately 45 days after the end of the breeding season to determine if pregnancy has occurred. The operator should sell open cows rather than putting further feed and other costs in them. Failing to check cows for pregnancy only contributes to higher annual cow costs, lower calf crop percentages, and higher break-even costs of the calves produced.

STOCKER-YEARLING PRODUCTION

Stocker-yearling producers handle cattle that are fed and managed for growth prior to going into a feedlot for finishing to slaughter weight. Replacement heifers that are intended to go into the breeding herd are typically included in the stocker-yearling category. However, discussion here will relate to steers and heifers being grown prior to going into the feedlot for finishing.

There are several alternate stocker-yearling production programs as identified in Figure 19-1. In some programs, one operator owns the calves from birth through the feedlot finishing phase, and the cattle are raised on the same farm or ranch. In some programs, one operator retains ownership, but the cattle are custom fed during the growing and finishing phases. In other programs, the cattle are bought and sold once, and still in other programs, they are bought and sold several times.

The primary basis of the stocker-yearling operation is to market available forage and high roughage feeds such as grass, crop residues (e.g., corn stalks, grain stubble, and beet tops), wheat pasture, and silage. Stocker-yearling operations also exist to use grazing areas that are usable only for summer grazing and are not adaptable for the production of supplemental feed for the winter.

The primary factors affecting the costs and returns of stocker-yearling operations are marketing (both purchasing and selling the cattle), the gaining ability of the cattle, amount and quality of the available forage or roughage, and the health of the cattle.

Stocker-yearling producers need to be aware of current market prices both for cattle they might purchase or sell. They also need to understand the loss of weight of the cattle from the time of purchase until the cattle are delivered to their farm or ranch. This loss in weight is called shrink and can sometimes reflect the difference in the profit or loss of the stocker-yearling operation. Therefore it is common for calves and yearlings to shrink 6% to 12% from purchased weight to delivered weight. For example, yearlings purchased at 700 lbs that shrink 8% will have a delivered weight of 644 lbs. It typically takes 2 to 3 weeks to recover the weight loss.

The gaining ability of most stocker-yearling cattle is estimated visually. Cattle that are light weight for their age, thin but healthy, with a relatively large skeletal frame size, are those considered as having a high gain potential. Cattle that are light for their age are typically most profitable for the stocker-yearling operator, whereas cattle that are heavy for their age are most profitable for the cow-calf producer.

Stocker-yearling cattle that are purchased and sold several times encounter stress situations of fatigue, hunger, thirst, and exposure to many disease organisms. The more common diseases are shipping fever complex and other respiratory diseases. These stress conditions make it necessary for stocker-yearling producers to have effective health programs for newly purchased cattle. Producers who have poor herd health programs will typically experience higher costs of gain and higher death losses.

The primary objective of the stocker-yearling operation is to obtain the most pounds of cattle gain within economic reason, while having assurance that high-quality forage yields can be obtained consistently each year. Forage management to obtain efficient production and consumption of nutritious feed is another essential ingredient of a successful stocker-yearling operation. Time of grazing and intensity of grazing (number of animals per acre) are important considerations if maximum forage production and utilization are to be maintained.

FEEDLOT CATTLE PRODUCTION

Feedlot cattle are those cattle being fed for slaughter in small pens or areas where harvested feed is brought to them. Some cattle are finished to slaughter weights on pasture, but they represent only 10% to 15% of the slaughter steers and heifers. They are sometimes referred to as nonfed cattle as they are fed little, if any, grain or concentrate feeds.

The cattle feeding areas in the United States are shown in Figure 19-9. These areas correspond to the primary feed-producing areas where cultivated grains and roughages are grown. These locations are determined primarily by soil type, growing season, and amount of rainfall or irrigation water. Figure 19-10 shows where the approximately 25 million feedlot cattle are fed in the various states. Numbers are shown for only 23 states because very few cattle are fed in the other states. Figure 19-11 identifies the states that have shown increases and decreases in fed cattle numbers over the past 10 years. These changes give

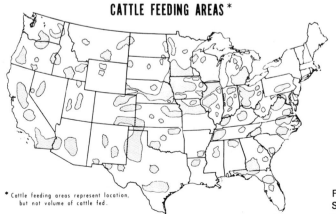

CATTLE FEEDING AREAS *

* Cattle feeding areas represent location, but not volume of cattle fed.

Figure 19-9. Cattle feeding areas in the United States. The areas represent location, but not volume of cattle fed.

Figure 19-10. Fed cattle marketings 1980 (1,000 head). For example, Kansas marketed 1,125,000 more cattle in 1980 than in 1970. This is the largest increase in the United States.

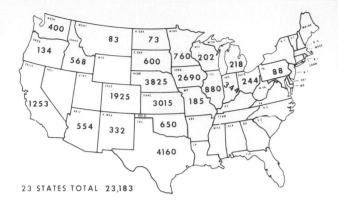

23 STATES TOTAL 23,183

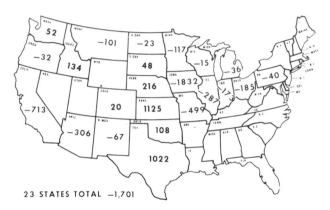

23 STATES TOTAL —1,701

Figure 19-11. Change in fed cattle marketings 1970 to 1980 (1,000 head). For example, Kansas marketed 1,125,000 more cattle in 1980 than in 1970. This is the largest increase in the United States.

evidence that the Plains states are becoming increasingly influential as the cattle feeding area of the United States.

TYPES OF CATTLE FEEDING OPERATIONS

Two basic types of cattle feeding operations are (1) commercial feeders and (2) farmer-feeders. These two types are generally distinguished by type of ownership and size of feedlot (Figure 19-12 and Figure 19-13). The farmer-feeder operation is usually owned and operated by an individual or family. The commercial feedlot may be owned by an individual or partnership, but quite often a corporation owns it, especially as feedlot size increases.

The two types of feeding operations are usually referred to as those over 1,000 head feedlot capacity (commercial) and those under 1,000 head (farmer-feeder). Approximately 70% of the fed cattle are fed in feedlots with over 1,000 head capacity, while 30% of the cattle are fed in feedlots under 1,000 head capacity. A number of commercial feedlots have capacities of 40,000 head or higher, and a few have capacities of over 100,000 head. Some of the commercial feedlots custom-feed cattle where someone else owns the cattle and the commercial feedlot provides the feed and feeding service.

Each of the two types of feeding operations has different advantages and disadvantages. The larger commercial feedlots usually have some economic advantages associated

Figure 19-12. A large commercial feedlot where thousands of cattle can be easily fed in fenceline bunks using feed trucks.

Figure 19-13. A farmer feedlot. In this operation the feed is fed in troughs located within the pen.

with size and have more professional expertise in nutrition, health, marketing, and financing. The farmer-feeder has the advantages of distributing labor over several enterprises using high-roughage feeds effectively, creating a market of home-grown feeds through cattle, and more easily closing down the feeding operation during times of unprofitable returns.

MANAGING A FEEDLOT OPERATION

The primary factors needed to analyze and manage a feedlot operation properly are the investment in facilities, cost of feeder cattle, feed cost per pound of gain, nonfeed costs per pound of gain, and marketing of the cattle. A more detailed analysis of these factors is shown in Table 19-2.

The investment in facilities varies with the type and location of the feedlot. The larger commercial feedlots are quite similar regardless of where they are located in the United

Table 19-2. The major component parts of a feedlot business analysis

Major component with primary factors influencing it				
Investment in facilities	Cost of feeder cattle	Feed cost per pound of gain	Nonfeed cost per pound of gain	Total dollars received
Land	Grade	Ration	Death loss	Market choice
Pens	Weight	Rate of gain	Labor	Transportation
Equipment	Shrink	Feed efficiency	Taxes	Shrink
Feed mill	Transportation	Length of feeding time	Insurance	Dressing percentage
Office	Gain potential		Utilities	Quality grade
			Veterinary expenses	Yield grade
			Repairs	Manure value

States. The general layout is an open lot of dirt pens with pen capacities varying from 100 to 500 head. The pens are sometimes mounded in the center to provide a dry resting area for cattle. The fences are pole, cable, or pipe. A feedmill, to process grains and other feeds, is usually a part of the feedlot. Special trucks distribute feed to fenceline feedbunks where cattle stand and eat inside the pens. Bunker trench silos hold corn silage and other roughages. Grains may be stored in these silos; however, they are more often stored in steel bins above the ground. The investment cost per head for this type of feedlot is approximately $100.

Feedlots for farmer-feeder operations vary from unpaved, wood-fenced pens to paved lots with windbreaks, sheds, or total confinement buildings. The latter may have manure collection pits located under the cattle, which stand on slotted floors. The feed may be stored in airtight structures. In most farmer-feeder operations, however, feeds are stored in upright silos and grain bins, particularly where rainfall is high. Feeds are typically processed on the farm and distributed with tractor-powered equipment to feedbunks located either inside or outside the pens. Investment costs for these feedlots will vary from $200 to $500 per head.

Before buying feeder cattle, the feedlot operator first estimates anticipated feed costs and the price the fed slaughter cattle will bring. These figures are then used to project the **cost of the feeder cattle** or what the operator can afford to pay for them. Feeder cattle are priced according to weight, sex, **fill** (content of the digestive tract), skeletal size, thickness, and body condition. Most commercial feeders prefer to buy cattle with **compensatory** gain. These are cattle that are thin and relatively old for their weight. They have usually been grown out on a relatively low level of feed. When placed on feedlot rations, they gain very rapidly and compensate for their previous lack of feed.

The feeder cattle buyer typically projects a high gain potential in cattle that have a large skeletal frame and very little finish or body condition. However, not all cattle of this type will gain fast.

Heifers are usually priced a few cents a pound under steers of similar weight. The primary reason is that heifers gain more slowly and the cost per pound of gain is higher.

Feeder cattle of the same weight, sex, frame size, and body condition can vary several cents a pound in cost. This value difference is usually due to differences in fill. The differences in fill can amount to 10 to 40 lbs in the live weight of the feeder cattle. Feed and water consumed prior to weighing, distance and time of shipping, temperature, and the manner in which the cattle are loaded and transported are some of the major factors affecting the amount of shrink. Shrink results primarily from loss of fill, but weight losses can occur in other parts of the body as well.

Feed costs per pound of gain form the major costs of putting additional weight on the feeder cattle. Typically, feed costs will be 60% to 75% of the total costs of gain.

Feed costs per pound of gain are influenced by several factors and the knowledge of these factors is important for proper management decisions. The choice of feed ingredients and how they are processed and fed are key decisions that affect feed costs.

Cattle that can gain more rapidly and efficiently on the same feeding program will have lower feed costs. Some of these differences are genetic and can sometimes be identified with the specific producer of the cattle. Most feeder cattle receive feed additives (e.g., rumensin) and ear implants (e.g., Ralgro® or Synovex®) that improve gain and efficiency and eventually the cost of gain (Figures 19-14 and 19-15).

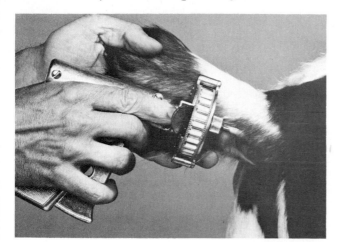

Figure 19-14. Implanting a steer with the growth stimulant Ralgro®.

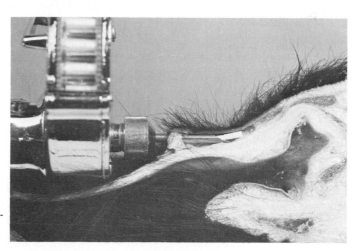

Figure 19-15. Cut-away section of the ear showing the location of three Ralgro® pellets. Note that they are located approximately one inch from the base of the ear, just under the skin.

Feed cost per pound of gain gets progressively higher as days on feed increase. Therefore, cattle feeders should be careful to avoid feeding cattle beyond their optimum combination of slaughter weight, quality grade, and yield grade.

Nonfeed cost per pound of gain is sometimes referred to as yardage cost. Yardage cost includes all the other costs of gain other than feed. These costs can be expressed as either the cost per pound of gain or the cost per head per day. Obviously, cattle that can gain faster will move in and out of the feedlot sooner and accumulate fewer total dollar yardage costs.

Death loss and veterinary expenses caused by feeder cattle health problems can increase the nonfeed costs significantly. Most cattle feeders prefer to feed yearlings rather than calves beause the death loss and health problems in yearlings are significantly lower than in calves.

Total dollars received for the slaughter cattle emphasize the need for the cattle feeder to be aware of marketing alternatives and the kind of carcasses the cattle will produce. Most slaughter cattle from the feedlots are sold directly to the packer, through a terminal market when professionals make the marketing transaction, or through an auction where the cattle are sold to the highest bidder. Nearly 80% of the fed cattle are sold direct to a packer. This marketing alternative requires the cattle feeder to be aware of current market prices for the weight and grade of cattle being sold.

Many of the slaughter cattle at large commercial feedlots are sold on a standard shrink of 4%, with the cattle being weighed at the feedlot without being fed the morning of weigh day. Feeders who ship their cattle some distance before a sale weight is taken should manage their cattle to minimize the shrink.

Most of the cattle are sold on a live weight basis with the buyer estimating the carcass weight, quality grade, and yield grade. Slaughter cattle of yield grades 4 and 5 are usually discounted quite heavily in price. The price spread between the Good and Choice quality grades can vary considerably over time. Some marketing alternatives will not show a price differential between Good and Choice if the cattle have been well fed for a minimum number of days (120 days for yearling feeder cattle).

Cattle feeders who manage their cattle consistently for profitable returns know how to purchase high-performing cattle at a reasonable purchase price. These cattle feeders will formulate rations that will maximize performance at the lowest possible feed costs. Also their cattle have minimum health problems with a low death loss. The feeder feeds the cattle the minimum number of days required to assure carcass acceptability and palatability and develops a marketing plan that will yield the maximum financial returns.

STUDY QUESTIONS

1. What is meant by a break-even price for a cow-calf producer? Show how a break-even price is calculated.
2. Why is calf crop percentage such an important trait for the cow-calf producer?
3. Tell how crossbreeding can improve both calf crop percentage and weaning weights.
4. What role does a stocker-yearling operator play in producing beef cattle?
5. Name some of the most apparent differences between a commercial feeder and a farmer-feeder.
6. What is meant by a yardage cost for the feeder? What are some of the primary factors influencing yardage cost?
7. What can a cow-calf producer do to keep calving difficulty at a relatively low level and yet produce calves that can weigh heavy at weaning time?

8. Explain how nutrition affects calf crop percent, weaning weight, and annual cow cost.
9. What is meant by shrink? Discuss how it can affect the cow-calf producer, yearling-stocker operator, and feeder. What are some of the main factors affecting shrink?
10. Identify the areas in the United States where cow-calf production and feedlots are most heavily concentrated.

SELECTED REFERENCES

Dyer, I. A. and O'Mary, C. C. 1977. *The Feedlot.* 2nd edition. Philadelphia: Lea and Febiger.

Ensminger, M. E. 1976. *Beef Cattle Science.* 5th edition. Danville, Illinois: Interstate Printers and Publishers.

Lasley, J. F. 1981. *Beef Cattle Production.* Englewood Cliffs, New Jersey: Prentice-Hall, Inc.

Neumann, A. L. 1977. *Beef Cattle.* 7th edition. New York: John Wiley and Sons.

O'Mary, C. C. and Dyer, I. A. 1972. *Commercial Beef Cattle Production.* Philadelphia: Lea and Febiger.

Preston, T. R. and Willis, M. B. 1974. *Intensive Beef Production.* 2nd edition. New York: Pergamon Press.

Price, D. P. 1981. *Beef Production.* Dalhart, Texas: Southwest Scientific.

20

Dairy Cattle Breeds and Breeding

The dairy cow could be considered a foster mother because many human babies have started early life by consuming cows' milk from a bottle. Milk is probably the most important food in nutrition, particularly for young people. Production of milk per cow has been increased markedly in the past 50 years by improvements in breeding, feeding, sanitation, and management.

The number of milking cows is shown by states for 1978 in Figure 20-1. It can be seen from Figure 20-1 that Wisconsin has more dairy cows than any other state and that New York, California, Minnesota and Pennsylvania are also important dairy states.

CHARACTERISTICS OF BREEDS

Six major breeds of dairy cattle (Holstein, Ayrshire, Brown Swiss, Guernsey, Jersey, and Shorthorn) are used for milk production in the United States.

Figure 20-1. The number of milking cows by states (× 1,000).

Table 20-1 summarizes the production characteristics of these six breeds and gives other information about them. The breeds are shown in Figure 20-2 and the external parts of a dairy cow are shown in Figure 20-3.

The average productive life of a dairy cow is short (approximately 3 years), because many cows must be culled due to low milk yield, udder breakdown, feet and leg weaknesses and disease. Dairy cows and goats to be culled because of low milk production should be culled during or following the first lactation.

Holstein cows produce extremely large amounts of milk—up to 100 lbs per day at the time of peak lactation. Thus, in high milk producing cows, great stress is placed on the udder, which can, as a result, cause the ligaments to no longer hold up the udder, called breakdown. If an udder breaks down, it is more susceptable to injury and disease, often necessitating culling the cow.

Ayrshires and Brown Swiss produce milk for more years than Holsteins, but their production levels are lower.

In fact, efficiency of milk production per 100 lb body weight is about the same for the major breeds of dairy cows.

Guernsey and Jersey cows produce milk having high percentages of milk fat and solids-not-fat, but the total amount of milk produced is relatively low. The efficiency of energy production of the different breeds varies less among the breeds than do either total quantity of milk produced or percentage of fat in the milk.

IMPROVEMENT IN MILK PRODUCTION

Great strides have been made over the past 50 years in improving milk production through improved management and breeding. For example, the amount of milk produced in the United States in 1979 was about the same as that produced in 1945, yet the number of dairy cows in 1979 was less than half the number in 1945. Thus, the average production per cow in 1979 was more than double that in 1945. In the past five years alone, the annual milk production by the major breeds of dairy cattle in the United States has increased by a total of 54 lbs per cow per year.

SELECTION OF DAIRY CATTLE

In the past, dairy operators have placed great emphasis on "dairy type" (a cow of ideal dairy type would have refinement, well-attached udders, prominent milk veins, and strong feet and legs), but recent studies (White and Vinson, 1976) indicate that some of the components of dairy type are negatively related to amount of milk produced. Therefore, selection for dairy type may have little or no value and might even be deleterious if overstressed in selection. However, a properly attached udder and strong feet and legs are the best indicators that a cow will remain a high producer for a long time. When used to replace cows that are **culled** for low production, heifers whose ancestral records indicate they will be high producers will raise the level of production in the herd significantly. Such heifers should also be used to replace cows that leave the herd because of infertility, mastitis, or death, although the improvement gained thereby is generally modest.

The Sire Index, computed by the U.S. Department of Agriculture, is based on comparing daughters of a given sire with their contemporary herd mates. Each sire is assigned a

Table 20-1. Characteristics of the breeds of dairy cattle

	Holstein	Aryshire	Brown Swiss	Guernsey	Jersey	Shorthorn
Origin	Holland	Scotland	Switzerland	Guernsey Island	Jersey Island	England
Weight						
Male	2,200 lb 1,000 kg	1,850 lb 839 kg	2,000 lb 908 kg	1,600 lb 728 kg	1,500 lb 681 kg	2,000 lb 908 kg
Female	1,500 lb 681 kg	1,200 lb 545 kg	1,400 lb 636 kg	1,100 lb 499 kg	1,000 lb 454 kg	1,400 lb 636 kg
Color	Black and white	Mahogany and white spotted, may have pigmented legs	Solid blackish; hairs dark with light tips	Light red and white, yellow skin	Blackish hairs have white tips to give gray color, or red tips to give fawn color; also can be solid black or white spotted	Red, roan, or white, or red and white or roan and white
Yearly milk yield (1979) for DHIA official records	Very high 15,014 lb	Intermediate 11,839 lb	Intermediate 12,368 lb	Low 10,858 lb	Low 10,231 lb	Low 10,451 lb
Percent fat	Very low (3.7%)	Intermediate (4.2%)	Intermediate (4.2%)	High (4.9%)	Very high (5.4%)	Low to intermediate (3.9%)
Udders	Very large	Large, strong	Large, strong	Medium size, strong	Small, strong	Large, strong

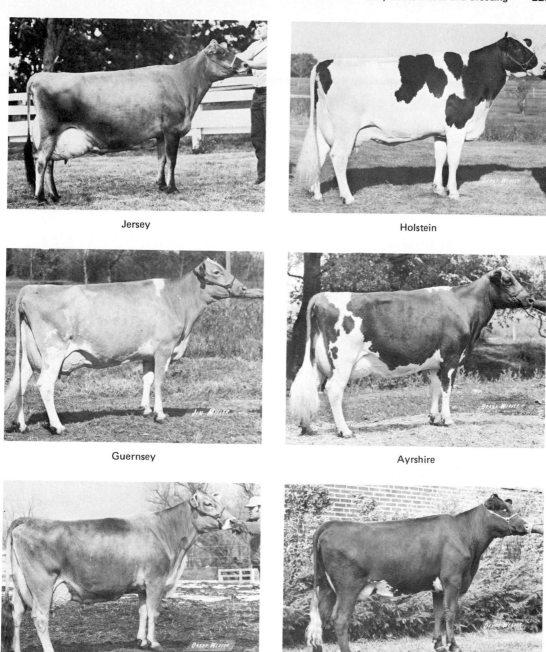

Figure 20-2. Major breeds of dairy cows used for milk production in the United States.

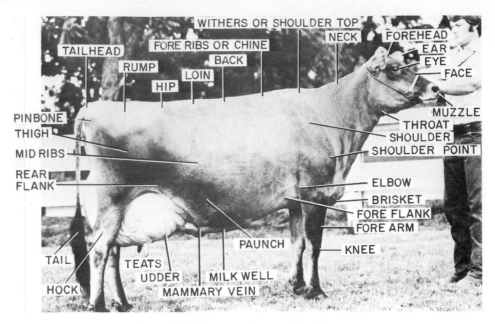

Figure 20-3. The external parts of the dairy cow.

predicted difference between himself and other sires based on the extent to which his daughters are superior or inferior to their herd mates. Many sires have daughters in 50 to 100 different herds, so the predicted difference is generally highly reliable and provides a sound basis for the selection of semen to be used for inseminating cows. The USDA publishes predicted differences between sires semiannually.

The best basis for evaluating a dairy cow is the quality (total solids) and quantity of milk that she produces. A good cow produces 10 to 15 times her body weight in one lactation. The four sources of information used in evaluating breeding dairy animals are the animal itself, its ancestors, its collateral relatives, and its progeny. The estimates of breeding value for yield are available through the USDA Dairy Herd Improvement Association for all cows enrolled in the official dairy herd improvement (DHI) and dairy herd improvement record (DHIR) testing plans and for bulls with sufficient daughters for a Dairy Herd Improvement Association (USDA-DHIA) sire summary. The most accurate estimate of transmitting ability of bulls is the USDA-DHIA Modified Contemporary Comparison (MCC). The second most accurate method is herd-mate comparisons of daughters. The dam-daughter comparison is third in accuracy and it is followed by the average of the daughters with no selection of information. The best cow indices for milk production are the USDA-DHIA cow index, followed by deviations from herd mates, daughter-dam comparisons, the cow's own lactation record, and selected records such as the highest production record. Certainly as more dairy cows are housed on hard floors, difficulties in maintaining healthy feet and legs may be expected to increase unless selection for strength in feet and legs is practiced.

A registry association for purebred animals exists for each of the breeds of dairy cattle in the United States. In addition, certain selective registries honor cows with outstanding production performance and bulls with daughters that are outstanding in production. The Holstein selective registry, for example, has the "gold medal dam" and the "gold medal sire" awards. A gold medal cow must produce 100,000 lbs of milk or an average of 12,500 lbs of milk and 500 lbs of milk fat per year during her productive life; she must also score 80 points as a 2- or 3-year-old, 81 as a 4- or 5-year-old, and 82 at 6 years of age or older on the basis of 100 points for perfection, when points are accumulated on a score card for scores assigned to the various parts of the animal. The gold medal cow should also have three daughters that meet the requirements of milk production mentioned and three male or female offspring that meet the classification (score) standards. A gold medal sire must have 10 or more daughters that produce an average of at least 13,000 lbs of milk and 585 lbs of milk fat or more per year during their productive lives. Each of these daughters must score at least 82 points in classification. Other breed associations have selective registries as a means of encouraging improvement in production.

BREEDING DAIRY CATTLE

Milk production, milk composition, efficiency of production, and characteristics that indicate that a cow will likely remain productive for several years are all highly important in selecting dairy cows. These traits are usually emphasized according to their relative heritability and economic importance. Milk and milk fat production are 20% to 30% heritable, and udder attachment is 30% heritable. Fertility is extremely important but low in heritability, so marked improvement in this trait is more likely to be accomplished environmentally through good nutrition and management than through selection.

Most dairy cattle in the United States are straightbred, because crossing breeds has failed to stimulate improvements. No combination of breeds, for example, equals the Holstein straightbred in total milk production. Because milk production is controlled by many genes and the effect of each of the genes is unknown, it is impossible to manipulate genes that control milk production by the same method that one can use in a case of simple inheritance. Also, milk production is a sex-limited trait expressed only in the female. Furthermore, milk production is highly influenced by the environment. To improve milk production, the environment must be standardized among all animals present to ensure, insofar as possible, that differences between animals are due to inheritance rather than environment.

In addition to a high level of milk production, the characteristics of longevity, regularity of breeding, ease of milking, and quiet disposition are important in dairy cows. Although it is difficult to predict whether or not a young cow will have longevity of production, such longevity can possibly be enhanced by replacing culled cows with heifers born to dams that are highly productive when mature. Milk production, milk composition, efficiency of production, and characteristics that indicate that a cow will likely remain productive for several years are all highly important in selecting dairy cows.

Bulls are evaluated for their ability to transmit the characteristic of high-level milk production both by considering the production level of their ancestors (pedigree) and by considering the production level of their daughters (progeny testing). An index is used as a predictive evaluation of the bull's ability to transmit the characteristic of high-level milk production.

TESTING DAIRY COWS

For a dairy operator to know which cows are good producers and which should be culled, a record of milk production is essential.

DHIA programs in which the USDA and University Extension Service personnel work with dairy operators help improve milk production. Records obtained are analyzed so that the dairy operator knows how each cow compares with all cows in the herd and how the herd compares with other herds in the area. One can use several testing plans; some are official and others are unofficial, but any of them may be useful to a dairy operator interested in improving production in a herd. Six important types of testing plans follow.

Official DHIA is the most common testing plan. All cows must be properly identified and all DHI rules enforced. The results of official records may be published and are used by the USDA to evaluate sires.

Official DHIR uses the same rules and procedures as the DHIA. The main difference is that DHIR records are sent to the offices of the breed clubs and extra surprise tests may be conducted. The DHIR program is for registered cattle only, whereas the DHIA program is for all cattle.

Owner-sampler records form a program in which the herd owner, rather than the DHIA supervisor, records the milk weights and takes the samples. The information recorded is the same as in official tests, but the records are for private use and are not published.

Tester-sampler records are similar to official records in that the DHIA supervisor samples and weighs the milk. The records are unofficial, however, and the enforcement of cow identification and other DHIA rules is less rigid than in official tests. These records are not for publication.

"A.M.-P.M." records are also unofficial. In this plan the herd is tested each month throughout the year, but the supervisor takes only the morning (A.M.) milking for one month and the afternoon (P.M.) milking for the second month. Each daily milk weight is doubled to determine the daily milk weight for calculation by the test interval method. The daily milk weights printed on the herd report are the average daily milk weights for the last two consecutive tests (one A.M. and one P.M.).

WADAM ("weigh a day a month") records are unofficial records in which only milk weights are used; no tests are made. Each cow's record and the herd summary are based only on the amount of milk produced. Unofficial records may also be combined in that owner-sampler records can be A.M.-P.M. or WADAM. It is also possible to combine WADAM and A.M.-P.M.

The dairy industry (breed associations, National Artificial Breeders Association, and the USDA) now use the "Best Linear Unbiased Prediction" (BLUP) method for estimating predicted differences among sires. This method was developed largely at Cornell University by Dr. C. R. Henderson. The **BLUP** procedure accounts for genetic competition among bulls within a herd, genetic progress of the breed over generations, pedigree information available on young bulls, and differing numbers of herdmate's sires, and partially accounts for the differential culling of daughters among sires. In the BLUP method direct comparisons are made among bulls that have daughters in the same herd. Indirect comparisons are made by use of bulls that have daughters in two or more herds. An example presented by the Holstein Association (1980) depicts the use of three different sires in two herds:

Herd 1

Daughters of Sire A

Daughters of Sire C

Herd 2

Daughters of Sire B

Daughters of Sire C

Direct comparisons can be made between sires B and C in herd 2. Indirect comparisons between sires A and B can be made by using the common sire C as a basis for comparison.

Pedigree information is used in the BLUP procedure to group bulls based on a pedigree index of one-half of the sire-predicted difference plus one-fourth of the maternal-grandsire predicted difference. The pedigree information is especially valuable for evaluating bulls that have few progeny, and whose evaluations are therefore of low repeatability.

The BLUP method seems to correct the problem of the herdmate comparison method, which overestimates bulls with low predicted differences and underestimates bulls with high predicted differences. This improvement in accuracy of prediction results largely from the BLUP procedure of accounting for genetic progress of a herd over generations.

The Total Performance Index (**TPI**) combines the predicted difference for milk production, the predicted difference for fat percentage, and the predicted difference for type into a single value. Since milk production and fat percent are negatively related, it is important that fat percentage be included. The Holstein Association combines these three traits by use of a ratio of 3:1:1 of predicted difference for milk, predicted difference for fat percentage, and predicted difference for type (that is, 3 times the predicted difference for milk/1 predicted difference for fat percentage/1 predicted difference for type). Using this ratio, fat percentage is maintained while milk production and type are improved.

Type has value from a sales standpoint. Although most of the components of type score are negatively related to milk production (White and Vinson, 1976), type is important as a measure of the likelihood that a cow will sustain a high level of production over several years. Feet and legs and mammary systems, including attachments, support, and quality, contribute to the ability of cows to stay in the herd.

In addition to BLUP procedures for estimating predicted differences in milk production, fat percentage, and type, the Holstein Association identifies sires that are carriers of seven undesirable genes: **bulldog** (a lethal trait characterized by large bulging head, and thick shoulders), **dwarfism**, hairless, imperfect skin, mule-foot (having one instead of two toes), **pink tooth** (congenital porphyria, pink-gray teeth, susceptibility to sunburn), and prolonged gestation.

The heritability estimate for milk production is 20% to 30%. Fat, protein and solids-not-fat have heritability estimates ranging from 20% to 50%. Scores for the components of type range from 10% to 50%, with heel depth low and upstandingness high in heritability. The heritability estimates for udder characteristics range from 5% for udder quality to 30% for rear to fore teat spacing.

STUDY QUESTIONS

1. Compare United States milk production and consumption in 1945 with figures for production and consumption in 1979.
2. Why has the crossing of breeds of dairy cattle been used only to a small extent in commercial dairying?

3.   a.   In 1945 there were 28 million dairy cows in the United States. How many are there today? What has happened to total milk production in the United States? What change has occurred in the average amount of milk produced per cow?

       b.   Of what importance is dairy type as far as milk producton is concerned?

4.   How is the Total Performance Index (TPI) computed by the Holstein Association?

5.   How do breeds of dairy cows compare with respect to efficiency of production per 100 lbs of body weight in cows?

6.   For each of the six breeds of dairy cows give annual milk production, % fat of the milk, and weight of the cows.

7.   What is meant by BLUP? What are its major advantages in providing useful information for selecting dairy bulls?

8.   Described the official DHIA program for obtaining production records on dairy cows.

SELECTED REFERENCES

Bath, D. L., Dickinson, F. N., Tucker, H. A. and Appleman, R. D. 1978. *Dairy Cattle: Principles, Practices, Problems, Profits.* Philadelphia, Pennsylvania: Lea and Febiger.

Campbell, J. R. and Marshall, R. T. 1975. *The Science of Providing Milk for Man.* New York: McGraw-Hill Book Co.

Holstein Friesian Association of America. 1980. *New Developments in Holstein Type Evaluations.* BLUP. Best Linear Unbiased Prediction, TPI-Summaries. Battleboro, Vermont.

White, J. M. and Vinson, W. E. 1976. *Type in the Holstein Friesian Breeding Program.* A Holstein Science Report. Battleboro, Vermont. Holstein Friesian Association of America.

21

Managing Dairy Cattle

The dairy industry in the United States has changed greatly from the days of the family milk cow to a highly specialized industry that includes production, processing, and distribution of milk. A large investment in cows, machinery, barn, and milking parlor is necessary. Dairy operators who produce their own feed need additional money for land on which to grow the feed and machinery to produce, harvest, and process it.

FEEDING DAIRY CATTLE

The dairy operator must provide feed to keep all animals healthy and, in addition, lactating cows must receive ample feed of good quality for milk production; however, these feeds must be provided at a low cost.

Lactating cows. The average milk production in a lactation period of 305 days is 12,000 lbs for all cows on DHIR, but some cows produce several times this amount (records of 50,000 lbs of milk in a lactation are known). Thus, some lactating cows may produce more than 147 lbs of milk and more than 5 lbs of milk fat per day. A great need for energy is created by lactation. For example, a cow weighing 1,400 lbs that produces 40 lbs milk daily needs 1.25 times as much energy for lactation as for maintenance. If this cow is producing 80 lbs of milk daily, she will need 2.5 times the energy for milk production than is needed for maintenance.

Some dairy operators allow lactating cows a free-choice of good-quality legume hay and corn silage and feed concentrates according to the amount of milk each cow gives. Chopped hay and silage can be delivered to mangers on each side of an open alleyway (Figure 21-1). In fact, some dairy operators feed concentrates in the milking parlor, where the concentrates are released automatically in relation to the amount of milk a cow gives. Some economic advantage lies in feeding concentrates in relation to quantity of milk produced. However, this method tends to feed a cow that is declining in production too generously and to feed a cow that is increasing in production inadequately.

235

Figure 21-1. Arrangement for feeding chopped hay and silage to dairy cows. The feed is delivered to the feeding area from a truck as the truck proceeds along the open alleyway.

It is a poor practice economically to allow cows in their last two months of lactation to have all the concentrates that they could consume because heavy feeding at this stage of lactation does not result in increased milk.

Young (2-year old) cows that are genetically high-producing animals should be fed large amounts of concentrates to provide the nutrition they need to grow and produce milk. If young cows are provided inadequate nutrition, their subsequent breeding and lactation can be hindered.

Forages such as alfalfa hay are high in calcium and low in phosphorus, whereas grains are high in phosphorus. If the concentrates that are fed include oil meals, the balance of calcium and phosphorus is adequate, but if most of the protein is supplied with alfalfa hay, some additional phosphorus will be needed. Corn is low in manganese, therefore, if corn is used as the primary energy source along with corn silage as the roughage, manganese should be added to the diet. There is usually no other specific need except that salt should be available to the cows at all times.

Feeding pregnant dry cows and heifers. Good summer pasture provides all of the nutritional needs of pregnant **dry cows** and heifers until the last one-third of pregnancy. Some concentrates should then be given, especially to heifers. One can run dairy cows on pasture or keep them in dry lot and bring forage feeds to them. Allowing cows to graze on pastures eliminates the labor costs of harvesting the forage, but much of the forage is wasted because cows trample it, soil it with manure, and push it into moist soil by lying on it. It is estimated that cows waste about 30% of the forage in a pasture. For these reasons, some dairy operators prefer to keep cows in dry lot and feed them harvested forage. In winter, dry pregnant animals can be fed good-quality hay and silage to provide their nutritional needs until they are in the last one-third of pregnancy, after which they should be fed concentrates at 0.5% of their body weight. The concentrate feeding should be increased to 1% of their body weight during the last three weeks of pregnancy to provide for high milk production after **freshening** (giving birth to the calf). By the time that a cow is ready to bear a calf, she weighs 125 to 200 lbs more than when she is not pregnant due to the weight of the calf and associated fluids and membranes.

Growing out young heifers. Young heifers 5 months of age and older on good pastures can probably obtain all of the feed that they need for normal growth. If the pasture is poor, they may need to be provided with good-quality legume hay and a small amount of grain. In winter, good-quality legume hay and silage usually provide all of the necessary nutrition. If the hay is grass hay and the silage is grass silage, young heifers may need a limited amount of grain to keep them growing well.

Ideally, heifers should be large enough to breed at 13 to 15 months of age so they can calve as 2-year-old animals. Guernsey and Jersey heifers should weigh 600 lbs when they are first bred; Holstein and Brown Swiss heifers, 800 lbs. If their respective weights are considerably less then these figures at 13 to 15 months of age, first breeding should be delayed until they are sufficiently large.

Feeding and care of bulls. Young bulls should be fed sufficiently to grow without becoming fat. A good pasture provides most of the nutritional needs, but it may be necessary to feed limited amounts of concentrates when the pasture is less than ideal. Bulls being used heavily for breeding should be fed legume hay, silage, and limited concentrates.

BREEDING DAIRY COWS

Proper management of dairy cows at the time of breeding, and at calving, the milking operations, and in the control of diseases is essential for a profitable dairy operation.

The time of breeding is an important phase in the management of dairy cattle. Because they are milked each day, dairy cows are more closely observed than beef cows and therefore more easily detected when in heat. When in heat, dairy cows may show restlessness, enlarged vulvas, and a decline in milk production; also, other cows tend to ride them.

Dairy cows are generally hand mated (each cow is taken to the bull for breeding) or inseminated artificially. Bulls are never run with a string of milking cows, because bulls are often dangerous. Technicians are available to artificially inseminate cattle, but individual operators who have thoroughly mastered the procedure by participating in an **artificial insemination** short course can also do it. The use of semen from outstanding, proven sires is highly desirable, even though this semen may cost more than semen from an ordinary bull. Considering the additional milk production that can be expected from heifers sired by a good bull, the extra investment can return high dividends. Semen from outstanding bulls at $50 to $100 per ampule is a better investment than semen from ordinary bulls at $10 per ampule.

Cows need a period of two months between lactations to give their bodies an opportunity to replenish fat and protein that the previous lactation may have depleted. It is advisable to dry the cow (stop milking her) two months before she is due to have her next calf.

CALVING OPERATIONS

Dairy cows that are almost ready to deliver their calves should be separated from the other cows, and each should be placed in a maternity stall that has been thoroughly cleaned and bedded with clean straw. The cow in a maternity stall can be fed there and must have access to water. A cow that delivers her **calf** without difficulty should not be disturbed, but when abnormal presentations occur, assistance is necessary. Difficulties in delivering a calf require the services of a veterinarian who should be called as early as possible.

As soon as the calf arrives, it should be wiped dry. Any membranes covering its mouth or nostrils should be removed, and its **navel** should be dipped in a tincture of iodine solution to deter infection. One should be sure newborn calves have some of the first milk (colostrum) produced by the dam, because colostrum contains antibodies to help the calf resist any invading microorganisms that might cause illness. The cow should be milked to encourage her to produce much milk.

Many commercial dairy operators dispose of bull calves shortly after the calves are born. However, some bull calves are kept for veal, and some are castrated and grown out for beef. Heifer calves are usually all grown out and most of them are kept for one lactation to determine how much milk they produce. Dairy calves that have been allowed to nurse from their dams are removed at about 3 days of age and given either milk that is not salable or milk substitutes. Grain and leafy hay are provided to young calves to encourage them to start eating dry feeds and to encourage rumen development. As soon as the calves are able to consume dry feeds (45 to 60 days) milk is removed from the diet because of its high cost.

Calves can be fed by bottle once or twice daily by allowing them only a certain amount of milk, or they may be allowed to nurse at will if the milk in their feeders is kept cold and heat lamps are provided. Whereas calves fed warm milk consume so much that digestive disorders result, those fed cold milk become chilled after drinking moderate amounts and go to the heat lamp for warmth. Thus, they do not overeat or gorge themselves.

MILKING OPERATIONS

The modern milking parlor is designed to speed the milking operation and thereby reduce labor. Cows are trained to come to the proper location in the parlor so that the milking machine cups can be applied to their teats (Figure 21-2). Prior to milking, the udder and teats are cleaned and dried, and after milking is complete, the teats are dipped into a weak iodine solution.

Modern milking machines (Figure 21-3) are designed to milk cows speedily and comfortably. Some even have a take-off device to remove the milking cup once the rate of flow drops below a designated level. Because of potential damage to the udder, under no circumstances should suction be continued after milking is completed.

It is of the utmost importance that the milking parlor and its equipment be kept sanitary. Milk is piped from the milking area into a refrigerated tank so that it can be cooled quickly to help limit bacterial growth. The pipes must be cleaned and drained well between milkings.

Regular milking times must be established. The time between milkings must be spaced equally during each 24-hour period. Many cows are milked twice per day but cows produce more milk if they are milked more often than twice per day. The reason for milking twice per day rather than more frequently is to minimize labor costs. Those persons doing the milking should be kind to the animals and should be quiet so that they do not excite the cows. Some people provide continuous music to muffle disturbing noises.

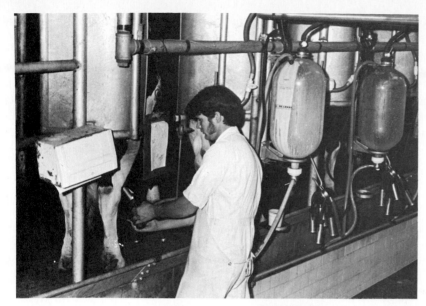

Figure 21-2. The suction cups are applied to the teats of the cow so that she can be milked by machine after she has come to her place in the milking parlor.

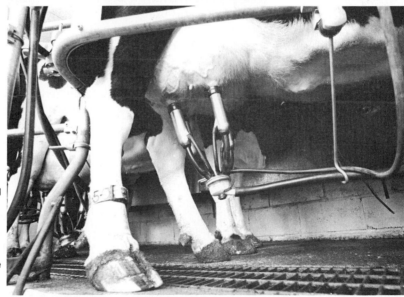

Figure 21-3. Modern milker on cow. The personnel in charge of milking observe closely to be certain that the suction cups are removed as soon as the cow has been milked unless the machine used automatically reduces its suction at that time. Injury to the udder can result if suction continues for some time after milking has been completed.

CONTROLLING DISEASES

Several diseases, such as tuberculosis and brucellosis, can affect dairy cattle. Cattle can be tested for these diseases; all animals that are reactors, that is, which show swelling at the place where killed bacteria were injected under the skin are removed.

Requirements for the production of grade A milk are that the herd must be checked regularly with the ring test for tuberculosis and must be bled and tested for brucellosis.

Milk that is low in bacteria count, has not been altered by adding water to increase the volume and comes from cows that are checked by the ring test for brucellosis and tested for tuberculosis and found to be free from both of these diseases is sold as grade A milk. For milk to grade A, the milking parlor and barn must meet specific sanitary requirements also. Grade A milk sells at a higher price than lower grades of milk; therefore, most dairy producers strive to produce grade A milk.

Perhaps the most troublesome condition in dairy cattle is mastitis, the inflammation of the udders or mammary glands that destroys tissue and impedes milk production. This disease cost the U. S. dairy industry $1.5 billion in 1979 alone for treatment expenses, reduced milk sales, and the costs of replacing afflicted cows.

In its early stage of development, mastitis causes small white clots to occur in the milk. To locate a cow with insipient mastitis, milk is collected using a strip cup that has a fine screen on a dark background so that the white clots are collected and can be seen. As mastitis progresses, the milk shows white strings; this milk is called "ropy" milk (Figure 21-4). The signs of this stage of mastitis are obvious; the udder is swollen, red, and hot, and gives pain to the cow when touched. In the final stages of mastitis, the ropy strings become dark and are present in a watery fluid. At this stage the udder has been so badly damaged that it no longer functions properly.

Susceptibility to mastitis is genetically controlled, but many environmental factors such as bruises, improper milking, and unsanitary conditions, can trigger its development. Unfortunately, high-producing animals are more likely to injure their udders than low-producing animals. Also, cows with **pendulous** udders (Figure 21-5) are more likely to develop mastitis.

The best approach to controlling mastitis is using good management techniques to prevent its outbreak. If mastitis is suspected, well established tests are available to test for it. The California mastitis test can be used on milk at the barn to determine beginning stages of mastitis. It is based on a reaction that causes some ropiness of the milk if the mastitis infection exists. The Wisconsin mastitis test is more quantitative than the California test in

 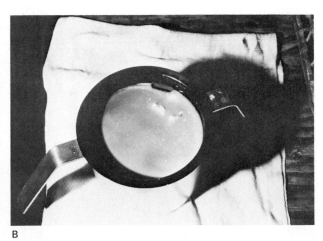

A   B

Figure 21-4. **(A)** Normal milk and **(B)** Ropey milk in strip cups.

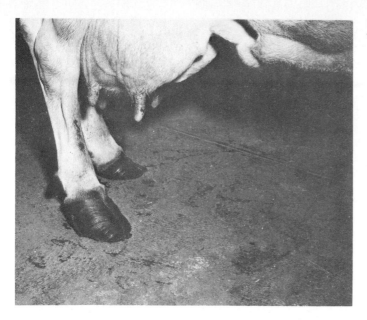

Figure 21-5. A pendulous udder afflicted with mastitis.

that it measures the agglutination that occurs in the milk when the test is applied, but it cannot be used as a barn test. Perhaps the somatic cell count, in which cells in the milk are counted, is the more effective test. Milk always contains some cells, but the number of **leucocytes** (white blood cells) increases when the udder is infected.

When mastitis is diagnosed, treatment must be started immediately if the udder is to be saved. The udder can be treated by inserting a small tube through the teat into the udder cistern and depositing **antibiotics** there, but it is best to employ the services of a veterinarian for this because injury can occur if not done properly. Milk from a cow treated with antibiotics must not be sold immediately for human consumption. Depending on the antibiotics used, milk should not be sold for human use for one to two weeks after the last treatment with antibiotics. Treating cows while they are not lactating is highly recommended.

STUDY QUESTIONS

1. What should one do as soon as a baby calf is born to assure that it has a good chance for survival?
2. How can one bottle feed baby calves on a self-service system?
3. At what age should heifers that are well developed be bred? What should heifers of each of the breeds weigh at the time that they are bred?
4. Young producing cows are still growing as well as lactating. What are the requirements for feeding such animals properly?
5. Should one feed concentrates in relation to amount of milk produced, or should cows in the peak of lactation be given all the concentrates they will consume as a means of evaluating their milk-producing ability?
6. State three important considerations that must be met in a milking operation.
7. What two things most often cause cows to go out of the herd?
8. What is the most troublesome disease with which the dairy operator must contend? Describe the tests that are used to diagnose it.

SELECTED REFERENCES

Bath, D. L., Dickinson, F. N., Tucker, H. A. and Appleman, R. D. 1978. *Dairy Cattle: Principles, Practices, Problems, Profits.* Philadelphia: Lea and Febiger.

Campbell, J. R., and Marshall, R. T. 1975. *The Science of Providing Milk for Man.* New York: McGraw-Hill Book Co.

Foley, R. C., Bath, D. L., Dickinson, F. N., and Tucker, H. A. 1972. *Dairy Cattle: Principles, Practices, Problems, Profits.* Philadelphia: Lea and Febiger.

Miller, W. J. 1979. *Dairy Cattle Feeding and Nutrition.* New York: Academic Press, Inc.

Schmidt, G. H., and Van Vleck, L. D. 1974. *Principles of Dairy Science.* San Francisco: W. H. Freeman and Co.

Wilcox, C. J., Van Horn, H. H., Harris, Jr. B., Head, H. H., Marshall, S. P., Thatcher, W. W., Webb, D. W. and Wing, J. M. 1979. *Large Dairy Herd Management.* Gainesville: University Presses of Florida.

22

Swine Breeds and Breeding

The breeds of swine are the genetic resources available to the swine producer. Most of the market pigs in the United States today are crossbreds, but these crossbreds have their base and continued perpetuation from the seedstock herds that maintain the pure breeds of swine. A knowledge of the breeds, their productive characteristics, and their crossing abilities is an important part of swine production.

CHARACTERISTICS OF SWINE BREEDS

In the early formation of the breeds of swine, breeders distinguished one breed from another through visual physical characteristics. The primary visual distinguishing characteristics for swine were, and continue to be, hair color and erect or drooping ears. Figure 22-1 shows the major breeds of swine and their primary distinguishing characteristics.

The importance of the breeds of swine can be indicated by pig registrations made by the different breed associations (Table 22-1). Although the registration numbers are strictly for purebred pigs, the numbers do reflect the demand for the breeding stock by commercial producers who produce market swine primarily by crossbreeding the purebreds. It is interesting to note how the popularity of the breeds has changed with time. This change is due to using objective measurement of productivity of the breeds and the breeds' ability to respond to the swine producer's demands and selection for higher productivity and profitability.

TRAITS AND THEIR MEASUREMENTS

Sow productivity, which is of extremely high economic importance, is measured by litter size, number reared per litter, weight at 21 or 56 days, and number of litters per sow per year. A combination of number of pigs born alive and the litter weight at 21 days best reflects sow productivity (Figure 22-2). A weight at 21 days is used to measure milk

Landrace
(white, large
drooped ears)

Chester White
(white, drooped
ears)

Yorkshire
(white, erect
ears)

Hampshire
(black with
white belt;
erect ears)

Poland China
(black with
white on legs,
snout, and tail;
drooped ears)

Berkshire
(black with
white on legs,
snout, and tail;
erect ears)

Duroc
(red, drooped
ears)

Spotted
(black and
white spots;
drooped ears)

Boar Power
(hybrids are not considered to be purebred
breeds but are used similarly to purebred
breeds. Foundation breeding of hybrids
comes from several breeds and hybrids are
bred as pure lines; no consistent color or
ear set.)

Figure 22-1. The major breeds of swine in the United States.

Table 22-1. Registration numbers of the major breeds of swine in the United States

Breed	Registration numbers			
	1955	1970-74	1978	1980
Berkshire	22,892	6,610	14,175	16,953
Chester White	20,449	19,034	55,596	58,818
Duroc	96,234	71,560	178,595	248,052
Hampshire	68,701	64,370	129,300	120,659
Landrace	5,580	6,896	39,415	82,326
Poland China	24,688	10,347	15,270	17,893
Spotted	21,912	15,714	111,205	123,248
Yorkshire	25,594	59,563	193,443	220,747

Source: National Society of Livestock Record Assn., Annual Reports.

production because the young pigs have consumed very little supplemental feed prior to that time. Sow productivity traits are lowly heritable, which means that the greatest improvement in these traits can be made through environmental changes rather than through selective breeding. Eliminating sows low in productivity while keeping highly productive sows is still a logical practice for economic reasons but not to make genetic changes. It is extremely important for commercial producers to incorporate into a cross-breeding program breeds that rank high in the maternal or sow productivity traits.

Figure 22-2. This crossbred sow farrowed 17 pigs and raised 15 of them. The number of pigs raised per sow and the weights of the litter at weaning effectively measures sow productivity.

Growth rate is economically important to most swine enterprises and has a heritability of sufficient magnitude (35%) to be included in a selection program. Seedstock producers need a scale to measure weight in relation to age (Figure 22-3). Growth rate can be expressed in several ways and typically is adjusted to some constant basis such as days required for swine to reach 220 or 230 lbs.

Feed efficiency measures the pounds of feed required per 100 lbs of gain. This trait is economically important because feed costs account for 60 to 70% of the total production costs for the commercial producer. Obtaining feed efficiency records requires keeping individual or group feed records. However, feed efficiency can be improved without keeping feed records by testing and selecting individual pigs for gain and backfat. This improvement occurs because fast-gaining pigs and lean pigs tend to be more efficient in their feed use. Obtaining feed efficiency records is done in some central boar testing stations; however, costs and other factors would need to be considered if similar records were to be taken on the farm.

Carcass traits are typically measured by ham-loin percentages, lean cut percentages, or both. Fortunately, these measures of carcass composition can be predicted reliably by backfat measurements taken on live pigs by either a metal ruler or ultrasonics (Figure 22-4).

Figure 22-3. Using the scales to measure growth rate of individual boars is an essential part of a sound breeding program.

Figure 22-4. A metal ruler is being used to probe this pig. The amount of backfat, measured in inches, will allow the swine breeder to predict the carcass desirability of a boar's offspring.

The metal probe is considerably less expensive and as accurate as the ultrasonic machines. Experienced individuals can visually predict significant differences in lean-to-fat composition in live pigs with a fairly high degree of accuracy.

Structural soundness, which with regard to swine is the capacity of breeding and slaughter animals to withstand the rigors of confinement rearing and breeding, is one of the vital needs of the swine industry today. Breeders commonly consider unsoundness as one of the results of confinement, but in reality, confinement rearing only makes unsoundness more noticeable. Some seedstock producers raise their breeding stock in confinement (similar to the manner in which most commercial producers raise their offspring) and then cull the unsound ones.

Recent studies report unsoundness to be medium in heritability; therefore, improvement can be made through selection. Soundness can be improved through visual selection if breeders decide to cull restricted, too straight legged, unsound boars lacking the proper flex at the hock, set of the shoulder, even size of toes or proper curvature and cushion to the forearm and pastern.

Inherited defects and abnormalities generally occur with a low degree of frequency across the swine population. It is important to know they do exist and that certain herds may experience a relatively high incidence. Certain structural unsoundness characteristics might be categorized as inherited defects. **Cryptorchidism** (retention of one or both testicles in the abdomen), umbilical and scrotal **hernias** (rupture) and inverted nipples (nipples go up into the teats instead of protruding), pale, soft, exuditive (PSE) carcasses are considered the most common occurring genetic defects and abonormalities. The PSE condition and the porcine stress syndrome (PSS) have occurred in recent years as a result of selection for extremes in muscling. Most of the important genetic abnormalities are inherited as a simple recessive pair of genes.

SELECTING REPLACEMENT FEMALES

The productivity of the sow herd is the foundation of commercial pork production. The sow herd also contributes half of the genetic composition of growing-finishing pigs. These factors together indicate the importance of careful selection of replacement gilts and wise decisions on their retention in the sow herd. The fastest growing, leanest gilts that are sound and from large litters should be selected for sow herd replacements (Figure 22-5). Among sows that have farrowed and will rebreed, those that have physical problems, bad dispositions, extremely small litters (two pigs below herd average), and poor mothering records should be culled.

A balance between sow culling and gilt selection needs to be established. Replacement gilts need to be available in sufficient numbers to replace culled sows. Gilts replacing sows represent the major opportunities for genetic change in the sow herd. Since sows generally produce larger litters of heavier pigs than gilts do, replacing sows with gilts may reduce production levels. This production differential and the low relationship between the performance of successive litters argue for low rates of culling based on sow performance to maintain high levels of production.

The higher productive rate of sows over gilts must be balanced against the genetic change made possible by bringing gilts into production. A total gilt replacement level of 20% to 25% is suggested for each farrowing.

Figure 22-5. This gilt has outstanding performance records for genetic superiority in the economically important traits.

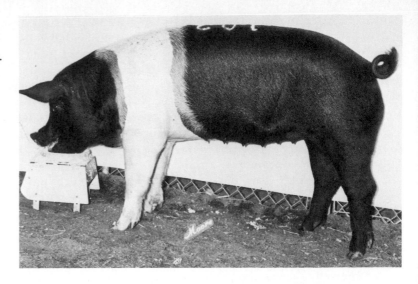

Pork producers may find economic advantages in timing the culling of sows to take advantage of high prices for sows. Consideration may also be given to possible tax savings through shifting income to capital gains by marketing young sows. Selling young sows makes a higher percentage of the hogs sold eligible for capital gains because more cull sows would be sold. Some producers choose to sell all sows after only one litter to maximize capital gains deductions.

The gilt selection/sow culling scheme suggested assumes that there are no major genetic anatagonisms between litter size and maternal performance on the one hand and rate of gain and low-backfat thickness on the other hand. There is some evidence that the so-called "very meaty gilt" does not make a good sow. There is, however, no documented evidence that selecting fast-growing, low-backfat gilts will adversely affect sow performance. Guidelines for gilt selection by age and weight are given in Table 22-2.

BOAR SELECTION

Boar selection is extremely important in making genetic change in a swine herd. Over several generations the boars selected will contribute 80 to 90% of the genetic composition of the herd. This contribution does not diminish the importance of female selection. Boars should have dams that are highly productive sows. The productivity of the replacement gilts is highly dependent on the level of sow productivity passed on by the boars.

The review of performance records is a must in selecting boars effectively (Figure 22-6). The economic importance of some of their records typically available is shown in Table 22- 3. Based on this comparison of Boar A with Boar B given in Table 22-3, if Boar A sired 700 pigs, he would return to the producer an extra $903 ($1.29 x 700). The superiority of replacement gilts retained in the herd would increase Boar A's value even more. This comparison shows the value of genetically superior boars and the justification for paying more for them.

Boars should be selected from the top 50% of the test group regardless of whether the selected boars come from a central test station or from the farm. Boars meeting the standards shown in Table 22-4 should receive serious consideration as potential herd sires.

Table 22-2. Gilt selection calendar

When	What
Birth	Identify gilts born in large litters. Hernias, cryptorchids, and other abnormalities should disqualify all gilts in a litter from which replacement gilts are to be selected.
	Record birth dates, litter size, identification.
	Equalize litter size by moving boar pigs from large litters to sows with small litters. Pigs should nurse before moving.
	Keep notes on sow behavior at time of farrowing and check on: (a) disposition, (b) length of farrow, (c) any drugs such as oxytocin administered, (d) condition of udder, (e) extended fever.
3 to 5 wk	Wean litters. Feed balanced, well-fortified diets for maximum growth and development.
	Screen gilts identified at birth by examining underlines and reject those with fewer than 12 well-spaced teats. If possible, at this time select and identify about two to three times the number needed for replacement.
180 to 200 lbs	Weigh and backfat-probe potential replacement gilts. Evaluate for soundness.
	Select for replacements the fastest growing, leanest gilts that are sound and from large litters. Save 25% to 30% more than needed for breeding.
	Remove selected gilts from market hogs. Place on restricted feed.
	Allow gilts to have exposure to a boar along the fenceline that separates them.
	Observe gilts for sexual maturity. If records are kept, give advantage to those gilts that have had several heat cycles prior to final selection.
Breeding time	Make final cull when the breeding season begins and keep sufficient extra gilts to offset the percentage of nonconception in your herd.
	Make sure all sows and gilts are ear-tagged or otherwise identified.

Source: Pork Industry Handbook, PIH-27, Guidelines for Choosing Replacement Females.

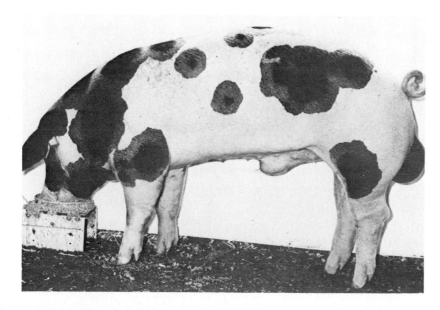

Figure 22-6. This boar has outstanding records for growth, feed efficiency, and backfat thickness. He is out of a sow that has produced several large litters of heavy pigs at weaning time.

Table 22-3. Economic value of production records of boars

	Daily gain	Feed efficiency	Adjusted backfat
	(lb/day)	(lb F/G)	(in.)
Heritability	0.30	0.40	0.50
Economic value/unit change	$4.00/lb/day	$12.00/lb F/G	$3.50/in.
Records			
Boar A	2.26	2.53	0.89
Group average	2.06	2.71	1.00
Boar B	1.86	2.89	1.11
Superiority of A over B	+0.40	-0.36	-0.22

Value of Boar A over Boar B

Trait	(Superiority) ×	(Heritability) ×	(Genetic Influence) ×	(Economic value) =	Added value/ 230-lb hog
Daily gain	0.40	0.30	0.50	$ 4.00	$0.24
Feed efficiency	0.36	0.40	0.50	12.00	0.86
Backfat	0.22	0.50	0.50	3.50	0.19
					$1.29

Source: Pork Industry Handbook, PIH-9, Boar Selection Guidelines for Commercial Producers.

Table 22-4. Suggested selection standards for replacement boars

Trait	Standard
Litter size	10 or more, farrowed, and 8 or more, weaned.
Underline	12 or more fully developed, well-spaced teats.
Feet and legs	Wide stance both front and rear, free in movement, good cushion to both front and rear feet, equal-sized toes.
Age at 230 lbs	155 days or less.
Feed/gain, boar basis (60 to 230 lbs)	275 lbs of feed/hundred pounds (cwt) of gain.
Daily gain (60 to 230 lbs)	2.00 lbs/day or more.
Backfat probe (adj. 230 lbs)	1.0 in. or less.

Source: Pork Industry Handbook, PIH-9, Boar Selection for Commercial Production.

EFFECTIVE USE OF PERFORMANCE RECORDS

Records are valuable when used in the breeding program, whereas those obtained only for promotion are eventually self-defeating. The primary reason for obtaining records should be to improve accuracy of selection. To be of value, records of different pigs must be obtained under comparable environments. The breeder who gives a small group of pigs preferential treatment is deceiving only himself or herself. If this procedure is used, the breeder can no longer compare accurately even the individuals in his or her own herd, and records thus obtained will yield false conclusions.

The ultimate objective in an improvement program is to predict an animal's breeding value. Breeding value is a measure of the animal's ability to transmit desired genetic traits to the resulting offspring. Proper records on the individual and on his or her relatives can help a great deal in predicting breeding value.

Heritability estimates are medium for performance traits and high for carcass traits. Genetic principles indicate that the most rapid rate of improvement for these traits is through measurement and subsequent selection based on the performance of the individuals being considered for selection. Thus, the breeding value of an animal for most traits can be predicted at the young age of 5 to 6 months and will generally result in more rapid genetic progress than a selection scheme based on sib or progeny tests or on pedigree information. Regardless of the types of facilities available, individual performance tests can be conducted on the farm if all animals are treated similarly. On-the-farm testing is essential to a program of rapid genetic improvement because it permits testing of a larger sample of the potential breeding population than is possible in central testing stations.

The principle of using on-the-farm records to identify the genetically superior animals is quite simple. It is a matter of standardizing environment and then measuring the traits to provide an estimate of each animal's potential breeding value. If an animal is better because of environment, this animal will breed worse than its record would suggest. By contrast, an animal that has received a poorer than average environment will probably breed better than its record would suggest. Therefore, it is important to determine if an animal is better because of environment or because of genetics. In comparing animals in different herds, this determination is difficult to make.

To compare animals within a group more objectively, indexes have been developed for use in testing stations and farm programs. The index is a single numerical value that is determined from a formula that combines the values for several traits. One such index, currently in use, combines the performance of daily gain, feed efficiency, and backfat into one numerical value for individual boars. In the formula which follows, $\overline{\text{ADG}}$ (average daily gain), $\overline{\text{F/G}}$ (feed per pound of gain), and $\overline{\text{BF}}$ (backfat) represent the averages for all the pigs in the group that are tested together. The index is as follows:

I   (index) = 100 + 60 (average daily gain of individual − $\overline{\text{ADG}}$)
          − 75 (feed efficiency of individual or group of litter mates − $\overline{\text{FG}}$)
          − 70 (backfat probe or sonoray of individual − $\overline{\text{BF}}$)

An example that shows the use of this index would be the index of boar A in Table 22-3. This index shows that boar A is 33 points superior for a combination of daily gain, feed efficiency, and backfat when compared to the average of the other boars in the test group. The index of the average boar is 100. About 20% of the boars are expected to exceed 120 index points, and 20% are expected to fall below the recommended minimum culling level of 80.

CROSSBREEDING FOR COMMERCIAL SWINE PRODUCERS

The primary function of the purebred or seedstock breeder is to provide breeding stock for the commercial producer. The breeding program of the seedstock producer will be designed to make genetic improvement in the economically important traits. Within a specific breed, the rate of improvement will be dependent on the heritability of the traits, how much selection is practiced, and how quickly generation turnover occurs. The most rapid genetic changes that have occurred over the past few decades are growth rate and leanness.

The commercial producer will also practice direct selection for the same economically important traits as does the seedstock producer. The commercial producer has an added advantage in making genetic change because crossing the various breeds of swine has given increased productivity. The extent of this genetic phenomenon of heterosis or hybrid vigor is shown in Table 22-5. The 41% improvement in total litter market weight of the crossbreds is a cumulative effect of larger litters, higher pig survival, and increased growth rate of the individual pigs. This marked increase in productivity answers the question as to why 90% of the market hogs in commercial production are crossbreds. Crossbreeding allows genetic improvement in some traits with low heritabilities such as

**Table 22-5. Effect of crossbreeding on
swine productivity (expressed as percent of purebred breeds)**

Trait	Purebred	First cross (two breeds)	Multiple cross (crossbred sow bred to purebred boar)
Number of pigs born alive	100	101	108
Litter size at weaning	100	110	124
21-day litter weight	100	110	127
Days to 230 lbs	100	107	107
Total litter market weight	100	122	141

Table 22-6. Relative reproductive performance of female breeds[a]

Trait	Breed					
	Berkshire	Chester White	Duroc	Hampshire	Landrace	Yorkshire
Number of pigs born alive	87	100	89	84	94	100
Litter size weaned	83	100	86	84	89	100
21-day litter weight	74	100	84	86	96	100
21-day litter weight/female exposed	84	100	78	94	89	94

[a]Best breed performance is given 100 and compared to each of the other breeds.
Source: Adapted from Pork Industry Handbook, PIH-39, and the *National Hog Farmer*, July 15, 1979.

sow productivity. There is little hybrid vigor for traits with high heritabilities, for example, carcass traits.

An effective crossbreeding program takes advantage of using hybrid vigor and selecting genetically superior breeding animals but also of knowing which breeds best complement one another. A producer needs to know the relative strengths and weaknesses of the available breeds. This information is shown in Tables 22-6 and 22-7. These breed comparisons can change over time as a result of genetic change made in the individual breeds. Thus, breed comparisons need to be made on a continuing basis.

The crossbred female is an integral and important part of an effective crossbreeding program. Breed cross rankings for sow productivity are shown in Table 22-8. Even though crossbreeding provides an opportunity to reap the benefits of many genetic sources, an unplanned crossing program will not yield success or profit for the pork producer. A crossbreeding system must be selected that will capitalize on heterosis, take advantage of breed strengths, and fit the individual producer's management program.

Two basic crossbreeding systems may be considered, namely, the rotational cross and the terminal cross. The rotational cross system combines two or more breeds where a different breed of boar is mated to the replacement crossbred females produced the previous generation (Figures 22-7 and 22-8). In a terminal cross system, a two-breed single or rotational cross female is mated to a boar of a third breed. Thus, all the pigs produced are sired by the same breed of boar with all offspring marketed. The female stock is usually purchased through a system primarily emphasizing reproductive performance. This latter system will maximize heterosis; however, purchasing replacement females increases the risk of introducing disease problems into the herd. A two- or three-breed rotational crossbreeding system may better fit a producer's management program even though heterosis is less than a terminal cross system. A two-breed rotation results in 67% of the heterosis of a

Table 22-7. Sire breed influence on various reproduction, production, and carcass traits.[a]

Trait	Breed							
	Berkshire	Chester White	Duroc	Hampshire	Landrace	Poland China	Spotted	Yorkshire
Number of pigs born alive	—	100	96	98	—	—	—	100
Litter size weaned	—	100	96	92	—	—	—	100
21-day litter weight	—	100	90	90	—	—	—	100
21-day litter weight per female exposed	—	—	91	81	—	—	—	100
Daily gain	96	92	100	98	93	95	98	98
Feed efficiency	94	96	100	98	90	95	96	99
Backfat thickness	87	88	88	100	78	95	86	87
Carcass length	96	95	97	98	99	94	97	100

[a]Best breed performance given 100 and compared to each breed where adequate data are available.

Source: Adapted from Pork Industry Handbook, PIH-39, and the *National Hog Farmer*, July 15, 1979.

Table 22-8. Relative reproductive performance of crossbred females[a]

Female breed cross	Litter size born alive	Litter size weaned	Litter size 21-day wt	Litter size 21-day wt/or per female exposed
Chester x Duroc	83	79	86	—
Chester x Hampshire	92	81	77	—
Chester x Yorkshire	100	100	100	100
Chester x Landrace	100	100	100	100
Hampshire x Landrace	100	95	95	88
Hampshire x Yorkshire	91	87	87	81
Berkshire x Yorkshire	90	85	83	93
Berkshire x Landrace	92	90	87	91
Berkshire x Hampshire	81	77	76	74
Duroc x Yorkshire	93	85	82	85
Duroc x Landrace	92	93	86	79
Duroc x Hampshire	86	82	79	76
Duroc x Berkshire	93	82	79	77

[a]Best breed performance is given 100 and compared to each of the other breeds.
Source: Adapted from Pork Industry Handbook, PIH-39, and the *National Hog Farmer*, July 15, 1979.

terminal cross, while a three-breed rotation gives 86% of the heterosis of a terminal crossing system.

The three-breed rotational cross is probably the most popular crossbreeding system. It combines the strong traits of a third breed not available in the other two breeds. Sires from three breeds are systematically rotated each generation and replacement crossbred females are selected each generation. These females are mated to the sire breed furthest removed in the pedigree. Because reproductive performance is to be stressed in the initial two-breed cross, growth, feed efficiency, and superior carcass composition may be lower than desired. These traits can be emphasized in the third breed choice and individual sire selection to meet the producer's overall production goals.

Although limited research information is available on the use of crossbred boars, indications are that crossbred boars are more aggressive breeders, have fewer problems in leg soundness and improve overall breeding efficiency in comparison to straightbred boars. A crossbred boar could combine those traits that may not be available in one straightbred breed.

Hybrid boars sold by some commercial companies should not be confused with crossbred boars sold by private breeder concerns. Hybrid boars are developed from specific line crosses. These lines have been selected and developed for specific traits. When specific crosses are made, the hybrid boar must be used on specific cross females to maximize heterosis in their offspring.

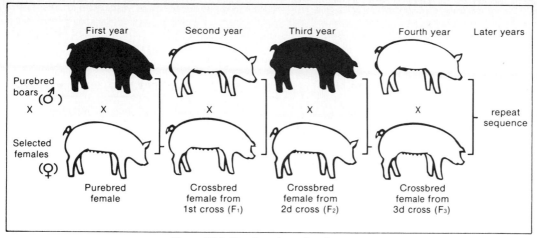

Figure 22-7. Two-breed rotational system.

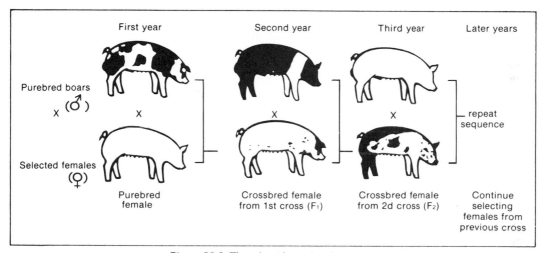

Figure 22-8. Three-breed rotational system.

The commercial pork producer has many selection tools, crossing systems, and genetic breeding stock sources for his use and evaluation. Producers must capitalize on heterosis and breed strengths and must require complete performance records on all selected breeding stock. Although there is no one best system, breed, or source of breeding stock, each producer must evaluate his total pork production program and integrate the most profitable combination of elements associated with a crossbreeding program.

STUDY QUESTIONS

1. Name the four most numerous breeds of swine based on registration numbers.
2. Why are so many of the market swine produced today crossbreds instead of purebreds?
3. Which breed of swine is superior for all of the economically important swine traits? Why?

4. Describe how you would select a boar that would make genetic improvement in the economically important traits.
5. How is sow productivity measured? How can this trait be improved in a swine herd?
6. Identify two breeds of swine that rank high for reproductive performance.
7. Identify two breeds of swine that rank high for the carcass traits.
8. Which breed of swine ranks highest for rate and efficiency of gain?
9. Identify the three main crossbreeding systems that are available to the commercial swine producer.
10. Plan a crossbreeding system considering an excellent combination of breeds and heterosis. Give justification to your plan.

SELECTED REFERENCES

Battaglia, R. A. and Mayrose, V. B. 1981. *Handbook of Livestock Management Techniques.* Minneapolis: Burgess Publishing Co.

Briggs, H. M. and Briggs, D. M. 1980. *Modern Breeds of Livestock.* New York: The MacMillan Co.

Johnson, R. K. 1981. Crossbreeding in swine: experimental results. *Journal of Animal Science* 52:906.

Lasley, J. F. 1978. *Genetics of Livestock Improvement.* Englewood Cliffs, New Jersey: Prentice-Hall.

National Hog Farmer, 1999 Shepard Road, St. Paul, Minnesota 55116.

Pond, W. G. and Maner, J. H. 1974. *Swine Production in Temperate and Tropical Environments.* San Francisco: W. H. Freeman and Co.

Warwick, E. J. and Legates, J. E. 1979. *Breeding and Improvement of Farm Animals.* New York: McGraw-Hill.

_____ *Pork Industry Handbook.* 1980. Cooperative Extension Service. Oklahoma State University, Stillwater, Oklahoma.

_____ 1981. *Guidelines for Uniform Swine Improvement Programs.* USDA Program Aid 1157.

23

Feeding and Managing Swine

Swine production in the United States is concentrated heavily in the region of the nation's midsection known as the Corn Belt. This area produces most of the nation's corn, which is the principal feed used for swine. The Corn Belt states of Iowa, Missouri, Illinois, Indiana, Ohio, northwest Kentucky, southern Wisconsin, southern Minnesota, eastern South Dakota, and eastern Nebraska produce nearly 75% of the nation's swine, with Iowa alone accounting for almost 25% (Figure 23-1).

Swine operations have been getting larger over the past 25 years as the number of farms selling hogs has dropped from 2.1 million to 450,000. Farms selling more than 1,000 hogs annually have increased more than 300% in the last 25 years, and these large operations now account for 25% to 30% of all hogs sold. Most swine producers farm 200 to 500 acres of land and have other livestock enterprises.

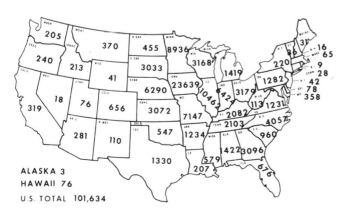

ALASKA 3
HAWAII 76
U.S. TOTAL 101,634

Figure 23-1. The United States 1980 pig crop (1,000 head). For example, Iowa is the leading state in number of pigs produced with 23,639,000 head.

257

TYPES OF SWINE OPERATIONS

There are four primary types of swine operations: (1) feeder pig production in which the producer maintains a breeder herd and produces feeder pigs for sale at an average weight of approximately 40 lbs, (2) feeder pig finishing in which feeder pigs are purchased and then fed to slaughter weight, (3) farrow-to-finish in which a breeding herd is maintained where pigs are produced and fed to slaughter weight on the same farm, and (4) purebred operations which are similar to farrow-to-finish except their salable product is primarily breeding boars and gilts.

Farrow-to-Finish Operations

Since farrow-to-finish is the major type of swine operation, primary emphasis will be given to the discussion of it and less on other types of operations. The information presented here is also pertinent to other operations, since most of them represent a certain phase of a farrow-to-finish operation.

Selection and breeding practices to produce highly productive swine are presented in Chapter 22. The primary objective in the farrow-to-finish production phase is to feed and manage sows, gilts, and boars economically to assure a large litter of healthy, vigorous pigs from each breeding female.

Boars should be purchased at least 60 days prior to the breeding season so they can adapt to their new environment. A boar should be purchased from a herd that has an excellent herd health program and then isolated from the new owner's swine for at least 30 days. During this time the boar can be treated for internal and external parasites if the past owner has not recently done this. The boar should be revaccinated for erysipelas and leptospirosis. Boars not purchased from a validated brucellosis-free herd should have passed a negative brucellosis test within 30 days of the purchase date. An additional 30-day period of fenceline exposure to the sow herd is recommended to develop immunities prior to the boar's use.

Younger boars (8 to 12 months of age) should be fed 5 to 6 lbs of a 14% balanced ration per day, while older boars should receive 4 to 5 lbs. The amount of feed should be adjusted to keep the boar in a body condition that is neither fat nor thin.

Young, untried boars should be test mated to a gilt or two prior to the breeding season as records show that approximately 1 out of 12 young boars are infertile or sterile. The breeding aggressiveness, mounting procedure, and ability to get the gilts pregnant should be evaluated.

Successful breeding of a group of females requires an assessment of the number of boars required and the breeding ability to each boar. Generally, a young boar can pen-breed 8 to 10 gilts during a 4-week period. A mature boar can breed 10 to 12 females. Handmating (producers individually mate the boar to each female) takes more labor but extends the breeding capacity of the boar. Young boars can be handmated once daily, and mature boars can be used twice daily or approximately 10 to 12 times a week. Handmating can be used to mate heavier, mature boars with the use of a breeding crate. The breeding crate takes most of the weight off the female. Breeding by handmating should correspond with time of ovulation (approximately 40 hours from the beginning of estrus) or litter size will be reduced.

Boars are to be considered dangerous at all times and handled with care. It is unsafe for a boar to have tusks as he may inflict injury on the handler or on other pigs. All tusks should be removed prior to the breeding season and every six months thereafter. Tusks can be removed with a bolt cutter.

Proper management of females in the breeding herd is necessary to achieve maximum reproductive efficiency. Gilts may start cycling as early as 5 months of age. It is recommended that breeding be avoided during their first heat cycle since the number of eggs ovulated is usually less than in subsequent heat cycles. Mixing pens of confinement-reared gilts and regrouping them with direct boar contact initiates early puberty.

Hot weather is detrimental to a high reproductive rate in the breeding females. High temperatures (above 85 °F) will delay or prevent the occurrence of heat, reduce ovulation rate, and increase early embryonic deaths. The animals will suffer heat stress when they are sick and have an elevated body temperature as well as when the environmental temperature is high.

Besides reducing litter size, diseases such as leptospirosis, pseudorabies, and those associated with stillbirths, mummified fetuses, and embryonic deaths may also increase the problem of getting the females to settle. Attention to a good herd health program will assure a higher reproductive performance over the years.

It is important that sows and gilts get the proper amount of nutrients during gestation to assure maximum reproduction. Feeding in excess is not only wasteful and costly but will likely increase embryonic mortality. A limited feeding system using balanced, fortified diets is recommended (Figure 23-2). It assures that each sow gets her daily requirements of nutrients without consuming excess energy. As a rule of thumb, 4 lbs of a balanced ration will provide adequate protein and energy; however, during cold weather an additional pound of feed may prove beneficial, especially for bred gilts. With limited feeding it is important that each sow gets her level of feed and no more. The individual feeding stall for each sow is best because it prevents the "boss sows" from taking feed from slower-eating or

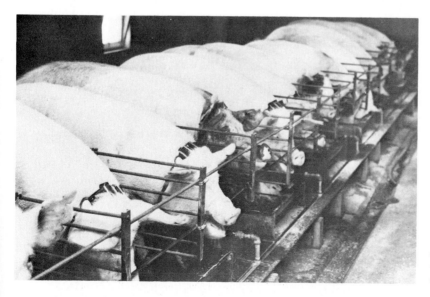

Figure 23-2. Sows in individual feeding units to control the feed intake for each sow.

timid sows. During gestation, gilts should be fed so they will gain about 70 to 80 lbs and sows should gain approximately 30 to 40 lbs.

MANAGEMENT OF THE SOW DURING FARROWING AND LACTATION

Farrowing is the process of the sow or gilt giving birth to her pigs. On the average, there are approximately 8.5 pigs farrowed per litter with approximately 7.2 pigs weaned per litter. The well-managed swine herds will be three to four pigs higher per litter than these average figures. Extremely large litters are usually undesirable because the pigs are small and weak at birth and the death losses are very high.

Proper care of the sow during farrowing and lactation is necessary to assure large litters of pigs at birth and at weaning. Sows should be dewormed about two weeks before they are moved to the farrowing facility. The sow should be treated for external parasites, at least twice, a few days prior to farrowing time. The farrowing facility should be cleaned completely of organic matter, disinfected, and left unused for five to seven days before a new group of sows is placed in the unit.

Before the sow is placed in the farrowing pen or stall, her belly and teats should be washed with a mild soap and warm water. This helps eliminate the bacteria that can cause nursing pigs to get diarrhea.

Sows should be in the farrowing facility at the right time. Breeding records and projected farrowing dates should be used to determine the proper time. If farrowing is to occur in a crate or pen, the sow should be placed there no later than the 110th day of gestation. This gives some assurance the sow will be in the facility at farrowing time since the normal gestation is 111 to 115 days in length. Farrowing time can be estimated by observing an enlarged abdominal area, swollen vulva, and filled teats. The presence of milk in the teats indicates that farrowing will usually occur within 24 hours.

Prior to farrowing, while the sows are in the farrowing facilities, they can be fed a ration similar to the one fed during gestation. A laxative ration can be helpful at this time to prevent constipation. The ration can be made bland or bulky by the addition of 10 lbs per ton of epsom salts, part of the protein as linseed meal, or the addition of oats, wheat bran, alfalfa meal, or beet pulp.

Sows need not be fed for 12 to 24 hours after farrowing, but water should be continuously available. Two to 3 lbs of a laxative feed may be fed at the first post-farrow feeding; the amount fed should be gradually increased until the maximum feed level is reached by 10 days after farrowing. Full feeding from the day of farrowing can be done successfully. Sows that are thin at farrowing may benefit from generous feeding early after farrowing. Sows nursing fewer than eight pigs may be fed a basic maintenance amount (6 lbs per day) with an added amount, such as 0.5 lb for each pig being nursed. It is unnecessary to reduce feed intake before weaning. Regardless of level of feed intake, milk secretion in the udder will cease when pressure reaches a certain threshold level.

BABY PIG MANAGEMENT FROM BIRTH TO WEANING

Most sows and gilts are farrowed in the farrowing stalls or pens to protect the baby pigs from the female lying on them (Figure 23-3). Producers who give attention at farrowing will decrease the number of pigs that die during birth or within the first few hours after

Figure 23.3 Sow and litter of baby pigs in a farrowing crate. The farrowing crate is a pen (approximately 5 ft × 7 ft) which confines the sow to a relatively small area. The panel, a few inches off the floor, separates an area for the baby pigs where the sow cannot lay on them. There is ample space for the pigs to nurse the sow.

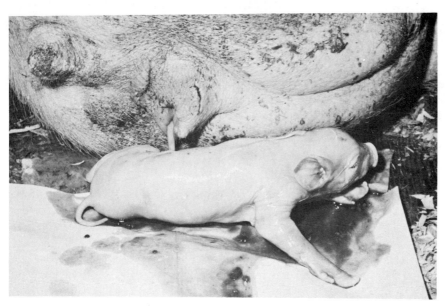

Figure 23-4. A baby pig just farrowed with its umbilical cord still extending into the vulva. Proper attention by the swine producer at farrowing time can assure a higher number of pigs weaned per litter.

birth. Pigs can be freed from membranes, weak pigs can be revived, and other care can be given that will reduce baby pig death loss (Figure 23-4). Manual assistance of the birth process should not be given unless obviously needed. Duration of labor ranges from 30 minutes to more than 5 hours, with pigs being born either head or feet first. The average interval between births of pigs is approximately 15 minutes but can vary from simultaneously to several hours in individual cases. An injection of oxytocin can be given to speed the rate of delivery after one or two pigs have been born.

Manual assistance, using a well-lubricated gloved hand, can be used to assist difficult deliveries. The hand and arm should be inserted into the reproductive tract as far as is needed to locate the pig; the pig should be grasped and gently but firmly pulled to assist delivery. Difficult births often enhance the occurrence of the symptoms of "MMA"—Mastitis (inflammation of the udder), Metritis (inflammation of the uterus), and Agalactin (inadequate milk supply). Antibacterial solutions, such as nitrofurazone, are infused into the reproductive tract after farrowing and sometimes intramuscular injections of antibiotics are used to decrease or prevent some of these infections.

It is very important that the newborn pig receive colostrum to give immediate and temporary protection against common bacterial infections. Pigs from extremely large litters can be transferred to sows having smaller litters to equalize the size of all litters. Care should be given to ensure that baby pigs receive adequate colostrum before the transfer takes place.

Air temperature in the farrowing facility should be 70 °F to 75 °F with the creep area for baby pigs equipped with zone heat (heat lamps, gas brooders, or floor heat) having a temperature of 85 °F to 95 °F. Otherwise the pigs will chill and die or become susceptible to serious health problems. The farrowing facility should also be controlled to remove moisture from the air without creating draft on the pigs.

Management of environmental temperature from birth to 2 weeks of age is critical. At 2 weeks of age, the baby pig has developed the ability to regulate its own body temperature.

Soon after birth, the navel cord should be cut to 3 to 4 in. from the body and then treated with iodine. This disinfects an area which could easily allow bacteria to enter the body and cause joints to swell and abscesses to occur.

The producer should also clip the 8 sharp needle teeth of the baby pig. This prevents the sow's udder from being injured and facial lacerations from occurring when the baby pigs fight one another. Approximately one-half of each tooth can be removed by using side-cutting pliers or toenail clippers.

Good records are important to the producer interested in obtaining maximum production efficiency. The basis of good production records is pig identification. Ear notching pigs at 1 to 3 days of age with a system shown in Figure 23-5 provides positive identification for the rest of the pig's life.

Baby pig management from 3 days to 3 weeks of age includes anemia and scour control, castration, and tail docking. Sow's milk is deficient in iron so iron dextran shots are given intramuscularly at 3 to 4 days of age and again at 2 weeks of age (Figure 23-6).

Baby pig scours are ongoing problems for the swine producer. Colostrum and a warm, dry, draft-free, and sanitary environment are the best preventative measures against scours. Orally-administered drugs determined to be effective against specific bacterial strains have been found to be the best control measures for scours. Serious diarrhea problems resulting from diseases such as transmissible gastroenteritis (TGE) and swine dysentery should be treated under the direction of a local veterinarian.

Castration is usually done before the pigs are 2 weeks old. Pigs castrated at this age are easier to handle, heal faster, and are not subjected to as much stress as those castrated later. The use of a clean, sharp instrument, low incisions to promote good drainage, and using antiseptic procedures make castration a simple operation.

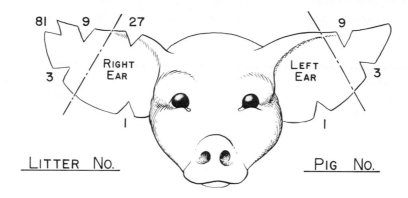

LITTER NO.   PIG NO.

EXAMPLES:

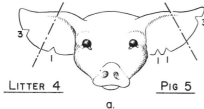

LITTER 4   PIG 5

a.

4-5 MEANS, LITTER 4, PIG 5

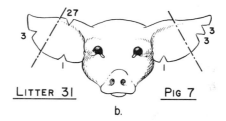

LITTER 31   PIG 7

b.

31-7 MEANS, LITTER 31, PIG 7

Figure 23-5. An ear notching identification system for swine. Individual pig identification is necessary for meaningful production and management records.

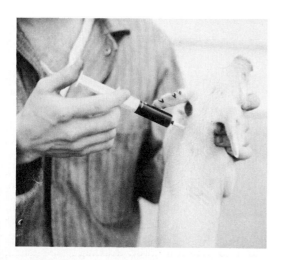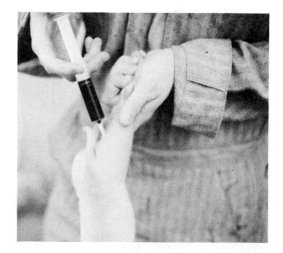

Figure 23-6. Iron shots being given to baby pigs. These injections can be given in either the ham or neck muscle.

Tail docking has become a common management practice to prevent tail biting during the confinement raising of the pigs. Tails are removed 1/4 to 1/2 in. from the body, and the stump should be disinfected along with the instrument after each pig is docked.

Pigs should be offered feed at 1 to 2 weeks of age in the creep area. These starter feeds can be placed on the floor or in a shallow pan. By three to four weeks after farrowing, the sow's milk production has likely peaked; however, the baby pigs should be eating the supplemental feed and growing rapidly (Figure 23-7).

Pigs can be weaned from their dams at a time that is consistent with the available facilities and the management of the producer. Pigs weaned at young ages have to receive a higher level of management. General weaning guidelines are: wean only pigs over 12 lbs, wean pigs over a 2- to 3-day period, wean the heavier pigs in the litter first, group pigs according to size in pens of 30 pigs or less, provide one feeder hole for 4 to 5 pigs and one waterer for each 20 to 25 pigs, and limit feed for 48 hours and use medicated water if scours develop.

FEEDING AND MANAGEMENT FROM WEANING TO MARKET

A dependable and economical source of feed is the backbone of a profitable swine operation. Since approximately 55 to 75% of the total cost of pork production is feed, the swine producer should be keenly aware of all aspects of swine nutrition.

The pig is an efficient converter of feed to meat. With today's nutritional knowledge, modern meat-type hogs can be produced with a feed efficiency of 3.3 lbs of feed per pound of gain from 40 lbs to market. To obtain maximum feed use, it is necessary to feed well-balanced rations designed for specific purposes.

Swine rations from growing-finishing market pigs, as well as rations for breeding stock, are formulated around cereal grains that are the largest component part of swine rations. The most common cereal grains that provide the basic energy source are corn, milo,

Figure 23-7. Baby pigs in a modern nursery unit. The pigs were weaned from the sow and placed into these pens when they weighed 15 to 20 lbs. The pigs will be moved into pen for growing and finishing when they weigh approximately 40 lbs.

barley, wheat, and their byproducts. Because of its abundance and readily available energy, corn is used as the base cereal for comparing the nutritive value of other cereal grains. Milo or grain sorghum are very similar in quality to corn and can completely replace corn in swine rations. Its energy value is about 95% the value of corn. Barley contains more protein and fiber than corn, but its relative feeding value is 85 to 95% of corn. It is less palatable than corn. Wheat is equal to corn in feeding value and is very palatable. However, for best feeding results, wheat should be mixed half and half with some other grain. Which grain or combination of grains to use can best be determined by availability and the relative costs.

Grinding the grains improves all the grains for feeding, especially those high in fiber, such as barley and oats. Pelleting the ration may increase gains and efficiency of gains from 5 to 10%. However, the advantages of pelleting are usually offset by the higher cost of pelleting the ration.

Cereal grains usually contain lesser quantities of proteins, minerals, and vitamins than swine require; therefore, the rations must be supplemented with other feeds to increase these nutrients to the recommended levels. Soybean meal has been proven to be the best single source of protein for pigs when palatability, uniformity, and economics are considered. Other feeds rich in protein (meat and bone meal or tankage) can compose as much as 25% of the supplemental protein if economics so dictate. Cottonseed meal is not recommended as a protein source unless the gossypol has been removed. Gossypol, in relatively high levels, is toxic to pigs. Table 23-1 shows the recommended protein levels for the different weights of pigs being grown and finished for slaughter.

If economics dictate, a producer can supplement specific amino acids in place of some of the protein supplement. Typically, lysine may be supplemented separately because lysine is the most deficient amino acid in swine rations.

Calcium and phosphorus are the minerals most likely to be deficient in swine rations, so care should be exercised to assure adequate levels in the ration. The standard ingredients for supplying supplemental calcium and phosphorus in the swine diet are limestone and either dicalcium phosphate or defluorinated rock phosphate. If an imbalance of calcium and zinc exists, a skin disease called parakeratosis is likely to occur.

Swine producers may formulate their swine rations several ways: (1) using grain and a complete swine supplement, (2) using grain and soybean meal plus a complete base mix

Table 23-1. Percent protein requirement for varying weights of pigs from weaning to slaughter

Period and weight	Percent protein in ration
Starter	
10 to 40 lbs	18 to 20
Grower	
40 to 100 lbs	16
Finisher I	
100 to 160 lbs	14
Finisher II	
160 to 240 lbs	13

carrying all necessary vitamins, minerals, and antibiotics, or (3) using grain, soybean meal, vitamin premix, trace mineral premix, calcium and phosphorus source, and antibiotic and salt mixed together. Which alternative one uses depends on the producer's expertise, the costs, and the availability of the ingredients. The trace mineral premix typically includes vitamin A, vitamin D, vitamin E, vitamin B_{12}, and other B vitamins (niacin, pantothenic acid, and choline).

Feed additives are used by most swine producers because the additives have the demonstrated ability to increase growth rate, improve feed efficiency, and reduce death and sickness from infectious organisms. The feed additives available to swine producers fall into three classifications: (1) antibiotics (compound coming from bacteria and molds that kill other microorganisms), (2) chemotherapeutics (compounds that are similar to antibiotics but are produced chemically rather than microbiologically), and (3) anthelmintics (dewormers).

Selection of the feed additive and the level to be used will vary with the existing farm environment, management conditions, and the stage of the production cycle. Disease and parasite levels will vary from farm to farm. Also, the first few weeks of a pig's life are far more critical in terms of health protection. Immunity from colostrum diminishes by 3 weeks of age, and the pig doesn't begin producing sufficient amounts of antibodies until 5 to 6 weeks of age. Also during this time, the pig is subjected to stress conditions of castration, vaccinations, and weaning. Therefore, at this age the additive level of the feed is much higher than when the pig is more than 75 lbs and progressing towards slaughter weight. Producers should be cautious because some of the additives have withdrawal times which means the additive should no longer be included in the ration. This is necessary to prevent the additives from occurring as residues in the meat.

MANAGEMENT OF PURCHASED FEEDER PIGS

Feeder pigs that are marketed to another producer to finish them to market weight are subjected to many more stresses than pigs produced in a farrow-to-finish operation. These stresses are fatigue, hunger, thirst, temperature changes, ration changes, different surroundings, and social problems. Almost every group has a shipping fever reaction. Care of the newly arrived pigs must be directed to relieving the stresses and treating the shipping fever correctly and promptly. The major management emphasis for newly purchased feeder pigs should be: (1) a dry, draft-free, well-bedded barn or shed with no more than 50 pigs per pen, (2) a specially formulated starter ration, having 12 to 14% protein, more fiber and a higher level of vitamins and antibiotics than a typical starter ration, (3) an adequate intake of medicated water containing sulpha products and electrolytes or water-soluble antibiotics and electrolytes, and (4) prompt and correct treatment of any sick pigs.

MARKET MANAGEMENT DECISIONS

The slaughter weight of the market hogs and the selected market can have a marked affect on the income and profitability of the swine enterprise. For many years the recommended weight to sell barrows and gilts was, in most instances, 200 to 220 lbs. The primary reasons for selling these animals at a maximum weight of 220 lbs were increased cost of

Figure 23-8. A group of pigs in a growing and finishing production unit. The building is totally enclosed and the floor is totally slatted. The manure from the pigs drops through the slats into a 36-inch deep pit. The waste is removed from the pit through a water flushing method.

gain and increased fat accumulations at heavier weights. Market discounts were common on pigs weighing over 220 lbs because of the increased amount of fat. Today swine producers can maximize income by marketing pigs weighing up to 250 lbs. These heavier pigs are leaner, and they are not subject to the same market discounts of the fatter pigs marketed several years ago (Figure 23-8).

Prices vary between markets and marketing costs, such as selling charges, transportation, and shrink (loss of live weight) can also vary. If more than one is available, a producer should occasionally patronize different markets as a check against his or her usual marketing program. No single market is continually and consistently the best market.

STUDY QUESTIONS

1. Where are most of the swine produced in the United States? Why?
2. What effect can high temperatures have on litter pig size? Why?
3. How can baby pigs be kept warm and yet not stay where the sow is likely to trample them?
4. What management practices can be used to give assurance that a young boar will be able to breed and settle gilts?
5. How should newly arrived feeder pigs be managed? Why?
6. Why are feed additives included in swine rations? At what ages of swine are these additives most useful? Why?
7. What are the basic component parts of a well-formulated swine ration for growing and finishing pigs?
8. How can groups of sows be limited fed a 4 lb daily ration per sow with assurance that all the sows will obtain their fair share?
9. What can a producer do to save more pigs per litter at farrowing time?
10. Why is it important that pigs receive an adequate supply of colostrum shortly after they are born?
11. Draw a diagram showing the ear notches of a 23-10 pig.
12. Why are needle teeth clipped, tails docked, and iron shots given as part of sound baby pig management?

SELECTED REFERENCES

Ensminger, M. E. 1970. *Swine Science*. Danville, Illinois: Interstate Printers and Publishers.

Krider, J. L., and Carroll, W. E. 1971. *Swine Production*. San Francisco: McGraw-Hill Book Co.

National Hog Farmer. 1976. Questions and Answers on Managing Swine: Hog Information Please, Vol. III, St. Paul: National Hog Farmer.

Pond, W. G. and Maner, J. H. 1974. *Swine Production in Temperate and Tropical Environments*. San Francisco: Freeman.

Van Arsdall, R. N. 1978. *Structural Characteristics of the United States Hog Production Industry*. Agricultural Economic Report No. 415. USDA

————. 1980. *Pork Industry Handbook*. Cooperative Extension Service, Oklahoma State University, Stillwater, Oklahoma.

24

Sheep Breeds and Breeding

Sheep provide meat and/or milk for human consumption in many parts of the world as well as wool for making clothing. Sheep have declined greatly in numbers in the United States in the past 40 years. Much of the decline has been in range sheep because of the difficulty in obtaining good herders. Farm-flock sheep occupy an important position because they can digest feed sources that are otherwise wasted. Sheep producers must increase rate and efficiency of production if they are to compete with producers of beef cattle.

Sheep have been bred in the past for three major purposes: the production of fine wool for making high-quality clothing; the production of long wool for making heavy clothing, upholstering, and rugs; and the production of mutton and lamb. In recent years, dual-purpose sheep breeds (which produce both wool and meat) have been developed by crossing fine-wool breeds with long-wool breeds and then selecting for both improved meat and wool production. Some breeds have been developed for specific purposes; examples are the Navajo breed, which is used to produce wool for making Navajo rugs, and the Karakul breed, which is used for skinning to obtain pelts for such clothing items as caps.

CHARACTERISTICS OF BREEDS

The breeds of sheep and their characteristics are listed in Table 24-1. The breeds of sheep commonly used in the United States are shown in Figures 24-1 and 24-2. All fine wool breeds and the dual-purpose breeds that possess fine-wool breeding have a herding instinct, so one sheepherder with dogs can look after a band of 1,000 ewes and their lambs on the summer range. Winter bands of 2,500 to 3,000 ewes are common. The meat breeds, by contrast, tend to scatter over the grazing area. They are highly adaptable to grazing fenced pastures where feed is abundant. Meat-type rams are often used for breeding ewes of fine-wool breeding on the range for the production of market lambs.

Table 24-1. Characteristics of breeds of sheep within each type

Breed	Size	Mutton conformation	Wool fineness	Wool length	Fleece weight	Color	Horns or polled	Other
Fine-wool breeds								
Merino	Small to medium	Poor	Very fine	Medium	Heavy	White	Rams horned, ewes polled	Good herding instinct
Rambouillet	Large	Medium	Fine	Medium	Heavy	White	Rams horned, ewes polled	Good herding instinct
Debouillet	Medium	Poor	Fine	Medium	Heavy	White	Rams horned, ewes polled	Good herding instinct
Long-wool breeds								
Romney	Medium large	Medium to good	Coarse	Long	Heavy	White	Polled	Spread in lambing date, lambs don't fatten well
Lincoln	Large	Good	Coarse	Long	Heavy	White	Polled	Lambs don't fatten at small size
Leicester	Large	Good	Coarse	Long	Heavy	White	Polled	Lambs don't fatten at small size
Cotswold	Large	Good	Coarse	Long	Heavy	White	Polled	Lambs don't fatten at small size
Mutton breeds								
Suffolk	Large	Excellent	Medium	Very short	Very light	White with black bare face and legs	Polled	Bare bellies, good milkers, black fibers
Hampshire	Large	Excellent	Medium	Medium	Medium	White with black face	Polled	Wool blindness, good milkers, black fibers
Shropshire	Medium to large	Good	Medium	Medium	Medium	White, dark face and legs	Polled	Wool blindness, excellent milkers

Table 24-1 continued.

Breed	Size	Mutton conformation	Wool fineness	Wool length	Fleece weight	Color	Horns or polled	Other
Southdown	Very small	Excellent	Medium	Short	Very light	White with brown face and legs	Polled	Used in hot house lamb production
Dorset	Medium	Good	Medium	Medium	Medium	White	Horned rams and ewes	Highly fertile, good milkers
Cheviot	Small	Excellent	Medium	Medium	Medium	White	Polled	Very rugged
Oxford	Large	Excellent	Medium	Medium	Medium to heavy	White, brown face and legs	Polled	
Tunis	Medium	Medium	Medium	Medium	Medium	Red or tan face	Polled or horned	
Dual-purpose breeds								
Corriedale	Medium to large	Good	Medium	Medium long	Heavy	White	Polled	Herding instinct
Columbia	Large	Good	Medium	Medium long	Heavy	White	Polled	Rugged, herding instinct
Targhee	Medium to large	Medium to good	Medium to fine	Medium	Heavy	White	Polled	Herding instinct
Panama	Large	Good	Medium	Medium long	Heavy	White	Polled	
Romeldale	Medium to large	Good	Medium	Medium	Heavy	White	Polled	
Montadale	Medium to large	Good	Medium	Medium	Medium	White	Polled	
Miscellaneous breeds								
Navajo	Medium	Poor	Coarse	Long	Medium	Variable	Polled or horned	Wool for making Navajo rugs
Karakul	Large	Poor	Coarse	Long	Medium	Black or brown	Rams horned, ewes polled	Used for pelts

Dorset

Montadale

Polled Dorset

Cheviot

Hampshire

Suffolk

Figure 24-1. Some breeds of sheep commonly used either as straightbreds or for crossbreeding to produce market lambs.

Rambouillet

Cotswold

Lincoln

Romney

Targhee

Finnsheep

Figure 24-2. Some breeds of sheep commonly used either as straightbreds or for crossbreeding to produce market lambs.

The Columbia and Targhee breeds were developed by the U.S. Sheep Breeding Laboratory, Dubois, Idaho. Both breeds have proven useful on western ranges because they have the herding instinct and they raise better market lambs when bred to a meat breed of ram than do the fine-wool ewes. The U.S. Sheep Breeding Laboratory has made great strides in improving the Rambouillet, Columbia, and Targhee breeds. These three breeds were faulted by having body folds, or wrinkles, that made shearing extremely difficult, and by "wool blindness" (a situation in which wool covering the face prevents the sheep from seeing). Wool blindness prevented the sheep from observing dangers and interfered with their ability to locate forage. Thus, woolblind ewes weaned lambs that averaged about 11 lbs less than those produced by open-faced ewes.

Selection against wool blindness and body folds had to be done by use of a subjective score because there was no way to measure these traits accurately. After several years of selection, the average score for these two undesirable traits was approximately the same as the average score prior to the selection program. However, photographs showed clearly that rapid progress was being made.

The most important trait in meat breeds of sheep is weight at 90 days of age, with some emphasis also given to conformation and finish of the lambs. Weight at 90 days of age can be calculated for the crosses in which lambs are weaned at younger or older ages than 90 days as follows:

$$\text{90-day weight} = \frac{\text{Weaning weight} - \text{Birth weight}}{\text{Age at weaning}} \times 90 + \text{Birth weight}$$

If a lamb weighs 10 lbs at birth and 100 lbs at 100 days, the 90-day weight is:

$$\frac{100 - 10}{100} \times 90 + 10 = 91 \text{ lbs}$$

If the birth weight was not recorded, an assumed birthweight of 8 lbs can be used.

Approximately 85% to 90% of the total income from sheep of the meat breeds is derived from the sale of lambs, with only 10% to 15% coming from the sale of wool. By contrast, the sale of wool accounts for 30% to 35% of the income derived from fine-wool and long-wool breeds. Even with wool breeds, the income from lambs produced constitutes the greater portion of total income.

The U.S. Sheep Experiment Station, Dubois, Idaho, is developing a new synthetic breed of sheep, called the Polypay, that is superior in lamb production and in carcass quality. Four breeds (Dorset, Targhee, Rambouillet and Finnsheep) provided the basic genetic material for the Polypay. Targhees were crossed with Dorsets and Rambouillets were crossed with Finnsheep, after which the two cross-bred groups were crossed. The offspring produced by crossing the two crossbred groups were intermated and rigid selection was practiced. This population has been closed. The Rambouillet and Targhee breeds contributed hardiness, herding instinct, size, a long breeding season and wool of high quality. The Dorset contributed good carcasses, high milking quality and a long breeding season. The Finnsheep contributed early puberty, early postpartum fertility, and high lambing rate.

This breed has shown outstanding performance in conventional once-a-year lambing under range conditions and superior performance in twice-a-year lambing when compared with other breeds or breed crosses.

BREEDING SHEEP

A large percentage of market lambs is of crossbred breeding (produced either from crossing breeds or from mating crossbred **ewes** with purebred rams), but the key to the success of crossbreeding is the improvement in production traits made by the purebred breeders in their selection programs.

The purebred breeder. The success of the purebred breeder in improving production traits of a flock depends heavily on the goal of the breeder, the standardization of the environment, and the accuracy of measuring the trait. Sheep breeders must know that their goal is sound and that it will be sound in the future. Breeders have made mistakes in the past. For example, Shropshire breeders at one time adopted a goal for their sheep of having "wool from the nose to the toes". This objective was achieved, but it resulted in wool-blind sheep that had little value under farm conditions.

Selecting replacement ewe lambs. Replacement ewe lambs should be those which are born in the early part of the lambing season, which have a high 90-day weight, and which are produced by ewes that have consistently produced outstanding lambs. Selection for twinning is effective so replacements (either single or twin) should be kept from ewes that have previously produced several sets of twins, in preference to even a twin from an older ewe that produced several singles.

In addition to a high 90-day weight, replacement ewe lambs should have a desirable body **conformation** and wool of high quality. Some traits such as conformation and degree of finish cannot be measured and must therefore be scored subjectively.

It is important to replace generations rapidly as a means of making genetic progess, although superbly producing ewes and their offspring should be retained in a flock. Also, rams that are outstanding as shown by records in their offspring should be used for another breeding season to increase their favorable selection differential. In general, emphasis is given to records on the individual for traits of relatively high heritability and to records on the offspring for traits of relatively low heritability.

Progress through selection. If 90-day weight is to be improved through selection, the progress that can be made depends largely on the selection pressure that is applied (**selection differential**) and the **heritability** of 90-day weight. Assume that the average 90-day weights of rams and ewes in a flock are:

90-day weight of rams = 100 lbs

90-day weight of ewes =   85 lbs

By selecting the relatively good ewes and rams from this flock, a breeding population might be obtained that averages:

90-day weight of rams = 120 lbs

90-day weight of ewes =   89 lbs

The selection pressure applied (selection differential) is the difference between the selected population and the original population:

120 lbs − 100 lbs = 20 lbs for rams

  89 lbs −   85 lbs =   4 lbs for ewes

To determine what the average improvement obtained equals when both sexes are considered together, the improvement in males and the improvement in females are summed and divided by 2. Thus, the average superiority obtained through selection in this example equals 12 lbs:

$$\frac{20 \text{ lbs} + 4 \text{ lbs}}{2} = \frac{24 \text{ lbs}}{2} = 12 \text{ lbs}$$

If the heritability of the 90-day weight is 0.50 lb, and if this same selection differential is reached each generation, the improvement expected in 90-day weight would be 6 lbs per generation. If the rams used average 2 years of age and the ewes average 4 years of age, $\frac{2+4}{2} = 3$ the generation interval is 3 years. Dividing the progress per generation of 6 lbs by the generation interval of 3 years, results in improvement of 90-day weight of 2 lbs per year. An annual increase of 2 lbs in 90-day weight throughout a 10-year period would result in lambs averaging 20 lbs more at 90 days than lambs in the original flock averaged prior to selection.

The preceding example shows that most of the progress from selection comes from selecting rams. Only about 5% of a breeder's ram lambs are needed as replacements, but about 35% of the ewe lambs are needed. Thus, about seven times the emphasis is placed on selecting rams as is given to selecting ewes. The selection differential can be increased by use of artificial insemination, because more than 1,000 ewes can be bred to a ram in a breeding season through artificial insemination. It is imperative that the rams, when semen is used for artificial insemination be truly outstanding, because any mistake made in selecting a ram is magnified about 30 times if artificial insemination is used rather than natural service. One problem related to artificial insemination of sheep today is that no good method for freezing ram semen is known.

Commercial sheep production. "Hot house" or Easter lambs (lambs born in December or January and sold as small, finished lambs for the Easter markets in the early spring) can be produced by breeding ewes of a highly fertile breed that produces large quantities of milk (such as Dorset) to a ram of a small breed that has excellent meat conformation (such as the Southdown). The lambs produced by such a mating receive ample milk to make rapid growth. The smaller size and excellent conformation contributed by the Southdown cause the lambs to have desirable conformation and to finish at a relatively small size. They should be sufficiently finished for slaughter at 70 to 80 lbs and make excellent small carcasses.

Most of the lambs produced in farm-flock operations are marketed at the time of weaning (when they are 100 to 130 days of age and weigh from 90 to 120 lbs), and are sufficiently finished by then to be graded as USDA Choice or Prime if they have been properly cared for and have good breeding. The best method for producing such lambs is to use a three-breed rotational crossbreeding system. One can use breeds such as Dorset, Suffolk, and Columbia in areas where relatively good grazing conditions exist. Under more rigorous conditions, the North Country Cheviot is preferable to the Columbia because of its ruggedness. On the average, larger lambs are produced if Columbia sheep are used in rotation, rather than Cheviots, but more slaughter lambs result from using Cheviots when a smaller sheep is needed. A good rotational program can be started by using what ewes one has and rams from highly selected flocks as follows:

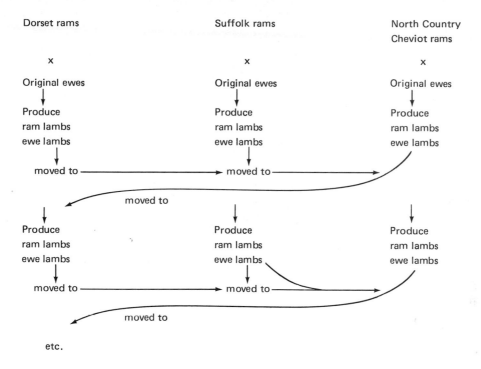

Dorset rams   Suffolk rams   North Country Cheviot rams

etc.

All rams and all cull ewe lambs are marketed.

Through this kind of program, one will eventually have crossbred ewes producing crossbred lambs (see Chapter 17). The crossbred ewes will be more fertile and will give more milk than straightbreds and will be able to adapt better to adversities. The crossbred lambs will be more vigorous at birth, will establish a higher level of milk production by their dams, and will grow more rapidly. The use of rams from highly selected flocks will contribute to better meat qualities and higher growth rates.

Terrill (1974) reviewed the studies that have been done on crossbreeding of sheep in the United States. **Crossbreeding** can be used to combine meat conformation of the sire breeds with lambing ability and wool characteristics of the ewe breeds and also to obtain the advantage of fast growth rate of lambs that results from heterosis. The additive as well as the heterotic effects are evident in crossbred lambs. Maximum **heterosis** is obtained in three- or four-breed crosses, either rotation or terminal. Crossing meat-type sheep with the Finnsheep offers a means for rapidly increasing efficiency of lamb meat production. The Finnsheep breed was introduced into the United States in 1968, and its numbers are increasing rapidly in this country. Finnsheep (Figure 24-3) are extremely fecund (two to six lambs per lambing), and studies show that crossbred ewes of 25% to 50% Finnsheep breeding produce more lambs than straightbred or crossbred ewes of the meat breeds (Dickerson, 1978). One method of breeding being used experimentally now is a three-breed rotational

Figure 24-3. One of the first Finnsheep ewes imported into the United States with her first lambs, illustrating the high productivity of Finnsheep.

crossbreeding system that uses rams of meat type and Finnsheep breeding. The results of this program are encouraging since number of lambs produced is greater but growth of lambs sired by Finnsheep is lower than those sired by rams of other breeds.

Evidence suggests that crossbred rams are more vigorous and more highly fertile than straightbred rams. Thus, crossbred rams might be used effectively in a rotational crossbreeding program.

Age to breed young ewes. Young ewes have traditionally been bred to lamb when they are 2 years old; however, ewe lambs that are properly cared for should be of sufficient size for breeding to lamb when they are a year of age. Crossbred ewe lambs generally reach sexual maturity at younger ages than straightbred animals, so are more likely to produce large crops of lambs. If young ewes are bred to a lamb at a year of age, they must be fed at a higher rate and given more attention than ewe lambs that are wintered in an open (not pregnant) condition.

Three-breed terminal crossbreeding. It is important in some production schemes to develop crossbred females that can be bred to males of a third breed as a terminal cross for the production of offspring that are to be marketed. Sheep producers cross Columbias with Dorset Horns to develop large ewes that have long, fairly fine fleece, are hardy, and produce much milk. These crossbred ewes are bred to good Suffolk or Hampshire rams. A high percentage of lambs from the latter cross are finished at weaning and weigh 90 to 100 lbs at about 90 to 100 days of age. The entire lamb crop from this terminal mating is marketed. All the two-breed crossbred ewes that are productive are kept until they reach an age at which their production of lambs declines. Replacements are produced by crossing Columbias and Dorset Horns. A producer carrying out such a program keeps a small flock of either straightbred Dorset Horns or Columbias to produce the two-breed crossbred ewes. The purebred rams are purchased as needed and they are used for breeding as long as they are serviceably sound.

INHERITED ABNORMALITIES OF SHEEP

Certain genetic abnormalities are important to guard against when managing sheep. Although exceptions occur, sheep showing obvious genetic defects are usually culled. Some of the inherited abnormalities follow.

Cryptorchidism is inherited as a simple recessive; therefore a ram with only one testis descended into the scrotum should never be used for breeding. In addition, one should cull rams that sire, or ewes that produce, lambs having cryptorchidism.

Dwarfism is inherited as a recessive and is lethal; therefore, ewes and rams having dwarf offspring should be culled.

The mode of inheritance of **entropion** (turned-in eyelids) has not been determined, but it is known to be under genetic control. A record should be made of any lamb having entropion so that it may be sent to market.

Sheep have lower front teeth but lack upper front teeth. They graze by closing the lower teeth against the dental pad of the upper jaw. If the lower jaw is either too short or too long, the teeth cannot close against the dental pad and grazing is difficult. The mode of inheritance of these jaw abnormalities is unknown, but these conditions are under genetic control. Sheep having abnormal jaws should be culled.

DETERMINING THE AGE OF A SHEEP BY ITS TEETH

The age of a sheep can be determined by its teeth. Lambs have four pairs of narrow lower incisors called milk teeth or baby teeth. At approximately a year of age the middle pair of milk teeth is replaced by a pair of larger, permanent teeth. At 2 years, a second is replaced. This process continues until, at 4 years of age, the sheep has all permanent incisors. The teeth start to spread apart and some are lost at about 6 to 7 years of age. When all the permanent incisors are lost, the sheep has difficulty grazing and should be marketed.

STUDY QUESTIONS

1. Give one highly desirable characteristic for each of two breeds of sheep for each of these classes: fine-wool, long-wool, dual-purpose, and mutton.
2. What breeding system would you use in a commercial farm-flock operation to maximize the production of slaughter lambs?
3. A lamb was born February 15 weighing 8 lbs. On June 1, this lamb weighed 88 lbs. Calculate the lamb's 90-day adjusted weight.
4. You have a flock of sheep with the following averages for 90-day weight: rams = 90 lbs, ewes = 78 lbs. By selecting from within this flock you obtain a breeding population that has the following averages for 90-day weight: rams = 108 lbs, ewes = 82 lbs. The average ages for your selected population are: rams = 1 year, ewes = 3 years. Calculate and label (a) selection differential, (b) progress expected per generation assuming a heritability of 0.50, (c) generation interval, and (d) progress per year.

SELECTED REFERENCES

Dickerson, G. E. 1978. *Crossbreeding Evaluation of Finnsheep and Some U.S. Breeds for Market Lamb Production.* North Central Regional Publication #246. Washington, D.C.: Agricultural Research Service, USDA and Lincoln, Nebraska: University of Nebraska.

Dolling, C. H. S. 1970. *Breeding Merinos.* Adelaide: Rigby.

Scott, G. 1975. *The Sheepman's Production Handbook,* 2nd edition. Sheep Industry Development Program, Inc.

Spurlock, G. M., Weir, W. C., Bradford, G. E., and Albaught, R. 1966. Production Practices for California Sheep. Davis, California: California Agricultural Exp. Sta. and Extension Service Manual 40.

Terrill, C. E. 1974. *Review and Application of Research on Crossbreeding of Sheep in North America.* Madrid, Spain: 1st World Congress on Genetics Applied to Livestock Production.

25

Feeding and Managing Sheep

The number of sheep in the United States reached a maximum of 56 million in 1942, after which ensued a steady decline to a level of 13 million in 1979. Although the price received for lambs increased by a spectacular 320% (from 14.7¢ per pound to 62.7¢ per pound) between 1961 and 1979, this increase has been partly offset by increased production costs, scarcity of good sheep herders, heavy losses from disease, predators, and adverse weather.

The numbers of ewes on farms (1980) and lambs produced (1979) are presented in Figure 25-1. It can be seen that Texas has more sheep than any other state and that California, Colorado, South Dakota and New Mexico are also important states in sheep numbers. The midwestern states generally raise more lambs per ewe than the western

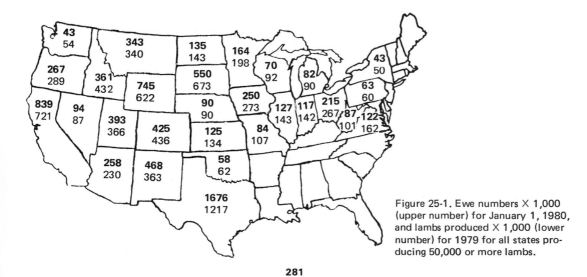

Figure 25-1. Ewe numbers × 1,000 (upper number) for January 1, 1980, and lambs produced × 1,000 (lower number) for 1979 for all states producing 50,000 or more lambs.

states, although Washington, Oregon, and Idaho resemble the midwestern states in this regard.

Even though much of the decline in the number of sheep in the United States has occurred in the West, this region, with its extensive public and private range lands, still produces 80% of the sheep raised in this country today. Most sheep growers in the United States have small flocks (50 or fewer sheep) and most raise sheep as a secondary endeavor. About 40% of the West's sheep producers maintain flocks of more than 50 sheep, and these flocks contain about 93% of the sheep in that region. Only about one-third of the operators in the West who have flocks of 50 or more sheep specialize in sheep, while the other two-thirds have diversified livestock operations.

Sheep are produced on farms where pastures are generally good and where meadow aftermaths, cereal grains, and grass grown for seed can provide nutritional needs at certain times. Sheep are also produced under range conditions where vegetation is sparse and much of the feed is **forbs** and **browse.** Farm-flock operations and range operations each have their own unique problems and methods.

About 23% of all sheep born in the western United States and 13% of those born in the north-central region of this country are lost before they are marketed. In the West, predators, especially coyotes, and weather are the most important causes of death. In the north-central region, the most important causes are weather and disease before docking, and disease and attacks by dogs after docking.

The range operator produces some **finished** lambs if good mountain range is available, but most range-produced lambs are feeders (lambs that must be fed to fatten). Feeder lambs are either fed by the producers, sold to feedlot operators, or fed on contract by feedlot operators. Colorado feeds more lambs for finishing than any other state. Some feedlots have a capacity for 10,000 or more lambs on feed at one time and many feedlots that have a capacity of 1,000 are in operation.

REQUIREMENTS FOR PRODUCTION OF FARM-FLOCK SHEEP

The objective of the farm-flock operator is to produce finished lambs ready for market at weaning time. Various provisions are required to care for farm flocks.

Pastures. Good pastures are essential to the typical farm-flock operator. Grass-legume mixtures such as rye grass and subterranean clover or orchard grass and lespedeza are ideal for sheep. Sheep can also graze on temporary pastures of such plants as Sudan grass or rape, which are often used to provide forage in the dry part of summer when permanent pastures may show no new growth. Crops of grain and grass seed may also provide pasture for sheep in the autumn and early spring. Sheep do not trample wet soil as severely as do cattle; pasturing sheep on grass-seed and small-grain crops in winter and early spring when the soil is wet does not cause serious damage to the sod.

Fencing. A woven-wire fence is necessary to contain sheep in a pasture. So that best use of the forage can be made and as a means of "rotating" (moving) sheep from one pasture to another to assist in the control of internal parasites, the pasture area should be subdivided into small pastures with woven-wire fencing. Some sheep operators use temporary fencing that they move to enclose particular areas when sheep are to be pastured. The fence should be tightly stretched and secured to either wooden or steel posts. Gates can be constructed from 1- x 4-in. lumber, with baling wire used both for hinges and for fastening the gate.

Metal gates are often used where permanent fencing is installed and a gateway is needed. Electric fences can also be used. It is usually necessary to use two electrified wires, one located low enough to prevent sheep from going underneath, and the other located high enough to prevent them from jumping over.

Corrals and chutes. It is occasionally necessary to put sheep in a small enclosure to sort them into different groups or to treat ailing individuals. Proper equipment is helpful, because sheep are often difficult to drive, particularly when ewes are being separated from their lambs. A cutting chute constructed so that sheep can be directed into various small lots as they proceed down it is needed (Figure 25-2). A well-designed **cutting chute** is sufficiently narrow to keep sheep from turning around and can be blocked so that sheep can be packed together closely for such purposes as treating diseases and reading ear tags. The chute is constructed of lumber that is nailed to wooden posts set into the ground and properly spaced so that the proper width (14 to 16 in.) is provided when the boards are nailed on the inside. The pens used to enclose small groups of sheep are constructed of a woven-wire fencing. A loading chute is used to place sheep onto a truck for hauling. A portable loading chute (Figure 25-3) is ideal because it can be taken along and used for unloading.

Figure 25-2. Sheep being directed into a special pen by use of a cutting chute.

Figure 25-3. Sheep being unloaded from a truck directly onto pasture by use of a portable loading chute.

Shelters. Sheep do not normally suffer from cold because they have a heavy wool covering; therefore, open sheds are excellent for housing and feeding wintering ewe lambs, pregnant ewes, and rams. Although the ewes can be lambed in these sheds, the newborn need an enclosed and heated room when the weather is cold.

Lambing equipment. Small pens, about 4 x 4 ft, can be constructed for holding ewes and their lambs until they are strong enough to be put with other ewes and lambs. Four-foot panels can be constructed from 1- x 4-in. lumber and the two panels can be hinged together. The lambing pens (called lambing jugs) can be made along a wall by wiring these hinged panels. Heat lamps are extremely valuable for keeping newborn lambs warm. A heat lamp above each **lambing jug** can be located at a height that provides a temperature of 90 °F at the lamb's level (Figure 25-4).

It is necessary to identify each lamb and to record which lambs belong to which mothers. Often, a ewe and her lambs are branded with a scourable paint to identify them. Numbered ear tags can be applied to the lamb at birth. A record book for keeping all necessary records is required. If newborn lambs are to be weighed, a dairy scale and a large bucket are needed. The lamb is placed in the bucket, which hangs from the scale, for weighing. It is advisable to immerse navel cords of newborn lambs in a tincture of iodine.

Feeding equipment. All sheep are given hay in the winter feeding period. Feeding mangers (Figure 25-5) should be provided, because much hay is wasted if it is fed on the ground. It may also be necessary to feed some sheep limited amounts of concentrates. The concentrates may be fed in the same bunk as hay if the hay bunk is properly constructed, or separate feed troughs (Figure 25-6) may be preferable.

Additional concentrates can be provided for lambs by placing the concentrate in a **creep** (Figure 25-7) that is constructed with openings large enough to allow the lambs to

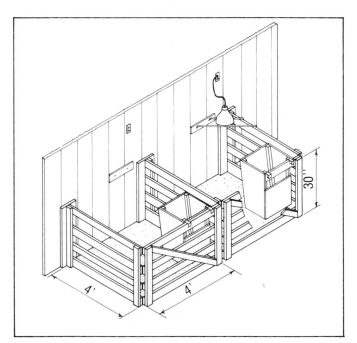

Figure 25-4. A lambing jug with heat lamp for ewes with newborn lambs.

Figure 25-5. Sheep eating
from hay bunk.

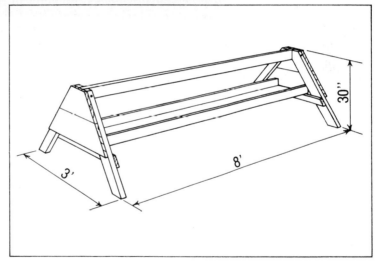

Figure 25-6. A feed trough for providing concentrates for sheep. Note
the top board that prevents sheep
from getting into the trough.

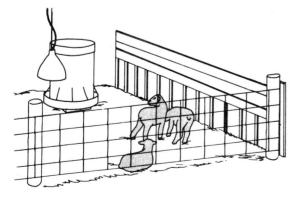

Figure 25-7. Creep with heat lamp for lambs.

enter but small enough to keep out the ewes. In addition to the concentrates, it is advisable to provide good-quality legume hay, a mineral mix containing calcium, phosphorus, and salt, and a heat lamp in the creep area.

Water is essential for all sheep at all times. It can be provided by tubs or automatic waterers. Tubs should be scrubbed clean once each week. Large buckets are usually used for watering ewes in lambing jugs.

A separate pen equipped with a milk feeder may be needed for orphan lambs. Milk or a milk replacer can be provided free-choice if it is kept cold. Heat lamps kept some distance from the milk feeder are provided.

Feed storage. Areas should be provided for storage of hay and concentrates so that feeds can be purchased in quantity or home-grown feeds can be stored where they will remain dry. An open shed is satisfactory for hay storage. Some operators prefer a feed bin for concentrates that is designed so that the feed can be put in at the top and removed from the bottom.

PRODUCTION SCHEMES

Some people raise purebred sheep and sell **rams** for breeding. Commercial producers whose pastures are good produce slaughter lambs on pasture, whereas those whose pastures are less desirable produce feeder lambs. Lambs that are neither sufficiently fat nor large for slaughter at weaning time usually go to feedlot operators where they are fed to slaughter weight and condition; however, most of the feedlot lambs are obtained from producers of range sheep.

Purebred breeder. Purebred sheep are usually fed more as they grow than commercial sheep because the purebred breeder is interested in growing sheep that have a desirable appearance. Records are essential to indicate which ram is bred to each of the ewes and which ewe is the mother of each lamb. All ram lambs are usually kept together to locate those that are desirable for sale. The ram lambs are usually retained by the purebred breeder until they are a year of age, at which time they are offered for sale. Buyers usually want to obtain rams from purebred breeders in early summer.

The purebred breeder has grave responsibilities to the sheep industry, because this person determines the productivity of commercial sheep. The purebred breeder should rigidly select the animals that are kept for breeding and should offer only those rams for sale that will contribute improved productivity for the commercial producer.

Slaughter lamb producer. The producer of slaughter lambs whose feed conditions and pasture conditions are favorable strives to raise lambs that finish at 90 to 120 days of age at a weight of about 100 lbs. each. Any lamb that is small or lacks finish requires additional feeding. Usually, the price decline that occurs from June through September offsets the improvement that is made in value of lambs by feeding, so that feeding the cull farm-flock lambs is generally unprofitable. Creep feeding of concentrates early in the life of lambs helps, but, because lambs may be disoriented when their creep is moved, they may not come to the creep after being put onto a new pasture. Therefore, creep feeding on pasture where pasture rotation is practiced may be ineffective. Keeping the sheep in good health and seeing that they have water at all times and on lush pasture helps bring about desirable early market lambs.

The commercial producer can do a great deal to assure raising lambs that are large and finished at weaning by using the proper kind of rams in a cross-breeding program. Rams that are themselves heavy at 90 days of age are more likely to sire lambs that are heavy at this age than are rams that are light in weight at 90 days of age. The breeds to use in a three-breed rotation should include one that is noted for milk-producing ability, one that is noted for rapid growth, and one that is noted for ruggedness. Perhaps the Dorset should be considered for milk production, either the Hampshire or Suffolk for rapid growth, and the Cheviot for its ruggedness. All of these breeds make desirable carcasses.

Commercial producers castrate all ram lambs. Young ewes that are selected to replace old ewes should be large and growthy and should be daughters of ewes that produce relatively many lambs. Ewes that have started to decline in production and poorly productive ewes should be culled.

Feeder lamb producer. Some commercial sheep are produced where pasture conditions are insufficient for growing the quantity or quality of feed needed for lambs to be large and finished at weaning. Lambs produced under such conditions may either be fed out in the summer or carried through the summer on pasture and finished in the fall. Sudan grass or rape can be seeded so that good pasture is available in the summer, and lambs can be finished on pasture by giving them some concentrates. If good summer pasture cannot be made available, it is advisable to wait until autumn, at which time the lambs are put on full feed in a feedlot.

Feedlot operator. Lambs that come to the feedlot for finishing are treated for internal **parasites** and for certain diseases, particularly **"overeating disease"**. They are provided water and hay initially, after which concentrate feeding is allowed and increased until they receive all the concentrates they will consume. Death losses may be high in unthrifty lambs.

The feedlot operator hopes to profit by increasing the value of the lambs per unit of weight and by increasing their weight. This operator must therefore purchase feeder lambs at a price below that of slaughter lambs because the cost of increasing the weight of lambs in the feedlot usually is greater than the sale price of the increased weight. Lambs on feed gain about 0.3 to 0.5 lb per head per day when they are gaining satisfactorily. Feeder buyers prefer feeder lambs weighing 70 to 80 lbs over larger lambs, because of the need to put 20 to 30 lbs of additional weight on the lambs to finish them which would make large feeders so heavy when finished that they would sell at a low price. A feeding period of 90 to 100 days should be sufficient for finishing thrifty lambs.

Feeding lambs is riskier than producing them because of death losses, price fluctuations of feed and sheep, and the necessity of making large investments.

FEEDS AND FEEDING

Mature, pregnant ewes probably need nothing more than good-quality legume hay during the first half of pregnancy, after which some concentrates should be fed. Any of the grains such as corn, barley, oats, milo, and wheat are satisfactory feeds. Sheep usually chew these grains sufficiently so that grinding or rolling is not essential. Rolled or cracked grains may be digested slightly more efficiently and may be more palatable, but finely ground grains are undesirable for sheep unless the grains are pelleted.

Rams should be fed to keep them thrifty but not fat. A small amount of grain along with good-quality hay satisfies their nutritional needs in winter. Bred, young ewes should be fed some grain along with all the legume hay they will consume because they must grow and also have some reserve for the subsequent drain of lactation. Ewe lambs should be grown out well but not fattened in the winter. Limited grain feeding along with legume hay satisfies their nutritional needs.

At lambing time, the grain allowance of ewes needs to be increased to assist them in producing a heavy flow of milk. Also, the lambs need to be fed a high-energy ration in the creep. Lambs obtain sufficient protein in the milk given by their mothers, but they need more energy. Rolled grains provided free-choice in the creep are palatable and provide the energy needed.

Lambs on full feed are allowed all the hay of good quality and concentrates they will consume. Some feedlot operators pellet the hay and concentrates, whereas others feed loose hay and grains. After the grasses and legumes start to grow in the spring, all sheep generally obtain their nutritional needs from the pasture.

CARE AND MANAGEMENT OF FARM-FLOCK SHEEP

The handling of sheep is extremely important. A sheep should never be caught by its wool. When a group of sheep is crowded into a small enclosure, the sheep will face away from the person who enters the enclosure. When the sheep's rear flank is grasped with one hand, the sheep starts walking backward; this allows one to reach out with the other hand and grasp the skin under the sheep's chin. When a sheep is being held, grasping the skin under the chin with one hand and grasping the top of the head with the other hand enables the holder to pull the sheep forward so its brisket is against the holder's knee. If a sheep is to be led, the skin under the chin can be grasped with one hand and the **dock** (the place where the tail was removed) can be grasped with the other hand for moving the sheep about. Putting pressure on the dock makes the sheep move forward while holding its chin with the other hand prevents it from escaping.

Certain times are especially important in a sheep operation, such as the times of breeding, lambing, castrating and docking, and shearing.

The breeding season. Sheep are bred in the late summer or early fall for winter or spring lambing. The gestation period is 147 to 152 days, depending on the breed.

Sheep may be hand mated or pasture bred. If they are **hand mated**, some **teaser rams** are needed to locate ewes that are in heat. Either a **vasectomized** ram (each vas deferens has been severed so sperm are prevented from moving from the testicles to become a part of the semen) can be used, or an apron can be put on the ram such that a strong cloth prevents copulation into the vagina of the ewe.

Prior to the breeding season, the rams to be used for breeding should be shorn and all wool removed from the scrotum to prevent summer, or heat, sterility. Fertility of the rams should be determined by examining the semen from two ejaculates collected 2 or 3 days apart.

Ewes that are found in heat are allowed one service each by the proper ram. It is generally best to keep these ewes separate from other ewes for one or two days so that the teaser ram does not spend his energies locating the same ewes that he has already located.

Ewes are checked twice daily for heat. Ewes normally stay in heat for 27 hours and ovulate near the end of heat; therefore, it is desirable to wait until the following morning to breed ewes that are found in heat in the late afternoon, and wait until late afternoon to breed ewes that are found in heat in the morning.

If ewes are to be pasture mated, they are sorted into groups according to the rams to which they are to be mated. It is best to have an empty pasture between breeding pastures so that rams do not have the urge to go through or over the fence to get to ewes in another breeding group.

Numbers can be put on sheep by using numbered branding irons to apply scourable paint. The brisket of the breeding ram can be painted with scourable paint so that he marks the ewe when he mounts her. If a light-colored paint is used initially, a dark color can be used after about 10 days so that any ewes that return in heat can be detected. An orange paint can be used initially followed, successively, with green, red, and black. The paint color can be changed each 10 days. Because the ewes are numbered, they can be observed daily and a record of the breeding date of each can be entered in the record book. Also, a ram that is not settling his ewes can be detected and replaced. Even though practice dictates that the semen of all rams be examined prior to the breeding season and only rams having semen that appears to be satisfactory are used, an occasional ram may be infertile or low in fertility.

The normal breeding season of 40 days provides a means of obtaining a lamb crop of uniform age and a means of locating ewes that have an inherent tendency for late lambing so that they can be culled.

Lambing operations. Prior to the time the ewes are due to lamb, the wool should be clipped from the dock, from the udder, and from the vulva region. Also all **dung** tags (small pieces of dung that stick to the wool) should be clipped from the rear and flank of the pregnant ewes. Young lambs will try to locate the teat of the ewe and may try to nurse a dung tag if it is present.

Ewes should be checked periodically to locate those that have lambed. The ewe and newborn lamb should be placed in a lambing jug. The ewe that is about ready to deliver can be watched carefully, but should not be interfered with if delivery is proceeding normally. In a normal presentation, the head and front feet of the lamb emerge first. If the rear legs emerge first (**breech presentation**), assistance may be needed if delivery is slow because the lamb can suffocate if deprived of oxygen too long. If the front feet are presented but not the head, the lamb should be pushed back enough to bring the head forward for presentation. The lamb can be identified by an ear tag, a tattoo, or both.

As soon as the lamb is born membranes or mucus that may interfere with its breathing should be removed. When the weather is cold, it may be necessary to dry the lamb by rubbing it with a dry cloth. If a lamb has become chilled, it can be immersed in warm water from the neck down to restore body temperature, after which it should be wiped dry. The lamb should be encouraged to nurse as soon as possible. A lamb that has nursed and is dry should survive without difficulty if a heat lamp is provided.

Some ewes may not want to claim their lambs. It may be necessary to tie a ewe with a rope halter so she cannot butt or trample her lamb.

Castrating and docking. Because birth is a period of stress for lambs, it is best to wait for three or four days to castrate and **dock** lambs. Castration and docking can be done by use of the elastrator (putting a tight rubber band around the scrotum above the testicles for castration or around the tail about an inch from the buttocks for docking) or the testicles and tail can be removed surgically. Because some losses are known to result when the elastrator is used, the surgical technique applied by qualified personnel is advisable. The emasculator is also useful for docking; the skin of the tail is pulled toward the lamb, the emasculator is applied about an inch from the lamb's buttocks and the tail is cut loose next to the emasculator. A fly repellent should be applied around any wound to lessen the possibility of **fly strike** (the laying of eggs in great numbers by flies).

Occasionally, ewes develop a vaginal or uterine **prolapse** (the reproductive tract protrudes to the outside through the vulva). This condition is extremely serious and leads to death if corrective measures are not taken by trained personnel. The tissue should be pushed back in place, even if it is necessary to hoist the ewe up by her hind legs as a means of reducing pressure that the ewe is applying to push the tract out. After the tract is in place, the ewe can be harnessed so that external pressure is applied on both sides of the vulva. In some cases, it may be necessary to suture the tract to make certain that it stays in place. Once a ewe has prolapsed her reproductive tract, the tract shows a weakness that is likely to recur; therefore, a record should be kept so that the ewe can be culled after she weans her lamb.

Shearing. Sheep are usually shorn in the spring prior to hot weather but after severely cold weather. Shearing is usually done by professional personnel, but classes are available to teach the operator. The usual method of shearing is to clip the fleece from the animal with power-driven shears, which leaves sufficient wool covering to protect the sheep's skin. Hand shearers are used where electricity is unavailable. In the shearing operation, the sheep is set on its rear and cradled between the shearer's legs, which are used to maneuver the sheep into the positions needed to make shearing easy.

The wool that is removed usually hangs together. It is spread with the clipped side up, rolled with the edges inside, and tied with paper twine. It is usually best to remove dung tags and coarse material that is clipped from the legs and put these items in a separate container. The tied fleece is put into a huge sack. When buyers examine the wool, they can obtain core samples of it from the sack. If undesirable material is obtained in the core sample, the price offered will be much lower than if only good wool is found.

If the wool is kept for some time before it is sold, it should be stored in a dry place and on a wooden or concrete floor to avoid damage from moisture.

FACILITIES FOR PRODUCTION OF RANGE SHEEP

Sheep differ from cattle in that they graze on weedy plants and brush as well as on grasses and legumes. They can eat such plants as tansy ragwort and halogeton, which are highly poisonous to cattle but only mildly so to sheep. These characteristics make sheep particularly well suited for range areas. Sheep of fine-wool breeding tend to stay together as they graze, which makes herding possible in range areas.

Range sheep are produced primarily in arid and semiarid regions. Range sheep are moved about either in trucks or by trailing so that they can consume available forage at various elevations. The requirements for the production of range sheep are quite different from those for farm-flock operations. One type of range sheep operation is described herein, but it must be noted that variations exist.

Range sheep are usually bred to lamb later than sheep in farm flocks; therefore, they can be lambed on the range. Few provisions are needed for lambing when sheep are lambed on the range, but under some conditions a tent or a lambing shed may be used to give range sheep protection from severe weather at lambing time. Temporary corrals can be constructed using snow fences and steel posts when it is necessary to contain the sheep at the lambing camp or at lambing sheds.

Sheep on range are usually wintered at relatively low elevations where little precipitation occurs. Wintering sheep are provided **feed bunks** if hay is to be fed and windbreaks to give protection from cold winds. Some producers of range sheep provide pelleted feed to supplement the winter forage. Pellets are usually placed on the ground, but are sometimes dispersed in grain troughs.

A chuck wagon in which are kept food for the attendant and supplies needed for caring for the sheep and the sheep dogs can be moved along as the sheep graze over the range area. Horses are used both for moving the chuck wagon and for carrying the herder. A sheep herder needs dogs to help look after and herd the sheep. The sheep are brought to a night bedding area each evening.

MANAGING RANGE SHEEP

Range sheep are grazed in three general areas: the winter headquarters, which is in relatively low and dry areas; the spring-fall range, which is somewhat higher and which receives more precipitation; and the summer grazing area, which is at high elevations in the mountains and which receives considerable precipitation that results in lush feeds.

The winter headquarters. The forage on the winter range, where there are usually fewer than 10 in. of precipitation annually, is composed of sagebrush and grasses. The grasses are cured on the ground from the growth of the previous summer; consequently, the winter forage is of low quality because the plants in the forage are mature and because they have lost nutrients. The soil in these areas is often alkaline and the water is sometimes alkaline.

Because the forage in the wintering area is of poor quality, supplemental feeding that provides needed protein, carotene, and minerals such as copper, cobalt, iodine, and selenium is necessary. A pelleted mixture made by mixing sun-cured alfalfa leaf meal, grain, solvent-extracted soybean or cottonseed meal, beet pulp, molasses, bone meal or dicalcium phosphate, and trace-mineralized salt is fed at the rate of 0.25 to 2.0 lbs per head per day, depending on the condition of the sheep. The feed may be mixed with salt to regulate intake so that feed can be before the sheep at all times. If the intake of feed is to be regulated through the use of salt, trace-mineralized salt should be avoided. Adequate water must also be provided at all times, because heavy salt intake is quite harmful if sheep do not have water for long periods of time.

The ewes are brought to the winter headquarters about the first of November. Rams are turned in with the ewes for breeding in November if lambing is to take place in sheds or if a spring range that is not likely to experience severe weather conditions is available for lambing. Otherwise, the rams are put with the ewes in December for breeding.

January, February, and March are critical months because the sheep are then in the process of exhausting their body stores and because severe snow-storms can occur. If sheep become snowbound, they should each be given 2 lbs of alfalfa hay plus 1 lb of pellets containing at least 12% protein. Adequate feeding of ewes while they are being bred and afterward results in at least a 30% increase in lambs produced and perhaps a 10% increase in wool produced. In addition, death losses are markedly reduced. Sheep that are stressed by inadequate nutrition either as a result of insufficient feed or a ration that is improperly balanced are highly susceptible to pneumonia. Heavy death losses can result.

The spring-fall range. Pregnant ewes are shorn at the winter headquarters (usually in April). They are then moved to the spring-fall range, where they are lambed. If they are lambed on the range, a protected area is necessary. An area having scrub oak or big sagebrush on the south slopes of foothills and ample feed and water is ideal for range lambing. Portable tents can be used if the weather is severe.

Ewes that have lambed are kept in the area where they lambed for about three days until the lambs become strong enough to travel. The ewes that have lambs are usually fed a pelleted ration that is high in protein and fortified with trace-mineralized salt, and either bone meal or dicalcium phosphate. Feeding at this time can help to prevent sheep from eating poisonous plants.

Ewes with lambs are kept separate from those yet to lamb until all ewes have borne their lambs. In addition, ewes that are almost ready to lamb are separated from those that will not lamb for some time yet. Thus, after lambing gets underway, three separate groups of ewes are usually present until lambing is completed.

If the ewes are bred to lamb earlier than is usual for range lambing and a crested wheatgrass pasture is available, ewes may be lambed in open sheds. If good pasture is unavailable, the ewes may be confined in yards around the lambing sheds, starting about a month prior to lambing. In this event, the ewes must be fed alfalfa or other legume hay and 0.50 to 0.75 lb of grain per head per day. The ewes should have access to a mixture of trace-mineralized salt and bone meal or decalcium phosphate.

Although shed lambing is more expensive than range lambing, increasing prices for lambs have made shed lambing advantageous. Fewer lambs are lost in shed lambing, and lambing can take place earlier in the year. The heavy market lambs that result produce enough income to more than offset the costs of shed lambing.

The summer range. Sheep are moved to the summer range shortly after lambing is completed if weather conditions have been such that snow has melted and lush growths are occurring. The sheep are put into bands of about 1,000 to 1,200 ewes and their lambs. In some large operations the general practice is to put ewes with single lambs in one band and ewes with twins in another. The ewes with twin lambs are given the best range area in an attempt to help the lambs grow well. Sheep are herded on the summer range to assist them in finding the best available forage.

In October, prior to the winter storms, the lambs are weaned. Lambs that carry sufficient finish are sent to slaughter and other lambs are sold to lamb feeders. It is the general practice among producers of Rambouillet, Columbia, or Targhee sheep to breed some of the most productive and best-wooled ewes to rams of the breed being used to raise replacement ewe lambs. Most of these ewe lambs are kept and grown out, and only the less desirable ones are culled. The remainder of the ewes are bred to meat-type rams, such as Suffolk or Hampshire, and all their lambs are marketed for slaughter, as feeders, or as stockers (animals used in the flock for breeding).

The fall range. As soon as the lambs are weaned, the ewes are moved to the spring-fall range. Later they go to the winter headquarters for wintering.

The number of ewes that can be bred per ram during the breeding period of about two months is 15 for ram lambs, 30 for yearling rams, and 35 for mature, but not aged, rams. These numbers are general and depend greatly on the type of conditions existing on the range.

CONTROLLING DISEASES AND PARASITES

Sheep raised by most producers are confronted with a few serious diseases, several serious internal parasites, and some external parasites. Some common diseases of sheep are the following:

Enterotoxemia ("overeating disease") is most often serious when sheep are in a high nutritional state (for example, lambs in the feedlot), but it can affect sheep that are on lush pastures. The disease can be prevented by administering Type D toxoid. Usually, three treatments are given; two about 4 weeks apart and a booster treatment 6 months later. Losses from this disease among young lambs can be prevented by vaccinating pregnant ewes.

Footrot is one of the most serious diseases of the sheep industry because of its common occurrence. The disease cannot be treated with systemic medication because the systemic medication cannot affect the bacteria causing rotting of tissue between the horny part of the hoof and the soft tissue below. It can be cured by severe trimming so that all affected parts are exposed, treating the diseased area with 10% formaldehyde (1 part of 38% formaldehyde to 19 parts of water), and then turning the sheep into a clean pasture so reinfection does not occur. A tilting squeeze is useful for restraining sheep that need treatment (Figure 25-8). Formaldehyde must be used with caution because the fumes are damaging to the respiratory system of both the sheep and the person applying it.

Once all sheep in the flock are free of footrot, it can best be prevented by making sure that it is not introduced into the flock again. Rams introduced for breeding should be

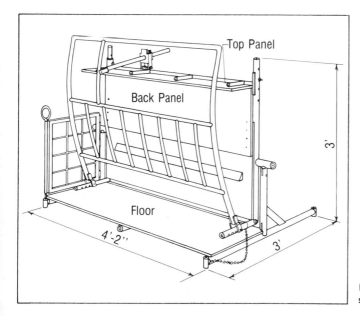

Figure 25-8. A tilting squeeze chute for restraining sheep.

isolated for 30 to 60 days to allow an opportunity for footrot to show and kept out if they show any signs of it. At present, much research is being conducted in an attempt to develop a vaccine against footrot. Some promising results indicate that the disease may be brought under control by this method.

Pneumonia usually occurs when animals have been stressed by other diseases, parasites, improper nutrition, or exposure to severe weather conditions. When pneumonia is recognized early, sulfonamides and antibiotics are usually effective against it. The afflicted animal should be given special care and kept warm and dry.

Ram epididymitis is an infection of the epididymis that reduces fertility. It can be controlled by vaccinating all rams one month before the breeding season. This should be done each year, even if a ram was vaccinated the previous year.

Sore mouth usually affects lambs rather than adult sheep. It is caused by a virus and can be contracted by humans. It can be controlled by vaccination, but sheep should not be vaccinated if this disease is not on the premises because this practice will introduce it.

Sheep are subject to a nutritional disease known as **white muscle disease.** To prevent it, pregnant ewes should be given an injection of selenium during the last one-third of pregnancy and the lambs should be given an injection of selenium at birth.

Sheep have internal and external parasites. The internal parasites are by far the more serious.

Liver flukes that infest sheep have snails as the intermediate host. After the infective stage of the fluke leaves the snail, it stays in water that sheep may drink, or, more frequently, encysts on grass that sheep may eat. Wet areas should be either drained or treated with copper sulfate to destroy snails. Sheep should not be provided water in ponds because snails are most likely to be present in such an area. Uninfested water can be piped from a fenced pond. Afflicted sheep can be treated by qualified persons with an anthelmintic, such as carbon tetrachloride. Great caution is needed in treating with carbon tetrachloride, because it is a dangerous drug that can kill the sheep.

Stomach worms of sheep constitute one of the most serious problems to producers. Stomach worms develop resistance to the **vermifuge** used; consequently, it is necessary to switch vermifuges. For example, sheep may be drenched prior to sending them to pasture with micronized (fine-particle) phenothiazine, after which the sheep are treated with mixtures of 1 part of phenothiazine to 9 parts of salt to hold down infestation in the summer. At the end of the pasture season the sheep should be drenched with tramisol. Sheep raised in drylot are unlikely to become infested with stomach worms. Some recent research by Dahmen in Idaho indicates that raising farm-flock sheep in drylot may be as economical as pasturing them on highly productive land.

Nodular worms cause knots, or nodules, in the cecum and colon of sheep. Treatments for control of stomach worms are also effective in controlling nodular worms.

Common external parasites of sheep include blowfly maggots, **keds** (sheep ticks), lice, mites, screw-worms, and sheep bots.

In wet periods when the weather is warm, **blowflies** may lay hundreds of eggs that hatch into maggots that irritate the skin of sheep. The area of the body where maggots are located should be shorn and the skin treated with an effective insecticide, by either rubbing or spraying the affected immediate areas with the treatment substance.

Keds are bloodsucking parasites. Sheep may be dipped, dusted, or sprayed using effective insecticides. Spray should be applied under high pressure (400 lbs) to ensure that it penetrates the wool.

Lice cause irritation to sheep and damage to wool. They are most likely to be serious when they infest unthrifty sheep. They are controlled by dipping, dusting, or spraying.

Mites spread mange, and they themselves are spread from sheep to sheep. Any sheep introduced into the flock should be examined carefully to be sure they do not have mange. Mites are controlled by dusting or spraying with lime sulfur.

All parasites that can be controlled by dipping, dusting, or spraying with an effective insecticide are controlled only if a systematic treatment schedule is followed. The insecticide kills only the adult parasites and does not destroy their eggs; consequently, two applications of the insecticide are necessary. The timing of the second application of insecticide is determined by the life cycle of the parasite. If the second dusting, spraying, or dipping is to be effective, the eggs of the parasite should have hatched but the adults should not have had time to lay more eggs. Guidance on timing of applications of insecticides can be obtained from an extension service or a veterinarian.

Screw-worms can attack any wound, so wounds such as those caused by castration or docking or by predatory animals should be protected by use of a fly repellent.

Sheep bot, or "grub in the head", results from a nasal fly that deposits larvae on the nostrils of sheep. These larvae crawl up the nasal passages and remain in the frontal sinuses.

The same general health program and treatments for diseases and parasites that have been discusssed for farm-flock sheep apply to range sheep. Range sheep are possibly more subject to nutritional deficiencies than are farm-flock sheep, and they can also be stressed more because of hauling or trailing. The herder who is attending 1,000 or more sheep on the range has grave responsibilities. The herder can see that the sheep are on good range at all times and keep them away from poisonous plants and predators. If the herder accomplishes these things, the incidence of diseases and parasites should be minor. Once a sheep on the range becomes very ill from a disease, death usually results because it is impossible to provide the individual attention necessary to assist it to recover.

STUDY QUESTIONS

1. Tell how you would control stomach worms in sheep.
2. If you were a commercial producer of farm-flock sheep interested in producing the maximum amount of slaughter lambs of high quality, what breeding system would you use? What breed or breeds would you use?
3. How can you assure that young lambs are prevented from chilling? If a lamb has chilled, how would you bring its temperature back to normal?
4. How would you control external parasites of sheep? Be specific.
5. At what age can young female sheep be bred? Upon what factors does the age at which a young ewe is first bred depend?
6. How can white muscle disease be prevented from occurring in lambs?
7. How would you provide for the nutritional needs of a flock of pregnant ewes from the time they are bred until the lambing season is completed?
8. Tell how you can identify each sheep in your flock. Be specific.
9. How would you control footrot in sheep?

10. If you were a farm-flock operator, would you raise feeder lambs or slaughter lambs? Why?
11. What factors have contributed to the decline in sheep numbers in the United States after 1942? Which factors are most important in the north-central states and which are more important in the Western states?

SELECTED REFERENCES

Battaglia, R. A. and Mayrose, V. B. 1981. *Handbook of Livestock Management Techniques.* Minneapolis: Burgess Publishing Co.

Chevelle, N. F. 1977. *Foot Rot of Sheep.* Washington, D.C.: Agricultural Research Service Farmers' Bulletin 2206.

Cole, H. H. and Roning, M. 1974. *Animal Agriculture. The Biology of Domestic Animals and Their Use by Man.* San Francisco: W. H. Freeman and Co.

Gee, C. K. 1979. *A New Look at Sheep for Colorado Ranchers and Farmers.* Ft. Collins: Colorado State University Experiment Station General Series 981.

Gee, C. K. and Madsen, A. G. 1974. *Structure and Operation of the Colorado Lamb Feeding Industry.* Ft. Collins: Colorado State University Experiment Station Technical Bulletin 121.

Gee, C. K. and Magleby, R. S. 1978. *Characteristics of Sheep Production in the Western Region.* Washington, D.C.: USDA Economic Research Service Agricultural Economic Report #345.

Gee, C. K., Magleby, R. S., Nielson, D. B. and Stevens, D. M. 1977. *Factors in the Decline of the Western Sheep Industry.* U.S. Department of Agriculture Economic Research Service Agricultural Economic Report #377.

Gee, C. K. and Van Arsdall, R. 1978. *Structural Characteristics and Costs of Producing Sheep in the North-Central States.* USDA Economics, Statistics and Cooperative Services SCS-19.

Michalk, D. L. 1979. *Sheep Production in the United States.* Wool Technology and Sheep Breeding. March/April 1979.

Pederson, J. H. 1974. *Sheep Handbook—Housing and Equipment.* MWP 53. Ames: Iowa State University.

Ulman, M. and Gee, C. K. 1975. *Prices and Demand for Lamb in the United States.* Fort Collins: Colorado State University Experiment Station Technical Bulletin 132.

26

The Poultry Industry

The production of domestic fowl, or poultry, is an extremely important enterprise in the United States. The estimated 294 million laying hens that produced 69.683 billion eggs worth $5.8 billion; 3.964 billion broilers worth $4.304 billion; and 165 million turkeys, worth $1.269 billion for a total of $11.373 billion for income from chickens and turkeys in 1980 is a sizeable industry. Figures for geese and ducks are not available but they contribute less to poultry income than turkeys. The operations that manage **poultry** are becoming increasingly efficient.

CHARACTERISTICS OF BREEDS

The term poultry applies to chickens, turkeys, geese, ducks, pigeons, peafowls, and guineas.

Chickens. Chickens are classified according to class, breed, and variety. A class is a group of birds that has been developed in the same geographical area. The four major classes of chickens are American, Asiatic, English, and Mediterranean. A *breed* is a subdivision of a class composed of birds of similar size and shape. Some important breeds are shown in Figure 26-1. The parts of a chicken are shown in Figure 26-2. A *variety* is a subdivision of a breed composed of birds of the same feather color and type of comb.

Factors such as egg numbers, eggshell color, egg size, efficiency of production, fertility, and hatchability are most important to commercial egg producers. Broiler producers consider such characteristics as plumage color, egg production, fertility, hatchability, growth, feed efficiency, livability, picking quality (no dark pin feathers so that a clean white carcass is obtained), and grade. Also, breeds that cross well with each other are important to broiler breeders.

The breeds of chickens listed in Table 26-1 were developed many years ago and are still used for breeding. Other breeds are exhibited at shows or propagated for specialty marketing, and are more a novelty than a part of the mainstream of the industry today.

White Plymouth Rock

S. C. White Leghorn (exhibition type)

White Cornish

New Hampshire

Barred Plymouth Rock

Hybrid Leghorn

Figure 26-1. Some breeds of chickens.

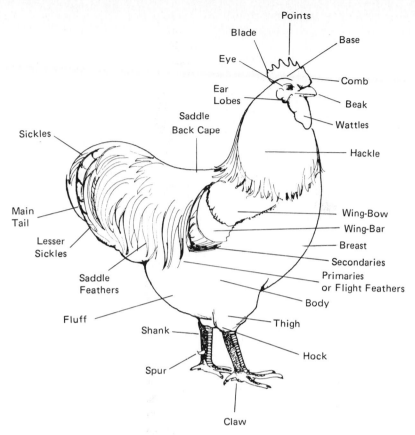

Figure 26-2. The external parts of the chicken.

Table 26-1. Certain breeds of chickens and their main characteristics

Breed	Purpose	Type of comb	Color of egg
American breeds			
White Plymouth Rock	Eggs and meat	Single	Brown
Wyandotte	Eggs	Rose	Brown
Rhode Island Red	Eggs	Single and Rose	Brown
New Hampshire	Eggs and meat	Single	Brown
Asiatic breeds			
Brahma (light)	Meat	Pea	Brown
Cochin	Meat	Single	Brown
English breeds			
Australorp	Eggs	Single	Brown
Cornish (white)	Meat	Pea	Brown
Orpington	Meat	Single	Brown
Mediterranean breed			
White Leghorn	Eggs	Single and Rose	White

Ducks. Many breeds of domestic ducks are known but only a few are of economic importance. The most popular breeds in the United States are White Pekin and Khaki Campbell. The Pekin is the most valuable for its meat and produces excellent carcasses in a rather short time of 7 to 8 weeks of age. The Khaki Campbell is used commercially in some countries for egg production.

Geese. Several breeds of geese are rather popular. The Embden, Toulouse, White Chinese, and Pilgrim are satisfactory for meat production. Birds of medium body size are preferred by a majority of commercial producers. According to Orr (1974), the characteristics that are desirable in geese are medium-sized carcass, good livability (death losses are low), rapid growth, and heavy coat of white or nearly white feathers. The Embden is a nearly white breed that meets these requirements. The Toulouse is gray. The Pilgrim is white. The White Chinese male is white and the female is light gray. These latter three breeds meet the requirements with the exception of feather color.

Turkeys. Two varieties of turkey are commercially important in the United States today: Beltsville White (small breed) and Broad White (large). Since the 1950s, the emphasis has been to produce more large, white turkeys (Figure 26-3). A corresponding decrease in the number of small whites has occurred.

Figure 26-3. Ideal Broad White turkey male.

CHANGES WITHIN THE POULTRY INDUSTRY

From 1900 until 1940, the primary concerns of the poultry industry in the United States were egg production by chickens and meat production by turkeys and waterfowl. Meat production by chickens was largely a byproduct of the egg-producing enterprises. The broiler industry, as we presently know it, was yet to be established. Egg production was well established near large population centers, but the quality of eggs was often low because of seasonal production, poor storage, and the lack of laws to control grading standards for eggs.

Prior to 1940, large numbers of small-farm flocks existed, and management practices, as applied today, were practically unknown. In the late 1950s, however, the modern mechanized poultry industry of the United States emerged. The number of poultry farms and hatcheries decreased, but the number of birds per installation dramatically increased. Larger cage-type layer operations appeared and egg-production units grew, with production geared to provide the consumer with eggs of uniform size and high quality. Huge broiler farms, which provided the consumer with fresh meat throughout the year were established, and large dressing plants capable of dressing 50,000 or more broilers per day were built. Thus, the poultry industry was completely revolutionized.

In this era of rapid expansion, the broiler industry shifted toward the southeastern United States—principally into Alabama, Arkansas, Georgia, Mississippi, and North Carolina. The main egg-production centers became located on the West Coast, in the Southeast, and in the Midwest (Table 26-2).

One dramatic change that occurred in the 1950s and 1960s was **integration**, which began in the broiler industry and is applied, to a lesser degree, in egg and turkey operations. Integration brings all phases of the enterprise under one head—corporate ownership of breeding flocks, hatcheries, feed mills, raising phase, dressing plants, services, and marketing and distribution of products. The actual raising of broilers is sometimes on a contract basis with a person who owns houses and equipment and who furnishes the necessary labor. The corporation furnishes the birds, feed, field service, dressing, and

Table 26-2. Egg production in the ten leading states, 1980

State	Number of eggs (millions)
California	8,713
Georgia	6,067
Arkansas	4,123
Pennsylvania	3,836
Indiana	3,536
Alabama	3,300
Florida	3,189
N. Carolina	3,155
Texas	2,795
Minnesota	2,183

Source: Economics, Statistics and Cooperatives Service. Poultry and Egg Situation. Washington, D.C.: U.S. Department of Agriculture. PES-305.

marketing. Payment for raising birds is generally made on the basis of a certain price per bird reared to market age. Bonuses are usually paid to those who have done a commendable job of raising birds. An important advantage to this system is that all phases are synchronized so that the utmost efficiency is realized.

Importance of the Poultry Industry

Eggs. About 69.70 billion eggs, which yield more than $5.80 billion in retail sales, are produced by chickens in the United States each year. The average American eats 300 eggs per year (this figure includes eggs used in all food preparations as well as fresh eggs), and the demand is increasing—Americans ate three more eggs per year per person in 1979 than in 1978.

The number of laying hens in the United States has actually declined over the years, but production is up because of a dramatic increase in the performance of the individual hen. In 1880, for example, the average laying hen produced 100 eggs per year; in 1950 the average was 175; and in 1980 it was 250! Furthermore, hens in large, efficient units often surpass the average.

The USDA estimates that 85% of the market eggs produced in the United States are from the country's 12,000 largest commercial producers (those that maintain one million or more birds, with the largest having 11 million). The 45 largest egg-producing companies in the United States have 97,276,600 layers, which is 35.5% of the nation's total.

Broilers. Broiler production in the United States has increased steadily since the inception of large-scale operations: about 3.9 billion broilers were produced in 1980, which is more than 100 times the 34 million produced in 1934.

Turkeys. Approximately 165 million pounds of live turkey were produced per month in the United States during 1980. The income from turkeys in 1980 was approximately $1.3 billion.

One of the most striking achievements in the poultry industry is the increased production of eggs and meat per hour of labor. One large commercial organization has developed a four-house, 8,000-hen complex that is managed by one employee. This labor reduction has been accomplished by automatic feeding, watering, egg collecting, egg packing, and manure removal.

BREEDING POULTRY

Chickens are used for the production of both meat and eggs while turkeys, geese, and ducks are used largely for production of meat. However, duck eggs are important in some parts of the world. Sophisticated selection and breeding methods have been developed in the United States for increased productivity of chickens for both eggs and meat. The discussion to follow is directed primarily at the improvement of the chickens; however, the methods described for chickens can be applied to improvement of meat production in turkeys, geese, and ducks.

Early poultry breeding and selection concentrated on qualitative traits which are, from a genetics standpoint, more predictable than quantitative traits. Qualitative traits, such as color, comb type, abnormalities and sex-linked characteristics are important; however, quantitative traits, such as egg production, egg characteristics, growth, fertility, and hatchability, are economically more important today.

Quantitative traits are more difficult to select for than are qualitative traits because the mode of inheritance is more complex and the role of the environment is greater. Quantitative traits differ greatly as to the amount of progress that can be attained through selection. For example, increase in body weight is much easier to attain than increase in egg production. Furthermore, a relationship usually exists between body size and egg size. Generally, if body size increases, a corresponding increase in egg size will occur. Most quantitative traits of chickens fall in the low to medium range of heritability, whereas qualitative traits fall in the high range.

Progress in selecting for egg production has been aided by the trapnest, a nest equipped with a door that allows a hen to enter but prevents her from leaving (Figure 26-4). The trapnest enables accurate determination of egg production of individual hens for any given period and helps to identify and help eliminate undesirable egg traits and identify clutch patterns and broodiness (the hen wanting to set on eggs to hatch them). Furthermore, it allowed the breeder to begin pedigree work within a flock.

The two most important types of selection applied primarily to chickens but to a lesser extent to other poultry today are mass selection and family selection. In mass selection, which is the older method, the best-performing males are mated to the best-performing females. This program is very effective in improving traits of high heritability. Family

Figure 26-4. Indoor turkey-breeding facility equipped with trapnests.

selection is a system whereby all offspring from a particular mating are designated as a family and selection and culling within a population of birds are done on the basis of the entire family. This type of selection is most adaptable to selection for traits of low heritability. This is not to say that progress for traits of high heritability cannot be made by using family selection—quite the opposite is true. However, mass selection is very effective for traits of high heritability and certainly easier.

Progress through selection is rather slow for chickens, especially for the traits of low heritability. Most breeding programs are geared to the improvement of more than one trait at a time. For example, to improve egg-laying lines, it might be necessary to attempt simultaneously to improve egg numbers, shell thickness, interior quality of the egg, and feed efficiency.

Outcrossing. Defined as the mating of unrelated breeds or strains, outcrossing is probably more adaptable to the modern broiler industry than to the egg-production industry, although it can be used to improve egg production. Numerous experiments have established fairly accurately which breeds or strains will cross well with each, and knowledge today is sophisticated to the extent that breeders know which strain or line to use as the male line and which to use as the female line.

Crossing inbred lines. The development of inbred lines for crossing is used largely for increasing egg production. Inbreeding is defined as the mating of related individuals or as the mating of individuals more closely related than the average of the flock from which they came. The mating of brother with sister is the most common system of inbreeding for poultry; however, mating parents with offspring gives the same results. Both of these types of inbreeding increase the degree of inbreeding at the same rate per generation.

Hybrid chickens are produced by developing inbred lines and then crossing them to produce chickens that exhibit hybrid vigor. The technique employed in the production of hybrid chickens is essentially the same as that used in the development of hybrid plants. The breeder works with many egg-laying strains or breeds while producing hybrids. The strains or breeds are inbred (mostly brother to sister) for a number of generations, preferably five or more. In the inbreeding phase, many undesirable factors are culled out—selection is most rigid at this stage.

In all phases of hybrid production, the strains are completely tested for egg production and for important egg-quality traits. After inbreeding the birds for the necessary number of generations, the second step, crossing the remaining inbred lines in all possible combinations, is initiated. These crossbreds are tested for egg production and for egg-quality traits. The crossbreds that show improvement in these traits are bred in the third phase, in which the best test crosses are crossed into three-way and four-way crosses in all possible combinations. Most breeders prefer the four-way cross. The offspring that results from three-way and four-way crosses are known as hybrid chickens.

Inbreeding is an extremely costly way to produce chickens that lay eggs of superior quality and quantity, because inbreeding takes an enormous economic toll from the strains that are inbred. Many strains do not survive the inbreeding phase. The events in this breeding scheme are timed so as to provide the commercial producer with chickens that will give maximum production at a definite point (usually after five or six generations of inbreeding). It is imperative that the breeder carry a control population for each strain to enable evaluation of the inbred lines and also to have birds from which to start new inbred lines as needed.

Strain crossing. Strain crossing is more easily done than inbreeding and crossing inbred lines because it entails only the crossing of two strains that possess similar egg-production traits. It is definitely a type of crossbreeding if the two strains are not related to each other.

Strain crossing has two definite advantages over inbreeding in that it is easy to do and is relatively inexpensive. In general, the day-old female chick produced through strain crossing costs less than the day-old hybrid chick.

The computer is used largely in chicken breeding to help select breeding stock. An overall merit index is computed using data on heritability of each trait considered, the relative economic importance of each trait, and the genetic correlations among the traits. Birds having the highest index are used for breeding.

In the production of broiler or layer hybrid chicks, the computer is used to determine which lines or strains are most likely to give chickens that are superior. In broiler hybrids, hatchability, growth rate, economy of feed use, and carcass desirability are important. In layer hybrids, hatchability, egg production, egg size, egg and shell quality, and livability are important.

Practically all line-cross layers are now free from **broodiness** because attempts have been made to eliminate genes for broodiness and records are available on which lines produce broody chicks when crossed. Lines that produce broody chicks in a line cross are no longer used for producing commercial layers.

STUDY QUESTIONS

1. With reference to poultry, define class, breed, and variety. List examples of classes and breeds of chickens.
2. What major changes have taken place in the poultry industry of the United States in the past 30 years?
3. How have modern poultry management practices improved the efficiency of production of poultry and poultry products?
4. What qualities are considered especially desirable in turkeys today? Why are turkey poults more expensive than young chickens?
5. What changes can be expected in the poultry industry in the foreseeable future?

SELECTED REFERENCES

Economics, Statistics and Cooperative Service. 1980. *Poultry and Egg Situation.* Washington, D.C: U.S. Department of Agriculture, PES-305.

Crop Reporting Board. Economics, Statistics and Cooperative Service. 1980. *Eggs, Chickens, and Turkeys.* Washington, D.C.: U.S. Department of Agriculture.

Plunkett, James. 1980. Four-house, 80,000-hen complex managed by a single employee. *Poultry Digest* 39:64, 68.

The Editors. 1980. Production companies have 35.5 percent of nation's layers. *Poultry Tribune* 86:10, 12.

Pedersen, John. 1980. Egg market outlook for 1980. *Poultry Tribune* 86:24,26,28.

Neshlim, M. C., Austic, R. E., and Card, L. E. 1979. *Poultry Production.* 12th edition. Philadelphia: Lea and Febiger.

27

Managing Poultry

The success of modern poultry operations depends on many factors. The hatchery operator must care for breeding stock properly so that eggs of good quality are available to the hatchery and must incubate eggs under environmental conditions that ensure the hatching of healthy and vigorous birds.

INCUBATION MANAGEMENT

Each type of poultry has an incubation period of definite length (Table 27-1) and incubation management practices are geared to the needs of the eggs throughout that time.

The discussion on incubation management applies specifically to chickens. Although the principles of incubation management apply to other types of poultry, specific information should be obtained from the extension service.

All modern commercial hatcheries have some form of forced-air incubation system. Today's commercial incubator has a forced-air system that creates a rather uniform environment inside the incubator. Forced-air incubators are available in many sizes. All sizes are

Table 27-1. Incubation times for various birds

Type	Incubation period (days)
Chickens	21
Turkeys	28
Ducks	28
Muscovy Duck	32 to 36
Geese	30
Pheasant	21 to 25
Bobwhite Quail	23
Coturnix Quail	16
Guineas	27

equipped with sophisticated systems that control temperature and humidity, turn eggs, and bring about an adequate exchange of air between the inside of the incubator and the outside. The egg-containing capacity of forced-air incubators ranges from several hundred to 100,000 or more (Figure 27-1).

It is vital that temperature, humidity, position of eggs, turning of eggs, oxygen content, carbon dioxide content, and sanitation be regulated.

Temperature. Proper temperature is probably the most critical requirement for successful incubation of chicken eggs. The usual beginning incubation temperature in the forced-air incubator is 99.5 °F to 100.0 °F. The temperature should usually be lowered slightly (by 0.25 °F to 0.5 °F) at the end of the fourth day and remain constant until the eggs are to be transferred to the hatching compartments (approximately three days before hatching). Three days before hatching, the temperature is again lowered, usually by about 1.0 °F to 1.5 °F. Just before hatching, the chicks change from embryonic respiration to normal respiration and give off considerable heat, which results in a higher incubator temperature.

With reference to temperature, there are two especially critical periods during incubation—the first through the fourth day, and the last portion of incubation. Higher-than-optimum temperatures usually speed the embryonic process and result in embryonic mortality or deformed chicks at hatching. Lower-than-optimum incubation temperatures usually slow the embryonic process and also cause embryonic mortality or deformed chicks.

Humidity. A relative humidity of 60% to 65% is needed for optimum **hatchability**. All modern commercial forced-air incubators are equipped with sensitive humidity controls. The relative humidity should usually be raised slightly in the last few days of incubation. It has been shown that when the relative humidity is close to 70% in the last few days of

Figure 27-1. Large commercial room-type incubator.

incubation, success of hatching is greater. Providing optimum relative humidity is essential in reducing evaporation from eggs during incubation.

Position of eggs. Eggs hatch best when they are incubated with the large end (the area of the space called the "air cell") up; however, good hatchability can also be achieved when eggs are set in a horizontal position. Eggs should never be set with the small end up, because a high percentage of the developing embryos will die before reaching the hatching stage. The head of the developing embryo should develop near the air cell. In the last few days of incubation, the eggs are transferred to a different type of tray—one in which the eggs are placed in a horizontal position.

Shortly before hatching, the beak of the chick penetrates the air cell. The chick is now able to receive an adequate supply of air for its normal respiratory processes. The horny point of the beak eventually weakens the eggshell until a small hole is opened. The egg is now said to be "pipped." The chick is typically out of the shell within a few hours. The hatching process varies considerably among different species.

Turning of eggs. Modern commercial incubators are equipped with time-controlled devices that permit eggs to be turned periodically. Eggs that are not turned enough during incubation have little or no chance of hatching because the embryo often becomes stuck to the shell membrane.

Commercial incubators are equipped with setting trays or compartments that allow eggs to be set in a vertical position. They also have mechanisms that rotate these trays so that the eggs rotate 90° each time they are turned. The number of times the eggs are turned daily is most important. If one were to rotate the eggs once or twice daily, the percent of hatchability would be much lower than if the eggs were rotated five or more times daily.

Oxygen content. The air surrounding incubating eggs should be 21% oxygen by volume. At high altitudes, however, the available oxygen in the air may be too low to sustain the physiological needs of developing chick embryos, many of which, therefore, die. Good hatchability can often be attained at high altitudes if supplemental oxygen is provided. This is necessary for commercial turkey hatching, but hatchability can be improved by selecting breeding birds that hatch well in environments in which the supply of oxygen is limited.

Carbon dioxide content. It is vital that the incubator be properly ventilated to prevent excessive accumulation of carbon dioxide (CO_2). The CO_2 content of the air in the incubator should never be allowed to exceed 0.5% by volume. Hatchability is lowered drastically if the CO_2 content of air in the incubator reaches 2.0%. Levels of more than 2.0% almost certainly reduce hatchability to near zero.

Sanitation. The incubator must be kept as free of disease-causing microorganisms as possible. The setting and hatching compartments must be thoroughly washed or steam-cleaned and fumigated between settings. In many cases, they should be fumigated more than once for each setting. An excellent schedule is to fumigate eggs immediately after they have been set and again as soon as they have been transferred to the hatching compartment. A good procedure for fumigation (Neshlim *et al.*, 1979) is:

1. Prepare 1.5 ml of 40% formalin for each cubic foot of incubator space.
2. Prepare 1.0 g of potassium permanganate for each cubic foot of incubator space.
3. Set temperature of setting compartment or hatching compartment at proper reading (99.5 °F to 100.0 °F, and 98.5 °F to 99.0 °F, respectively).

4. Turn on fan.
5. Close vents.
6. Set relative humidity at proper reading (65%).
7. Place potassium permanganate in open container in compartment.
8. Pour formalin into potassium permanganate container and close incubator door for approximately 20 minutes.
9. Open door for 5 to 10 minutes or turn on exhaust system to allow formaldehyde gas to escape.
10. After gas has escaped, remove potassium permanganate container, close door, and continue normal operating procedures.

Candling of eggs. In maintaining a healthy and germ-free environment, eggs must be candled (examined by shining a light through each egg to see if a chick embryo is developing) at least once during incubation so that infertile and dead-germ eggs (eggs containing dead embryos) can be removed. Many operators candle chicken eggs on the fourth or fifth day and turkey and waterfowl eggs on the seventh to tenth day. Some operators candle eggs a second time when transferring eggs to the hatching compartment.

MANAGING YOUNG POULTRY

The main objective in managing young poultry is to provide a clean and comfortable environment with sufficient feed and water (Figure 27-2).

House preparation. The brooder house should be thoroughly cleaned, **disinfected**, and dried several days before it receives young birds. All necessary brooding equipment

Figure 27-2. Brooding of chicks with a floor-type system. Note litter on floor, automatic feeders and waterers, and gas brooder stoves in a raised position.

should have been tested ahead of time and be in the proper place. Such advance preparation is probably as essential to success as any management practice applied to young birds. It is imperative that the disinfectant have no effect on the meat or eggs produced by birds. Federal Drug Administration regulations that govern the use of disinfectants should be strictly observed.

Litter. Litter should be placed in the house when equipment is checked. Some commonly used substances are planer shavings, sawdust, wood chips, peat moss, ground corn cobs, peanut hulls, rice hulls, and sugar cane fiber. The entire floor should be covered by at least 2 in. of litter, which must be perfectly dry when the birds enter.

The primary purpose of litter is to absorb moisture. Litter that gets extremely damp should be replaced. Some operators add small amounts of litter as the birds grow.

Floor space. The availability of too much or too little floor space can affect growth and efficiency of production adversely. If floor space is insufficient, young birds have difficulty finding adequate feed and water. This could lead to feather picking and actual cannibalism. Too much space can cause the birds to become bored, which can cause problems similar to those caused by overcrowding.

The amount of space required varies with the type and age of bird. Regardless of type, each bird should have at least 7 to 10 square inches of space around and under the hover (a metal canopy that covers most brooder stoves). Chicks of the egg-producing type and birds of similar size fare well for five to six weeks on 0.5 square feet of floor space. Between the time these birds are 6 and 10 weeks old, the space per bird must be increased to nearly 1 square foot.

Broilers, turkeys, and waterfowl should have an area of 0.75 to 1 square foot for at least their first eight weeks. In certain situations such birds could be allowed as much as 1 to 1.5 square feet before they are 8 weeks old.

Feeder space. One linear inch of feeder space per bird is sufficient for most species from the age of 1 day to 3 weeks; 2 in. for 3 to 6 weeks; and 3 in. beyond 6 weeks. Quail require less space, whereas turkeys, geese, and ducks require slightly more. Most feeders are designed so that birds can eat on either side of them.

The operator can determine how much feeder space is needed through observation. If relatively few birds are eating at one time, the feeding area is probably too large. If too many are eating at once, the area is probably too small. Young birds are usually fed automatically after the brooder stove enclosure is removed.

Water requirements. Generally, two one-gallon size water fountains are adequate for 100 birds that are one day old. More waterers can be added as necessary. Most commercial producers switch from water fountains to an automatic watering system when birds are put on automatic feeding. If trough-type waterers are used, allow at least one inch of water space per bird until 10 to 12 weeks of age. More space might be needed thereafter. Trough-type waterers designed so that birds can drink from both sides are available. A space of at least 0.5 inch per bird is needed if pan-type waterers are used.

Lighting requirements. The amount of light provided by the operator varies according to the type of bird being raised. Many broiler producers use a 24-hour light regime, whereas others use systems such as 20 hours of light alternating with 4 hours of darkness. Future layers do not require as much light per day as future broilers because the producer desires that future layers not reach sexual maturity too quickly. Generally, about 12 to 14 hours of light are sufficient for future layers until they are about 12 to 14 weeks old. However, in window-type housing without light-checks (baffles that can be opened and closed), these birds might receive more than 12 to 14 hours of light in long-day periods. Supplemental light can be provided in short-day periods.

A system of lighting that takes into account the age of the bird, the time of year in which the bird was hatched (seasonal effect), and the type of housing (windowed or windowless) should be used. Pullets should be on a constant day-length regime or on a decreasing light regime by the time they are 10 weeks old to prevent early sexual maturity. In most cases, some type of dim light is provided to prevent birds from piling up in periods of darkness. After birds are about a week old, some producers substitute a light of relatively low intensity (15 to 25 watts) for a bulb of higher intensity (40 watts is standard for the first week).

Other management factors. Young birds should have access to the proper feed as soon as they are placed in the brooder house. It is important that birds such as turkey poults start eating feed early. Some young poults die for lack of feed although feed is readily available because they have not learned to eat it. It is a good practice to teach young poults to drink water and to eat. This can be done by dipping their beaks into water or by using some type of dropper to insert a few drops into their mouths. They can be taught to eat by placing some food in their mouths or by pushing the beak into the feed. Other species of domestic birds more readily learn to use the feed and water provided them.

Young birds are commonly debeaked sometime between their first day of life and several weeks of age. **Debeaking** is done with a debeaking machine by searing off approximately half of the upper mandible and removing a small portion of the lower one. Care should be taken to avoid searing the bird's tongue while debeaking. Debeaking is done to prevent birds from **picking** feathers from other birds; it prevents **cannibalism** from starting.

Many commercial operators "dubb" chickens at the time of debeaking. Dubbing is cutting off the comb in single-comb varieties. The wattles (naked, fleshy processes of skin hanging from the chin or throat) are frequently removed in dubbing also. Dubbing can be delayed to a later age—up to 10 weeks.

Birds should be vaccinated according to a prescribed schedule, a veterinarian can advise what to use and when to use it.

MANAGING 10- to 20-WEEK-OLD POULTRY

Management of poultry from approximately 10 weeks to 20 weeks of age is quite different from managing younger birds. Improper practices in this most critical period could adversely affect subsequent production.

Confinement rearing is used by commercial producers of replacement birds, those birds that will be kept for egg production. In the three basic systems of confinement rearing, birds are raised on solid floors, slatted floors, or wire floors or cages.

Birds raised on floors in confinement should have from one to two square feet of floor space per individual. The amount of space required depends on the conditions of the surrounding environment and also on the condition of the house. Caged birds should be allowed 0.5 square feet to not more than 0.75 square feet.

Replacement chickens that are raised in confinement are generally fed a completely balanced growing, or developing, ration that is 15% to 18% protein. Most birds are kept on a full-feeding program, but in certain conditions some producers restrict the amount of feed provided. Restricted feeding can take different forms: total feed intake may be limited, protein intake may be limited, or energy intake may be limited. The main purposes of a restricted feeding program are to slow the rate of growth (thus delaying the onset of sexual maturity) and to lower the cost of feeding. Feed intake must not be restricted whenever a restricted lighting system (which is also used to delay sexual maturity) is in effect.

Automatic feeding and watering devices are used in most confinement operations. Watering- and feeding-space requirements are practically the same as those for birds that are 6 to 10 weeks old. As long as birds are not crowding the feeders and waterers, there is no particular need to increase the feeding and watering space.

The lighting regime used in confinement rearing is very important. Birds raised in window-type housing receive the normal light of long-day periods unless the house is equipped with some type of light-check. In short-day periods supplemental lighting can be used to meet the requirements of growing chickens. Replacement chickens should receive approximately 14 hours of light daily up to 12 weeks of age. To delay sexual maturity, the amount should then be reduced to about 8 to 9 hours daily until the birds have reached 20 to 22 weeks of age. The light is then either abruptly increased to 16 hours per day, or it is increased by 2 to 3 hours with weeky increments of 15 to 20 minutes then added until 16 hours of light per day are reached.

Regardless of which system replacement pullets are reared under, they should be placed in the laying house when they are approximately 20 weeks of age so that they can adjust to the house and its equipment before they begin to lay.

MANAGEMENT OF LAYING HENS

The requirements of laying hens for floor space vary from 1.5 to 2 square feet per bird for egg-production strains and from 2.5 to 3.5 square feet for dual-purpose and broiler strains. Turkey breeding hens require four to six square feet but less area is needed if hens are housed in cages. If turkey hens have access to an outside yard, four square feet of floor space is best. Game birds such as quail and pheasants require less floor space than egg-production hens. In general, three square feet or more of floor space is adequate for ducks, whereas geese need approximately five square feet.

Breeder hens are housed on litter or on slatted floors while birds used for commercial egg production are kept in cages (Figure 27-3).

Floor-type houses for breeder hens are usually equipped with an area where hens can roost during darkness. The hens will deposit much manure in the roosting or dropping pits. The manure can be removed manually or mechanically. The tops of the pits are covered with some type of wire (preferably welded). Roosting perches, which are typically 2 in. wide and 2 in. thick, are placed on top of the wire and are spaced approximately 15 to 18 in. apart. The height of the roosting area is usually about 15 to 18 in. apart. The height of the roosting area is usually about 15 to 18 in. Slatted floors are usually several feet above the base of the building and manure is often allowed to accumulate for a rather long time before being removed. Many slatted-floor houses are equipped with mechanical floor

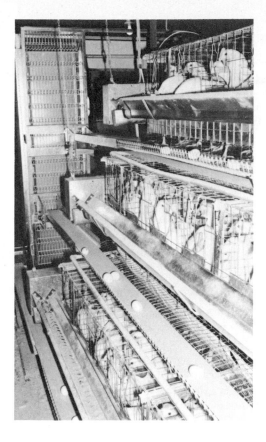

Figure 27-3. Composite of an automated commercial cage house for laying hens. Feeding, watering, egg gathering (vertical elevator at end of row), and manure removal are done automatically.

scrapers that remove the manure periodically. The frequent removal of manure lessens the chance that ammonia will accumulate.

Gathering of eggs in floor-type houses can be done automatically if some type of roll-away nesting equipment is used. In houses that are equipped with individual nests, the eggs are gathered manually. Some operators prefer colony nests (a group of open nests).

Some type of litter or nesting material must be placed in the bottom of individual and colony nests if eggs are to be gathered manually. Birds should not be allowed to roost in the nests in darkness because dirty nests result.

Cage operations are used by most commercial egg-producing farms. The basic type of cage operation is *individual* cage systems. Young chickens may be brooded in colony cages (Figure 27-4) while laying hens may be housed in triple-decked laying cages. The individual cages range in size from 8 to 12 in. in width, 16 to 20 in. in depth (front to back), and 12 to 15 in. in height. The number of birds housed per unit varies with the operator. The density of the cage population is an important factor in production.

Figure 27-4. Young chickens being brooded in colony cages.

All cages are equipped with feed troughs that usually extend the entire length of the cage. Some troughs are filled with feed manually, others automatically. It is advantageous to dub and remove the wattles from caged layers so they can obtain feed from the automatic feeders by reaching the head through the openings.

Two basic types of watering devices, troughs and individual "cup-type" waterers, are used in individual and colony cages. The cup-type might have a slight advantage in controlling disease because birds can only drink from their individual unit. The watering system should be equipped with a metering device that makes it possible to medicate the birds quickly when necessary by mixing the exact dosage of medicant required with the water.

The arrangement of cages within a house varies greatly. Some producers have a "step-up" arrangement, such as a double row of cages at a high position with a single row at a low position on either side. The droppings from birds caged in the double row fall free of the birds in the single row. An aisle approximately three feet wide is usually between each of the groups of cages. Rather than aisles, some systems have a movable ramp that can travel, above the birds, from one end of the house to the other. In other systems, several double cages are stacked on top of each other.

In cage operations, manure is typically collected into pits below the cages and removed by mechanical pit scrapers. Manure may also drop into pits that are slightly sloped from end to end and partially filled with water. The manure can be flushed out with the water into lagoons. This system minimizes the occurrence of flies and, if managed properly, renders the odor from ammonia negligible.

HOUSING POULTRY

Factors such as temperature, moisture, ventilation, and insulation are given careful consideration in the planning and management of poultry houses.

Temperature. Most poultry houses are built to prevent sudden changes in house temperature. A bird having an average body temperature of 106.5 °F usually loses heat to its environment except in extremely hot weather. Chickens perform well in temperatures between 35 °F and 85 °F, but 55 °F to 75 °F seems optimal.

House temperature can be influenced by such factors as the prevailing ambient temperature, solar radiation, wind velocity, and heat production of the birds. A four-and-a-half-pound laying hen can produce nearly 44 **British thermal units** (Btu) of heat per hour (one Btu is the quantity of heat required to raise the temperature of one pound of water 1 °F at or near 39 °F). The amount of Btu produced varies with activity, egg-production rate, and the amount of feed consumed. Heat production by birds must be considered in designing poultry houses (about 40 of the 44 Btu produced per hour by a laying hen are available for heating). Although heat produced by birds can be of great benefit in severe cold, the house must be designed so that excess heat produced by birds in hot weather can be dissipated.

Moisture. Excess house moisture, especially in cold weather, can create an environment that can be extremely uncomfortable, can lead to a drop in production by laying birds, and, if allowed to continue for long, can cause illness.

Much moisture in a poultry house comes from water in inflowing air, in water vapor from the birds themselves, and in their droppings. Every effort should be made, of course, to minimize spillage from watering equipment. The amount of moisture in a poultry house can be reduced by increasing the air temperature or by increasing the rate at which air is removed. Because air holds more moisture at high temperatures than at low temperatures (Table 27-2), raising the air temperature will cause moisture from the litter to enter the air and thus be dissipated. Air temperature in the house can be increased by retaining the heat produced by the birds themselves and by supplemental heat.

If incoming air is considerably colder than the air in the house, it must be warmed or it will fail to aid in moisture removal. In cold weather, most exhaust ventilation fans are run slower than normal, so most moisture removal is then accomplished by increasing the air temperature in the house.

Ventilation. A properly designed ventilation system provides adequate fresh air, aids in removing excess moisture, and is essential in maintaining a proper temperature within the house. The type and amount of insulation and the heat produced by the birds themselves must be considered in planning for ventilation.

Table 27-2. The water-holding capacity of air at various temperatures

Temperature (°F)	Lbs of water per 1,000 lbs of dry air
50	7.62
40	5.20
30	3.45
20	2.14
10	1.31
0	0.75
-10	0.45

Source: Meyer, V. M., and Walther, P. *Ventilate Your Poultry House: For Clean Eggs, Healthy Hens, More Profit.* Ames: Iowa State University Cooperative Extension Service Pamphlet 292. 1963.

Ventilation is accomplished by positive pressure in which air is forced into the house to create air turbulence or by negative pressure in which air is removed from the house by exhaust fans (Figure 27-5). The positive pressure system is accomplished by having fans in the attic so that air is forced through holes. The negative pressure system is accomplished by locating exhaust fans near the ceiling. Some houses are ventilated where the air is fairly warm and dry by having open walls. Also, when the outside air is cold and dry, air can enter at the low portions of the house and leave through vents near the ceiling as it warms. The heat created by the chickens will warm the colder air which will then take moisture from the house.

The two integral and necessary parts of the most common ventilation systems are the exhaust fan and the air-intake arrangement. The number of exhaust fans varies with the size of the house, the number of birds, and the capacity of the fans (Figure 27-5). Most fans are rated on the basis of their cubic feet per minute (cfm) capacity, that is, how many cubic feet of air they move per minute. Fans in laying houses should operate at 4 to 4.5 cfm per bird when the temperature is moderate. In summer, cfm's per bird could be as high as 10.

Fans also have a static pressure (water pressure expressed in inches) rating. A good fan should be rated at ⅛ in. static pressure.

Many fans have two speeds—a low speed for low house temperatures and a high speed for warm house temperatures. Others operate at only one speed—the amount of air removed is controlled by shutters that open so that more air can be removed when the temperature of the house exceeds the thermostatically fixed temperature. These shutters close at lower temperatures.

An air-intake area must be provided for the ventilation system. It is usually a slotted area in the ceiling or near the top of the walls. There should be at least 100 square inches of inlet space for each 400 cfm of fan capacity.

Figure 27-5. A battery of fans used to ventilate a large poultry house.

Insulation. Because energy needs are becoming ever more critical, the insulation of poultry houses of the future must be superior to that of the past. Sudden temperature changes inside the house must be avoided, especially for young birds. Houses in cold and windy areas require better insulation than those in milder climates, but good insulation is also essential in areas where outside temperatures are high. The insulating ability of any material is measured by its R value. This value is based on the material's ability to limit heat loss, as expressed in Btus. The better the insulating material, the higher its R value. For example, if the outside temperature is 11 °F cooler than the inside, a material having an R value of 11 loses approximately one Btu per hour to the outside for each square foot of area. A material having an R value of 20 loses one Btu per hour to the outside for each square foot of area when the outside temperature is 20 °F cooler than the inside.

By knowing how much heat the birds generate, the local variations in ambient temperature, expected wind velocities, and relative humidities experienced, the appropriate R value can be established. In many areas of the United States, especially in the cold regions, wall insulation in poultry houses should have an R value of 15; ceiling insulation, at least 20 (heat loss through the floor is negligible). In warmer areas, wall insulation should have an R value of 5 to 10; ceiling insulation, 10 to 15.

When computing the R value for a particular area of the house (walls, for example), the resistance of the outer surface, the insulation, the air space between the studding, the inner wall material, the resistance of the inner surface and, if windows are present, the presence of glass, must all be considered. Each of these factors has an R value and the total of these values establishes the R value for that area of the house. A building materials dealer can furnish the R value ratings of these factors.

Because too much moisture may accumulate in the house, especially in cold weather, some type of vapor barrier should be present in the walls and ceiling. This barrier can be a part of the insulation itself (foil backing) or separate. It should be placed between the insulation and the inside wall. The foil-back portion of the insulation should be placed next to the inside wall.

FEEDS AND FEEDING

Rations fed to poultry today are complex mixtures that should include all ingredients, in a balanced proportion, that have been found to be necessary to body maintenance, for maximum production of eggs and meat, and for optimum reproduction (**fertility** and **hatchability**). Poultry nutritionists are constantly searching for and testing new feedstuffs that, when incorporated into rations, will permit better efficiency of production. Feeding can be expensive, because an adequate intake of energy, protein, minerals, and vitamins is essential.

The computer is used in formulating least-cost rations that meet the nutritional requirements of poultry. Data input that must be provided the computer includes constraints that depend on age of birds and their productivity, cost of ingredients that are available, ingredients desired in the ration, composition of the nutritional materials, and requirements of the chickens to be fed.

The constraints set minimum and maximum percentages. For example, an upper limit on the amount of fiber that could be in the ration must be established because birds can not use fiber effectively. Likewise, a low or minimal level of yellow corn is important if it is to provide the carotene needs.

The large computers are expensive and only large poultry operations can afford them; however, the microcomputers are now designed and ready for use in computing least-cost rations. Most poultry producers can afford a microcomputer.

Energy requirements are supplied mainly by cereal grains, grain by-products, and fats. Some important grains are yellow corn, wheat, sorghum grains (milo), barley, and oats. Most rations contain rather high amounts of grain (60% or higher, depending on the type of ration). A ration containing a combination of grains is generally better than a ration having only one type. Animal fats and vegetable oils are excellent sources of energy. They are usually incorporated into broiler rations or any high-energy rations.

Protein is so highly essential that most commercial poultry feeds are sold on the basis of their protein content. The types of amino acids present determine the nutritional value of protein (see Chapter 13). Excellent protein can be derived from both plants and animals. Most rations contain both plant and animal protein so that each source can supply amino acids that the other source lacks. The most common sources of plant protein are soybean meal, cottonseed meal, peanut meal, alfalfa meal, and corn gluten meal. Cereal grains contain insufficient protein to meet the needs of birds. The best sources of animal protein are fish meal, milk byproducts, meat byproducts, tankage, blood meal, and feather meal. The protein requirements of birds vary according to the species, age, and purpose for which they are being raised. The protein requirements of certain birds are shown in Table 27-3. Where variable values are listed, the highest level is to be fed to the youngest individuals.

Whatever ration is adequate for turkeys is generally suitable for game birds. Some producers of game birds feed a turkey ration at all times. Others feed a complete game bird

Table 27-3. Protein requirements of poultry

Type	Age (weeks)	Percent protein required in diets
Chickens		
Broilers	0 to 9	20 to 23
Replacement pullets	0 to 14	16 to 20
Replacement pullets	14 to 20	12
Laying and breeding hens		15
Turkeys		
Starting	0 to 8	26 to 28
Growing	8 to 24	14 to 22
Breeders		14
Pheasant and Quail		
Starting and growing		28 to 30
Ducks		
Starting and growing		17

Source: National Academy of Sciences. Nutrient Requirements of Poultry. 6th ed. Washington, D.C.: National Academy of Sciences, National Research Council, 1971.

ration. The actual nutrient requirements of game birds are not as well known as are the requirements of chickens and turkeys.

A considerable number of minerals are essential, especially calcium, phosphorus, magnesium, manganese, iron, copper, zinc, and iodine. Calcium and phosphorus, along with vitamin D, are essential for proper bone formation. A deficiency of either of these elements can lead to a bone condition known as rickets. Calcium is also essential for proper eggshell formation.

The amount of calcium required varies somewhat with age, rate of egg production, and temperature. Chickens and turkeys up to 8 weeks of age require 1.0 to 1.2% of calcium in their diets. Calcium can be reduced to around 0.8% from 8 to 16 weeks of age. Laying hens require at least 2.75% calcium. The amounts of phosphorus required for chickens of different ages are as follows: 0 to 8 weeks, 0.7%; 8 to 18 weeks, 0.4%; and mature, 0.6%. Turkeys require from 0.7 to 0.8% phosphorus; game birds, approximately 1.0%. Requirements for other minerals vary greatly and, in many cases, are not completely known.

The vitamins that are most important to poultry are A, D, K, and E (fat-soluble), and thiamine, riboflavin, pantothenic acid, niacin, vitamin B_6, choline, biotin, folacin, and vitamin B_{12} (water-soluble).

STUDY QUESTIONS

1. List and discuss the factors that are essential for ideal incubation of eggs in still-air and forced-air incubators.
2. List five materials that make good litter for a poultry house. What is the most vital characteristic of a good litter?
3. Discuss the importance of providing birds in a poultry house with adequate floor space.
4. Why are the lighting requirements for chickens being raised as broilers different from those for chickens that are raised as layers or breeders?
5. Compare range-type and confinement systems of rearing poultry in terms of the management practices and facilities required. Which system seems most advantageous?
6. Discuss the importance of insulation in present and future poultry houses.
7. What management steps should be taken in order to provide an ideal environment inside a poultry house?
8. What constitutes a good poultry ration? From what sources do the ingredients in modern poultry feeds come?
9. List the protein requirements for young chickens, turkeys, and game birds. Why are protein requirements higher for young birds than adults?

SELECTED REFERENCES

Day, E. J. 1980. Microcomputers: Ready for least-cost ration formulation. *Feedstuffs* 52:1, 50,52.

Ensminger, M. E. 1971 *Poultry Science*. 1st edition. Danville, Illinois: Interstate Printers and Publishers.

Hicks, F. W. 1974. *Farm Flock Management Guide*. University Park: Pennsylvania State University Extension Bulletin 4-663.

Nakaue, H. S. 1975. *Fundamentals of Computer Ration Formulation*. Official Proceedings, 10th Annual Pacific Northwest Animal Nutrition Conference, pp. 31-36.

Neshlim, M. C., Austic, R. E., and Card, L. E. 1979. *Poultry Production*. 12th edition. Philadelphia: Lea and Febiger.

28

Horses and Donkeys

Horses, donkeys, and their crosses, mules and hinnies, have contributed significantly to civilization. In the past, all of these animals provided a swifter method for humans to move longer distances than was possible on foot, and pack (carrying) animals and draft (pulling) animals moved heavier loads than humans could. Horses and mules provided a means of transporting grains and livestock to marketing centers. The railroad and early road beds were developed by use of horses and mules as sources of power.

The external parts of a horse are shown in Figure 28-1. In the terminology applied to horses, a **stallion** is an intact male horse of breeding age. A **gelding** is a male horse that was castrated prior to reaching sexual maturity. The young male is called a **colt**. A female of breeding age is a **mare**. The young female is called a **filly**. Fillies and colts are collectively called **foals**. A mule is produced by crossing a male donkey (jack) with a mare. A hinny is produced by crossing a stallion with a female donkey (jenny). Horses are sold under such terms as "at the halter," which means that no guarantee exists that the horse is sound or usable; "sound," which means the horse is sound and healthy but may be mean or unbroken, or "fully guaranteed," which means the horse is sound and well trained.

ORIGIN AND DOMESTICATION OF THE HORSE

Fossil remains show that the *Eohippus*, a four-toed, small animal, was the beginning of the horse. It was originally a wet area inhabitant but through changes that occurred, it became larger and became the *Mesohippus,* an animal about the size of a collie dog. These animals also changed and became capable of foraging on the prairies in the Great Plains. The third toe grew longer and the other toes disappeared resulting in a foot that is characteristic of horses today.

Even though these horses were present some 50-60 million years ago, no horses were present on the continent of North America when Columbus arrived. All horses had completely disappeared. It is assumed that some crossed from Alaska into Siberia and it is from

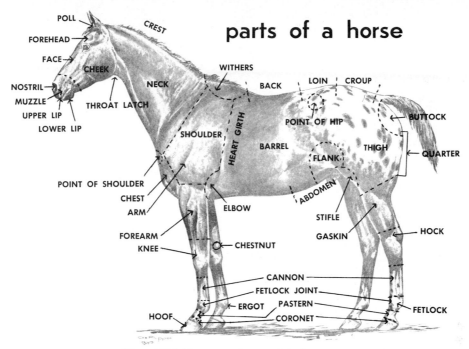

parts of a horse

POLL
CREST
FOREHEAD
FACE
CHEEK
WITHERS
NOSTRIL
NECK
BACK
LOIN
CROUP
MUZZLE
THROAT LATCH
UPPER LIP
LOWER LIP
POINT OF HIP
BUTTOCK
HEART GIRTH
SHOULDER
BARREL
THIGH
QUARTER
FLANK
POINT OF SHOULDER
CHEST
ABDOMEN
ARM
ELBOW
STIFLE
HOCK
FOREARM
GASKIN
KNEE
CHESTNUT
CANNON
FETLOCK JOINT
ERGOT
PASTERN
FETLOCK
HOOF
CORONET

Figure 28-1. The external parts of the horse.

these animals that horses evolved in Asia and Europe. The draft horses and Shetland ponies developed in Europe while the lighter, more agile horses developed in Asia and the Middle East.

The horse was one of the last of our farm animals to be domesticated which occurred about 5,000 years ago. Horses were first used as food, then for war and sports, and also for draft purposes. They were used for transporting people swiftly and for moving heavy loads. Also horses became important in farming, mining, and forestry.

The donkey was domesticated in Egypt earlier than when domestication of the horse occurred. They apparently descended from the wild ass of Africa.

The horse is in the kingdom Animalia, Phylum Chordata, Class Mammalia, Order Perissodactyla, Family Equidae, Genus Eques, and species equus caballus. Donkeys and zebras are in the same genus but are different species from horses. Horses mate with donkeys and zebras but the offspring that are produced are sterile.

Horses have been companions for people through most of their existence since domestication. They were important in wars, mail delivery, handling other livestock, farming, forest harvesting, and mining. The horse today is used in racing, shows, for handling livestock, and as companions for those who love horses. From young children to older adults, the horse has an appeal to people.

BREEDS OF HORSES

Whether a person keeps only one horse for a pet or operates a large ranch, care of horses is of vital concern to any owner. A knowledge of the breeds of horses and management techniques such as breeding, feeding, and control of diseases and parasites is important.

Horses have been used for so many special purposes that many breeds have been developed as a means of supplying the proper horse for a particular need. The major breeds of horses and the primary use each serves are listed in Table 28-1. It should be kept in mind that horses of most breeds may be used for several purposes. No attempt is made in Table 28-1 to indicate all the ways each of the breeds are used. For example, the Thoroughbred (Figure 28-2) is classified as a race horse primarily for long races, but it is used for several other functions.

CHARACTERISTICS OF HORSES

The "light" breeds of horses are those used to provide pleasure for their owners through such activities as racing, riding, and exhibition in shows and parades. Figure 28-3 shows some examples. Quarter Horses are excellent as cutting horses (horses that cut cattle out of herds) and for running short races. Ponies, such as the Pinto, are selected for their friendliness and safety with children. The Saddle Horse and the Tennessee Walking Horse have been selected for ease of riding and willingness to respond to the wishes of the rider. The Palomino, Appaloosa, and Paint breeds are display animals sometimes used in parades.

Beauty in color and markings is important in horses used for show purposes, and "color" breeds have been selected accordingly. The Appaloosa, for example, has color markings of two patterns, leopard and blanket. It appears that these two types are under different genetic controls. In Appaloosas, Pintos, and Palominos, coloration can be adversely affected or eliminated by certain other genes, such as the gene for roaning and the gene for gray. The gene for gray can eliminate both colors and markings, as shown by gray horses that turn white with age.

Another important group of horses is that of the draft horses, large and powerful animals that are used for heavy work. The Percheron, Shire, Clydesdale, Belgian and Suffolk are examples (Table 28-1).

Figure 28-2. Ruffian, a great thoroughbred racing filly. Note the beautiful conformation with strong muscling. Thoroughbreds excel as long-distance racehorses.

Table 28-1. Characteristics and uses of the breeds of horses

Breed	Color	Height in hands[a]	Weight in lbs and kg	Uses
Riding and harness horses				
Arabian	Bay, chestnut, brown, gray	14.2 to 15.2	850 to 1,000 lbs 386 to 454 kg	Pleasure
Thoroughbred	Bay, brown, gray, chestnut, black, roan	15.2 to 17.0	1,000 to 1,300 lbs 454 to 590 kg	Long races
Morgan	Bay, chestnut, brown, black	14.2 to 15.2	950 to 1,150 lbs 431 to 522 kg	Pleasure
Standardbred	Bay, chestnut, roan, brown, black, gray	14.2 to 16.2	850 to 1,200 lbs 386 to 545 kg	Harness racing
American Saddle Horse	Chestnut, bay, brown, black	15.0 to 15.3	1,000 to 1,150 lbs 454 to 522 kg	Gaited, saddle, show
Tennessee Walking Horse	All colors	15.0 to 16.0	1,000 to 1,200 lbs 454 to 545 kg	Riding for pleasure
Hackney	Bay, chestnut, black, brown	15.0 to 16.0	—	Heavy harness
American Quarter Horse	All colors	14.2 to 15.2	1,000 to 1,250 lbs 454 to 568 kg	Short races, stock horses
Ponies				
Pony of America Hackney	Appaloosa Bay, chestnut, black, brown	11.2 to 14.2	450 to 850 lbs 194 to 386 kg	Riding by children Harness pony
Welsh	Bay, chestnut, black, roan, gray	11.0 to 13.0	350 to 850 lbs 159 to 227 kg	Riding by children
Shetland	Bay, chestnut, brown, black, spotted, mouse	9.2 to 10.0	300 to 400 lbs 136 to 182 kg	Riding by children

(continued)

Table 28-1 continued.

Breed	Color	Height in hands[a]	Weight in lbs and kg	Uses
Draft[b]				
Percheron	Black, gray usually	15.2 to 17.0	1,600 to 2,200 lbs 728 to 999 kg	Heavy pulling
Clydesdale	Bay, brown, black	15.2 to 17.0	1,700 to 2,000 lbs 772 to 908 kg	Heavy pulling
Shire	Bay, brown, black	16.2 to 17.0	1,800 to 2,200 lbs 817 to 999 kg	Heavy pulling
Belgian	Chestnut, roan usually	15.2 to 17.0	1,900 to 2,400 lbs 863 to 1,090 kg	Heavy pulling
Suffolk	Chestnut	15.2 to 16.2	1,500 to 1,900 lbs 686 to 863 kg	Heavy pulling
Color Registries				
Palomino	Palomino	Any height		Parade use
Appaloosa	Leopard, blanket	Any height		Parade use
Paint	Tobiano, overo	Any height		Parade use

[a] Height is measured in inches but reported in "hands." A "hand" is 4 in.

[b] Draft horses are heavy horses used for pulling; other horses are called light horses and are used primarily as pleasure horses.

Quarter Horse

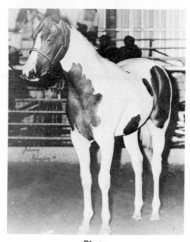

Pinto

Appaloosa Specks Red Dog

American White Horse R. R. Snow King

Palomino

United States Trotting Horse

Figure 28-3. Some breeds of pleasure horses.

BREEDING PROGRAM

Mares come into heat every 21 days if they do not become pregnant. Heat lasts for five to seven days. Ovulation occurs toward the end of heat. Mares usually produce only one foal. Although about 10% of all ovulations in mares are multiple ovulations, twinning occurs in only about 0.5% of the pregnancies that carry to term. The uterus of the mare apparently cannot support twin fetuses; consequently, most twin conceptions result in the loss of both embryos. The length of gestation is about 340 days (approximately 11 months). Because of the relatively long duration of heat and because ovulation occurs toward the end of heat, horse owners often delay breeding the mare for two days after she has first been observed in heat.

Mares of the light breeds reach sexual maturity at 12 to 18 months of age, whereas draft mares are 30 months of age when sexual maturity is reached. Mares usually come into heat five to seven days following foaling, and fertile matings occur at this heat if the mare has recovered from the previous delivery.

ABNORMALITIES OF HORSES

Two terms are used in denoting abnormal conditions in horses: *unsoundness* and *blemish*. An **unsoundness** is any defect that interferes with the usefulness of the horse. It may be caused by an injury or improper feeding, it may be inherited as such, or it may develop as a result of inherited abnormalities in conformation. A **blemish** is a defect that detracts from the appearance of the horse but does not interfere with its usefulness. A wire cut or saddle sore may cause a blemish without interfering with the usefulness of the horse.

Horses may have anatomical abnormalities that interfere with their usefulness. Many of these abnormalities are either inherited directly or develop because of an inherited condition. Abnormalities of the eyes, respiratory system, circulatory system, and conformation of the feet and legs are all important.

Cataract is inherited as a dominant trait that could be eliminated if it were not that horses produce several foals before the cataract develops.

Moon blindness is periodic ophthalmia in which the horse is blind for a short time, regains its sight, and then again becomes blind for a time. Periods of blindness may initially be spaced as much as six months apart. The periods of blindness become progressively closer together until the horse is continuously blind. This condition received the name "moon blindness" because the trait is first noticed as a reality when the periods of blindness occur about a month apart. It was originally thought that the periods of blindness were associated with changes in the moon.

Heaves is a respiratory defect in which the horse experiences difficulty in exhaling air. The horse can exhale a certain volume of air normally, after which an effort is exerted to complete exhalation. This condition is extremely serious when horses are placed under the stress of exercise.

Ruptured blood vessels is a defect of circulation in which the blood vessels are fragile and may rupture when the horse is put under the stress of exercising. Some racehorses have been lost due to hemorrhage from these fragile blood vessels.

Toeing in (Figure 28-4) refers to the turning in of the toes of the front feet, whereas *toeing out* (Figure 28-4) refers to the turning out of the toes of the front feet. These conditions influence the way in which the horse will move its feet when it travels.

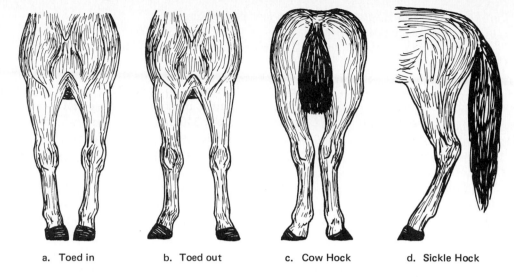

a. Toed in   b. Toed out   c. Cow Hock   d. Sickle Hock

Figure 28-4. Drawings illustrating toeing in, toeing out, cow hocks, and sickle hocks of horses. All four of these abnormalities affect the way horses travel and they may lead to the development of unsoundness.

The term *cow hocks* (Figure 28-4) indicates that the hocks are close together but the hind feet are far apart. Such hocks are greatly stressed when the horse is pulling, running, or jumping.

Sickle hocks (Figure 28-4) is a term used when the hock has too much set or bend. As a result, the hind feet are too far forward. The strain of pulling, jumping, or running is much more severe on a horse with sickle hocks than on a horse whose hocks are of normal conformation.

Contracted heels is inherited, with two dominant genes acting complementarily to produce the trait. Thus, genes A or C alone do not cause the trait to develop; it is expressed only when both genes A and C are present. Because of this mode of inheritance, horses that have contracted heels may produce offspring that do or do not have contracted heels, and horses that have normal heels can produce some offspring that have contracted heels. The foot of the horse is constructed with cartilage between the bone of the foot and the horny hoof. The heel can be spread apart because the cartilage of the heel bends. This provides a shock-absorbing mechanism for the foot. When the heel is contracted, it cannot spread; therefore, the pounding of the foot bruises the cartilage and its heals by ossification.

Sidebones is the abnormality that occurs when the lateral cartilages in the heel ossify. This can cause lameness.

Ringbone is a condition in which the cartilage all around the foot is ossified. Ringbone shows as a hard enlargement at the junction of bony hoof and hair areas.

Bog spavin is a soft swelling on the inner, anterior aspect of the hock.

Bone spavin is a bony enlargement on the inner aspect of the hock. Both bog and bone spavin arise when stresses are applied to horses that have improperly constructed hocks.

A *thoroughpin* is an enlargement between the large tendon (tendon of Achilles) of the hock and the fleshy portion of the hind leg.

Capped hock is a hard swelling at the point of the hock.

Curb is a hard swelling at the bottom-rear of the hock.

Fistula is a running sore on the withers.

Poll evil is a running sore on the poll of the head. These running sores develop after the head is bruised.

Several physiological abnormalities are known in horses. Some are due to inheritance. Others are due to faulty nutrition and management. An example of an inherited physiological abnormality in horses is hemolytic icterus. This condition is caused by a dominant antigen located on the red blood cells. Horses do not develop antibodies to their own antigens but they develop antibodies to the antigens of other individuals. If a stallion that is **RR** (**R** denotes antigen production) is mated to a mare that is **rr** (**r** denotes no antigen production), the foal is **Rr**. As this foal develops, it produces an antigen that may cross the extra-embryonic membranes and enter the circulation of the mare, who then produces antibodies in response to this **antigen**. After two or three matings of the **RR** stallion to the same **rr** mare, the mare will have produced a high titer (amount) of antibodies. This situation is very similar to the "Rh" condition in humans. One major difference between the mare and the human female is that the antigen or the antibodies must cross several maternal and placental cell layers in the mare. As a result, the antigen crosses these membranes and causes antibodies to be produced in the mare, but antibodies, which are larger molecules, do not enter the circulation of the foal. The foal is born normal, but the antibodies in the mother's milk can harm the foal. Because the gut wall of a newborn foal is quite porous, the antibodies in the mother's milk are absorbed from the gut into the foal's bloodstream. The antibodies react with the red blood cells, causing the red blood cells to clump and be destroyed.

Another condition that leads to death of the foal is the lack of iodine in the nutrition of the pregnant mare. If the fetus receives no iodine from the mare, it cannot produce thyroxine. The lack of thyroxine in the circulation of the fetus allows the anterior pituitary to produce large quantities of thyrotropic hormone. This hormone stimulates the development of the thyroid in the foal and results in the development of a goiter. A foal that is born with a goiter dies because a proper iodine-thyroxine balance cannot be established.

GAITS OF HORSES

The major gaits of horses, along with their modifications, are as follows:

1. *Walk* is a four-beat gait in which each of the four feet strikes the ground separately from the others.
2. *Trot* is a diagonal two-beat gait in which the right front and left rear feet hit the ground in unison, and the left front and right rear feet hit the ground in unison. The horse travels straight without weaving sideways when trotting.
3. *Pace* is a lateral two-beat gait in which the right front and rear feet hit the ground in unison and the left front and rear feet hit the ground in unison. There is a swaying from right to left when the horse paces.
4. *Gallop* is a three-beat gait. The hind feet hit the ground in unison. Both front feet hit the ground separately and at a different time than the hind feet.
5. *Canter* is a slow gallop.
6. *Rack* is a snappy four-beat gait in which the joints of the legs are highly flexed.
7. *Foxtrot* is a rhythmic trot.

ABNORMALITIES OF TRAVEL BY HORSES

A horse that toes out with its front feet tends to swing its feet inward when its legs are in action. Swinging the feet inward can cause the striding foot to strike the supporting leg so that **interference** to forward movement results. A horse that toes in tends to swing its front feet outward, giving a **paddling** action.

Some horses overreach with the hind leg and catch the heel of the front foot with the toe of the hind foot. This action, called **forging**, can cause the horse to stumble or fall. Some horses have a nervous condition called **stringhalt**, in which the foot is picked up normally but is put down with an abrupt jerk due to a sudden impulse. A horse may dislocate the hind leg at the stifle (see Figure 28-1). As a result, the horse drags the dislocated leg, because it cannot control its action. Setting the horse on its haunches by a sudden backward jerk on the bridle may snap the joint back into place.

DETERMINING THE AGE OF A HORSE BY ITS TEETH

The age of a horse can be estimated by its teeth (Figures 28-5, 28-6, and 28-7). A foal at 6 to 10 months of age has 24 so-called baby or milk-teeth—12 incisors and 12 molars. The incisors include three pairs of upper and three pairs of lower incisors.

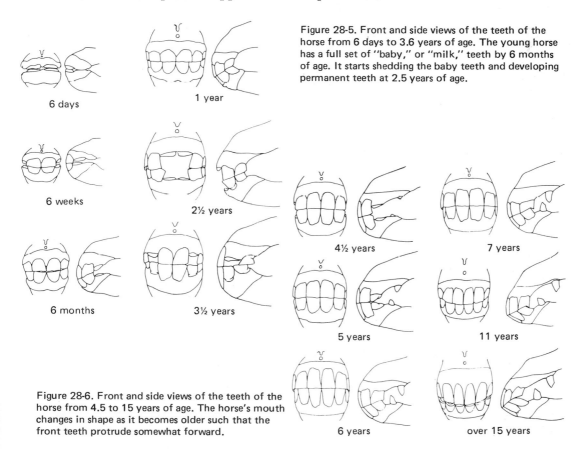

6 days

1 year

Figure 28-5. Front and side views of the teeth of the horse from 6 days to 3.6 years of age. The young horse has a full set of "baby," or "milk," teeth by 6 months of age. It starts shedding the baby teeth and developing permanent teeth at 2.5 years of age.

6 weeks

2½ years

4½ years

7 years

6 months

3½ years

5 years

11 years

Figure 28-6. Front and side views of the teeth of the horse from 4.5 to 15 years of age. The horse's mouth changes in shape as it becomes older such that the front teeth protrude somewhat forward.

6 years

over 15 years

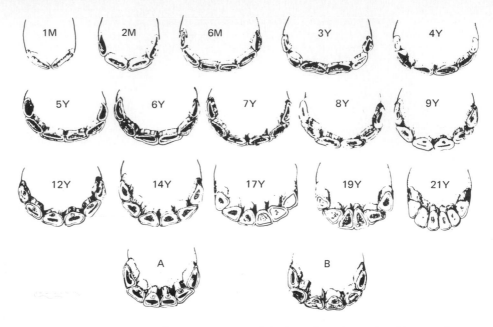

Figure 28-7. Table surfaces of the lower incisors of the horse from 1 month to 21 years of age.

Chewing causes the incisors to become worn. The wearing starts with the middle pair and continues laterally. At one year of age, the center incisors show wear; at 1.5 years, the intermediates show wear; and at two years, the outer, or lateral, incisors show wear. At 2.5 years, shedding of the baby teeth starts. The center incisors are shed first. Thus, at 2.5 years, the center incisors become permanent teeth; at 4 years, the intermediates are shed; at 5 years, the outer, or lateral, incisors are shed and replaced by permanent teeth.

A horse at 5 years of age is said to have a "full" mouth, because all the teeth are permanent. At 6 years, the center incisors show wear; at 7 years, the intermediates show wear; and at 8 years, the outer, or lateral, incisors show wear. Wearing is shown by a change from a deep groove to a rounded dental cup on the grinding surface of a tooth.

DONKEYS, MULES, AND HINNIES

Mules and hinnies have been used as draft animals in mining and farming operations, and as pack animals. When **mules** were needed for heavy loads, it was important to breed mammoth jacks to mares of one of the draft breeds. The crossing of smaller jacks with mares of medium size produced mules that were useful in mining and farming operations.

Donkeys, mules and hinnies have some characteristics that make them more useful than horses for certain purposes. Their sure-footedness makes them ideal pack animals for moving loads over rough areas. They are rugged and can endure abuse. In addition, they have the characteristic of taking care of themselves. For example, a mule that gets a leg caught in a barbed wire fence brays for someone to come to its rescue, whereas a horse might become excited and damage itself severely in an attempt to become free. Mules do not gorge themselves when grain feeds are before them; consequently, they do not normally founder (overeat). However, donkeys, mules, and hinnies do not respond to the wishes or commands of humans as readily as horses and are often quite contrary.

Recently, there has been a movement to use small donkeys (burros) for pets for children. Small mules from crosses between male burros and mare ponies have also been in demand as children's pets.

The primary reason why trucks and tractors replaced horses and mules in farming operations was that work could be done more rapidly and one person could do much more with tractors and trucks than with horses and mules. The high cost of oil products (gasoline and diesel fuels) and the possibility of scarcity of these items may result in more use of horses and mules in farming, logging, and mining operations, a partial return to former days. It is unlikely that horses and mules will be used to replace trucks and tractors, but certain operations on farms or in the timber industry may be done to advantage by horses and mules.

STUDY QUESTIONS

1. For the American Quarter Horse, Thoroughbred, Tennessee Walking Horse, and Welsh, give their approximate height and the use that is made of each.
2. Define or explain the following terms:
a.	walk	f.	ringbone
b.	trot	g.	thoroughpin
c.	bog spavin	h.	bone spavin
d.	pace	i.	sickle hocks
e.	curb	j.	interference
3. a. If a horse toes in, will its legs tend to paddle or to interfere?
   b. What do we call the striking of the heel of the front foot with the toe of the hind foot when a horse travels?
   c. When a pregnant mare has a deficiency of iodine, what happens to the foal? How can this condition be prevented?
   d. Explain why some foals develop hemolytic icterus. How can one prevent the development of this disease?
   e. What are three important characteristics that horses should possess to make them most useful as pleasure horses?

SELECTED REFERENCES

Barbalace, R. C. 1977. *An Introduction to Light Horse Management.* Minneapolis: Burgess Publishing Co.

Bone, J. F. 1975. *Animal Anatomy and Physiology.* 4th edition. Corvallis: Oregon State University Book Stores.

Evans, J. W., Borton, A., Hintz, H. F., and Van Vleck, L. D. 1977. *The Horse.* San Francisco: W. H. Freeman and Co.

Wagoner, D. M. Editor-Publisher. 1978. *Equine Genetics and Selection Procedures.* Dallas, Texas: Equine Research Publications.

29

Feeding and Managing Horses

Horses relate well to people and provide many forms of pleasure. Although many people enjoy horses, most own only a few; however, whether a person keeps only one horse for a pet or operates a large ranch, care of horses is of vital concern to any owner. A knowledge of the breeds of horses and management techniques such as breeding, feeding, and control of diseases and parasites is important.

Most breeders know that some weaknesses exist either in their entire herd or in some of the individuals of the herd. Breeders generally attempt to find a stallion that is particularly strong in the trait or traits that need strengthening in the herd. Obviously, it is also necessary to be sure that another weakness is not brought into the herd, so it is better to use a stallion with no undesirable traits, even if he is only average in the trait that needs correcting in the herd. To emphasize one trait only for correcting a weakness while bringing another weakness into the herd simply means that continued attempts must be made to correct weaknesses and that, even after 25 to 30 years of breeding, little or no improvement will be evident.

To have an effective breeding program, it is necessary to be completely objective. There is no place in a breeding program for sympathy toward an animal or for personal pets. Every attempt should be made to see that the environment is the same for all animals in a breeding program. A particularly appealing foal that is given special care and training may develop into a desirable animal. However, many foals less appealing in early life might also develop into desirable animals if they are given special care and training. A desirable animal that has had special care and training may transmit favorable hereditary traits no better than a less desirable animal that has not had this treatment. Selecting a horse for breeding that has had special care and training may in fact be selecting only for special care and training. Certainly, these environmentally produced differences in horses are not inherited and will not be transmitted. In fact, it is often wise to select animals that were developed under the type of environment in which they are expected to perform. If stock horses are being developed for herding cattle in rough, rugged country, selection under such conditions is more desirable than where conditions are less rigorous. Horses that possess inherited weaknesses tend to become unsound in rugged environment, and as a result are not used for breeding. Such animals might never show those inherited weaknesses in a less rugged environment.

The ideal environment for most horse-breeding programs is one in which there is plenty of forage of good quality covering an area that requires the horses to take considerable exercise over fairly rough land to obtain the forage. Where the land is highly productive and level, the animals may have to be forced to exercise a great deal. Forced exercise tends to keep the animals from becoming overfat and gives strength to the feed and legs. Horses need regular, not sporadic, exercise. Regular exercise, even if quite strenuous, is healthy for genetically sound animals and may reveal the weaknesses of those that are not. Strenuous exercise can, however, be harmful to animals that have not exercised for a considerable period of time.

An environment should be provided that is useful for distinguishing animals that are genetically the most capable for the purpose they are to fulfill. For example, horses that are being bred for endurance in traveling should be made to travel long distances every day to determine if they can do so without developing unsoundnesses; horses that are being bred for jumping ability should be made to jump as early in life as they can be trained so that those lacking the ability to jump or those that become unsound from jumping can be removed from the breeding program before they leave any offspring; draft horses (Figure 29-1) should be used for pulling heavy loads early (two or three years of age) in life to

Figure 29-1. A Clydesdale stallion.

determine their ability and willingness to pull and to see if heavy pulling causes them to become unsound. This information is needed before the animals are used in a breeding program. It is always best to know the capabilities of a horse prior to the time it is selected for breeding.

Any horse that is unsound should not be used for breeding regardless of the purposes for which the horse is being bred. Such abnormalities as toeing in or toeing out, sickle hocks, cow hocks, and contracted heels will likely lead to unsoundness and difficulties or lack of safety in traveling. Interference and forging actions are extremely objectionable because they can cause the horse to stumble or fall. Defects of the eyes, of the mouth, and of respiration should be selected against in all horses.

HOUSING AND EQUIPMENT

Barn. A barn constructed with a driveway between two rows of tie stalls provides for easy feeding, because feed can be placed in the mangers of the tie stalls as the feed truck goes down the hallway. If the tie stalls are such that they can be opened from the outside and cleaned with mechanical equipment, labor is saved. The manger in tie stalls should be such that hay can be fed in the manger and grain in a small feeder. Grain and hay should not be given in the manger at the same time, however, because the horse throws the hay out of the manger to get to the grain.

Box stall. Box stalls are used for foaling and for the mare and the foal when the weather is severe. It is difficult to construct box stalls that can be cleaned with mechanical equipment.

Feed. Feed should be stored in an area connected with or adjacent to the barn so that it can be moved horizontally. Overhead (loft) storage of feed requires a great deal of labor and expensive construction.

Fences. The lot fences should be constructed with wooden or steel posts and cable, or with wooden posts and two-inch lumber. Barbed wire should not be used for lot fences, because it can cut a horse severly.

Tack room. A tack room is needed to house riding equipment, including saddles, blankets, bridles, and halters. In addition, it may be advisable to have a wash room where horses can be washed.

Chute. A chute is essential for caring for horses. If the horse is in a chute with a bar behind its body, it is unlikely that the animal can kick the attendant when insemination is being done. The chute can also be used for "teasing" a mare with a stallion to determine if the mare is in heat. When the stallion approaches the front of the mare, the mare reacts violently against the stallion if not in heat, but squats and urinates with a winking of the vulva if in heat.

The chute is also useful when blood samples are being taken. Often, horses strike with a front foot when a needle is inserted into the jugular vein but the enclosed front end of the chute prevents the horse from injuring a person who is drawing blood.

When not in a chute, the animal can be prevented from kicking by placing ropes around its neck just in front of the shoulders, running the ropes between the front legs to the rear legs, and tying the ends to the rear legs (Figure 29-2). This **hobble** is often used when a mare is being bred either naturally or artificially to prevent the mare from kicking the stallion or the one doing the inseminating.

If a chute for confining the horse is unavailable, the upper lip of the horse can be squeezed tightly by means of a small rope that is twisted if the horse misbehaves (Figure 29-3). This action, called a **twitch**, subdues the animal.

FEEDS AND FEEDING

Horses eat roughages such as hay, straw, and stovers; they graze on grass and legume pastures and they eat grains of all kinds. Many horse owners consider oats as one of the best grain feeds for horses. Today, concentrate mixtures are prepared and sold by commercial feed companies, especially for persons who have only one or two horses.

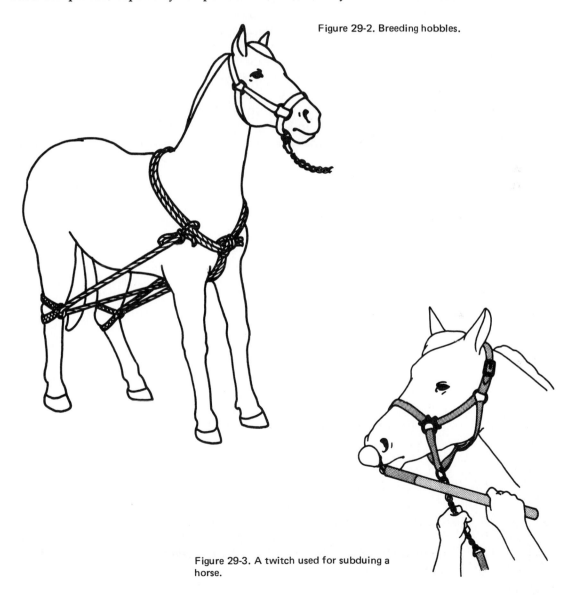

Figure 29-2. Breeding hobbles.

Figure 29-3. A twitch used for subduing a horse.

Although horses spend little time chewing, they grind such grains as oats, barley, and corn so that cracking or rolling these feeds beforehand is unnecessary. Wheat and milo, however, should be cracked to improve digestibility. The stomach is relatively small, composing only 10% of the total digestive capacity. Only a small amount of digestion takes place in the stomach and food moves rapidly through to the small intestine. From 60 to 70% of the protein and soluble carbohydrates are digested in the small intestine and about 80% of the fiber is digested in the cecum and colon. The large intestine has about 60% of the total digestive capacity with the colon being the large component. Bacteria that live in the cecum aid digestion there. Most fat, vitamins, calcium, and magnesium are digested in the small intestine.

Owners of pleasure horses are interested in having them make a desirable appearance and many feed their animals liberally. Perhaps more horses are kept too fat than are underfed. Also, many people want to be kind to their animals and keep them housed in a box stall when weather conditions are undesirable. This may not be best for the horse. Certainly if any deficiency exists in the feed provided, such a deficiency is much more likely to affect horses that are not running on pasture than those that are.

Young, growing foals should be fed to grow well following weaning but not to fatten. Quality of protein and amounts of protein and energy are important for young foals after they are weaned. Some milk powder in the concentrate mixture will provide the amino acids that might otherwise be missing or scarce in the ration.

Good-quality pasture or hay supplemented with a smaller amount of grain provide most of the nutrition needed by young horses. Trace-mineralized salt should be provided at all times and clean water is essential. The best method of providing water is to pipe it to a watering tank (Figure 29-4). The forage and grain mixture may be pelleted, in which case the grains in the mixture are usually cracked or rolled.

Figure 29-4. Horses drinking clean water from a watering tank.

Pregnant mares do well on good pastures or on good-quality hay supplemented with a small amount of grain. Oats make an excellent grain feed for pregnant mares. Pregnant mares should be in thrifty condition and not fat. Just prior to and immediately following delivery of the foal, a ration that has a slightly laxative effect is desirable. Some bran included in the grain provides this effect. Animals used for riding or working, whether they are pregnant or not, need more energy than those that are not working. Horses that are being exercised heavily should be fed ample amounts of concentrates.

Lactating mares require heavier grain feeding than do geldings or nonpregnant mares. If lactating mares are exercised heavily in addition to producing a heavy flow of milk, they must be fed grain and hay liberally. Stallions need to be fed at a rate that keeps them thrifty but not fat. Feeding good-quality hay with limited amounts of grain is desirable.

Feed companies provide properly balanced rations for feeding horses. An owner who has only one or two horses may find it highly advantageous to use such prepared feeds because it is difficult to prepare feeds for only a few animals, and the user of commercially prepared feeds can prevent errors.

MANAGING HORSES

Proper management of horses is essential at several critical times. These critical times are the breeding season, the time of foaling, the time of weaning of foals, the time of castration of colts, and the time that care is given to animals at hard work.

Breeding season. Mares bred naturally should have the vulva washed and dried and the tail wrapped prior to being served by the stallion (Figure 29-5). If the mare can be bred twice during heat without overusing the stallion, breeding two and four days after the mare is first noticed in heat is desirable. A mature stallion can serve twice daily over a short time and once per day over a period of one or two months. A young stallion should be used lightly at about three or fewer services per week.

Foaling time. Mares normally give birth to foals in early spring, at which time the weather can be unpleasant. A clean box stall that is bedded with fresh straw should be

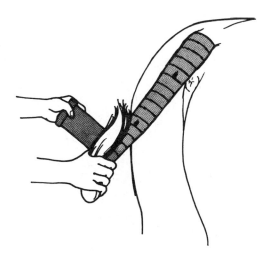

Figure 29-5. Tail of horse being wrapped prior to breeding.

available. If the weather is pleasant, mares can foal on clean pastures. When a mare starts to foal, she should be observed carefully, but she should not be disturbed unless assistance is necessary. If the head and front feet of the foal are being presented, it should be delivered without difficulty (Figure 29-6). If necessary, however, a qualified person can assist by pulling the foal as the mare labors. Do not pull when the mare is not laboring and do not use a tackle to pull the foal. If the front feet are presented but not the head (Figure 29-7), it may be necessary to push the foal back enough to get the head started along with the front feet. Breech presentations can endanger the foal if delivery is delayed; therefore, assistance should be given to help the mare make a rapid delivery if breech presentation occurs. If it appears that difficulties are likely to occur, a veterinarian should be called as soon as possible.

As soon as the foal is delivered, its mouth and nostrils should be cleared of membranes and mucus so it can breathe. If the weather is cold, the foal should be wiped dry and assisted in nursing. As soon as the foal nurses, its metabolic rate increases and helps it stay warm. The **umbilical cord** should be dipped in a tincture of iodine solution to prevent harmful microorganisms from invading the body through it.

In general, it is highly desirable to exercise pregnant mares up to the time of foaling. Mares that are exercised properly while pregnant are in better muscle tone and are likely to experience less difficulty when foaling than those that get little or no exercise while pregnant. Many mares used for plowing and similar work have foaled in the field without difficulty. The foal is usually left with the mare for a few days, after which the foal may be left in the box stall while the mare is ridden or used for other types of work. The mare and young foal do well when they are run together on a good, clean pasture.

Weaning the foal. When weaning time arrives, it is best to remove the mare and allow the foal to remain in the surroundings to which it is accustomed. Because the foal will make every attempt to escape in its effort to find its mother, it is left in a box stall or tightly fenced lot. Fencing other than barbed wire is used. The foal at weaning should be fed well by giving it good-quality hay and some grain.

Castrating colts. **Colts** that are not being kept for breeding can be castrated any time after they have recovered from the stress of weaning, but the stress of castration should not

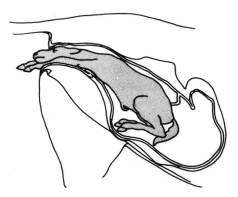

Figure 29-6. Correct presentation position of foal for delivery.

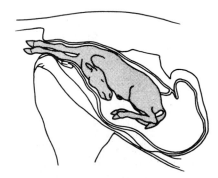

Figure 29-7. Malpresentation; head and neck back.

be imposed at weaning time. Some people prefer to delay castration until the colt has reached a year of age. The colt can be put in a turntable squeeze for castration. The scrotum is opened on each side, by a qualified person, and the membrane around each testicle is split to expose the testicle. The testicle is pulled somewhat so the cord can be clamped. The cord is crushed by the clamp and then severed. The crushing of the cord prevents excessive bleeding. It may be desirable to give the colt an injection of an antibiotic. If the scrotum was cleaned with a mild disinfectant prior to castration, the wound need not be washed with an antiseptic. Harsh disinfectants should not be applied to the wound. After the colt is castrated, it should be on clean pasture where it can be closely observed. If danger from fly strike exists, a fly repellant should be applied about the scrotal area.

Horses can be permanently identified by tattooing them on the inside of the upper lip. This does not disfigure the animals. The identification can easily be read by raising the upper lip.

Care of hard-working animals. Horses that are used for heavy work are hungry and thirsty when they are brought to the stable. If they are given grain, then hay, and then water, the grain is pushed into the large cecum and fermentation produces excessive amounts of gases. This can make the horse ill. The proper way to feed and water a hungry, thirsty horse is to give it water first, then hay, and then grain. This procedure allows the grain to be digested in the stomach and small intestine. If horses feed heavily on grain, they are likely to founder (overeat to the point of becoming ill). Foundering can cause death. The hooves of horses that survive foundering are likely to grow excessively.

CONTROLLING DISEASES AND PARASITES

Sanitation is of vital importance in controlling diseases and parasites of horses. Horses should have clean stables and should be groomed regularly. A horse that is to be introduced into the herd should be isolated for a month first so that any disease to which it may have been exposed prior to isolation is given time to express itself.

Horse manure is an excellent medium for microorganisms that cause **tetanus.** Horses should be given shots to prevent tetanus, which can develop if an injury allows tetanus-causing microorganisms to invade through the skin. Usually two shots are given to establish **immunity,** after which a booster shot is given each year. Because the same microorganisms that cause tetanus in horses also affect humans, those who work with horses should also have tetanus shots.

Strangles, also known as distemper, is a bacterial disease that affects the upper respiratory tract and associated lymph glands. High fever, nasal discharge, and swollen lymph glands suggest strangles. This disease is spread by contamination of feed and water. Afflicted horses must be isolated and provided clean water and feed. A strangles bacterin is available, but postvaccinal reactions limit its use to stables and ranches where the disease is endemic.

Brood mares are subject to many infectious agents that invade the uterus and cause **abortion;** examples include **Salmonella** and **Streptococcus** bacteria and the viruses of **rhinopneumonitis** and **arteritis.** Should abortion occur, obtain professional help to determine the specific cause and plan to prevent the problem in the future.

Sleeping sickness, or **equine encephalomyelitis,** is caused by viral infections that affect the brain of the horse. Different types, such as the eastern and western types, are known. Encephalomyelitis is transmitted by such vectors as mosquitoes. It can also be spread by horses rubbing noses together. This disease is also transmissable to humans. Vaccination against the disease consists of two intradermal injections spaced a week to 10 days apart. These injections should be given in April. The vaccination ensures immunity for only six months; therefore, the injections should be repeated each six months when the disease is prevalent.

Influenza is a common respiratory disease of horses. The virus that causes influenza is airborne, so exposure frequently occurs where horses congregate. The acute disease causes high fever and a severe cough when the horse is exercised. Rest and good nursing care for three weeks usually gives the horse an opportunity to recover. Severe aftereffects are rare when complete rest is provided. Those horse owners who plan for shows should vaccinate for influenza each spring. Two injections are required the first year, with one annual booster thereafter.

Horses can become infested with internal and external parasites. Control of internal parasites consists of rotating horses from one pasture to another, spreading manure from stables on land that horses do not graze, and treating infested animals.

Pinworms develop in the colon and rectum from eggs that are swallowed as the horse eats contaminated feed or drinks contaminated water. They irritate the anus, which causes the horse to rub the base of its tail against objects even to the point of wearing off hair and causing skin abrasions. Pinworms are controlled by oral administration of proper vermifuges.

Bots are the maggot stage of the bot fly. The female bot fly lays eggs on the hairs of the throat, front legs, and belly of the horse. The irritation of the bot fly causes the horse to lick itself. The eggs are then attached to the tongue and lips of the horse, where they hatch into larvae that burrow into the tissues. The larvae later migrate down the throat and attach to the lining of the stomach, where they remain for about six months and cause serious damage. Since the chemicals used in the control of bots can be injurious if not properly administered, or given at the proper dosages, professional assistance shou   be obtained for treatment.

Adult **strongyles** (bloodworms) are firmly attached to the walls of the large intestine. The adult female lays eggs that pass out with the feces. After the eggs hatch, the larvae climb blades of grass where they are swallowed by grazing horses. The larvae migrate to various organs and arteries where severe damage results. Blood clots form where arteries are damaged. Clots can break loose and plug an artery. Treatment of bloodworms consists of phenothiazine mixed with the feed or phenothiazine-piperazine mixture administered orally.

Adult **ascaris** worms are located in the small intestine. The adult female produces large numbers of eggs that pass out with the feces. The eggs become infective if they are swallowed when the horse eats them while grazing. The eggs hatch in the stomach and small intestine and the larvae migrate into the bloodstream and are carried to the liver and lungs. The small larvae are coughed up from the lungs and swallowed. When they reach the small intestine, they mature and produce eggs. The same chemicals used for the control of bots are effective in the control of ascarids.

STUDY QUESTIONS

1. Fertility is usually low in horses. What factors contribute to low fertility?
2. How would you satisfy the nutritional needs of a mare that is not at work from the time that she is bred in April until she foals?
3. When a horse that has been worked hard is brought to the barn for feed and water, in what order would you give grain, hay, and water? Why?
4. Name three diseases and three parasites of horses and tell how each is controlled and prevented.

SELECTED REFERENCES

Battaglia, R. A., and Mayrose, V. B. 1981. *Handbook of Livestock Management Techniques.* Minneapolis: Burgess Publishing Co.

Bone, J. F. 1975. *Animal Anatomy and Physiology.* 4th edition. Corvallis: Oregon State University Book Stores.

Cunha, T. J. 1980. *Horse Feeding and Nutrition.* New York: Academic Press.

Ensminger, M. E. 1966. *Horses and Horsemanship.* Danville, Illinois: Interstate Printers and Publishers.

Evans, J. W., Borton, A., Hintz, H. F., and Van Vleck., L. D. 1977. *The Horse.* San Francisco: W. H. Freeman and Co.

Wise, W. E., Drudge, J. H., and Lyons, E. T. 1972. Controlling Internal Parasites of the Horse. Lexington: University of Kentucky Cooperative Extension Service.

30

Goats

Goats have a large impact on the economy and food supply for people of the world. Worldwide, more goats' milk is consumed than cows' milk, and goats are an important source of meat in certain countries. In the United States, the consumption of both goat milk and meat is increasing, and the production of mohair is an important industry in Texas.

There are three classes of domesticated goats: the dairy goat, which is used largely for the production of milk; the Angora goat, which is used mainly for the production of mohair and meat; and the Pygmy goat, which is used chiefly as a laboratory animal and pet.

The milk produced by dairy goats differs from cows' milk in that all the carotene in goat milk is converted into vitamin A, and the type of curd formed when acids and enzymes act on goat milk is different from the curd of cow milk. Goat milk is more readily digested and assimilated by some people because of these differences. The dairy goat is a desirable animal for providing milk for the family because it is small and less expensive to feed than a cow and can consume large quantities of **browse** that is unsuitable for a cow.

Goats serve different functions in different places. In some parts of the world, goat meat production is extremely important and, in fact, meat goats outnumber dairy goats worldwide. Dairy goat numbers in the United States are modest because of the highly specialized and effective dairy industry built around the dairy cow. Yet in some parts of the world, large quantities of dairy products are produced from goats. In fact, there are probably more goat dairy products than dairy cow products worldwide.

CHARACTERISTICS OF BREEDS OF DAIRY GOATS

The external parts of a dairy goat are shown in Figure 30-1. The five major breeds registered in the United States are Toggenburg, Saanan, French Alpine, Nubian, and American LaMancha, all of which are capable of high productivity. The dairy goat and dairy cow are about equal in efficiency with which they convert feed into milk, even though the dairy goat produces much more milk in relation to its size than does the dairy

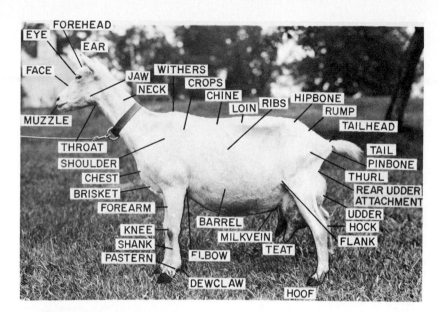

Figure 30-1. The external parts of the dairy goat.

cow. Feed needed for maintenance is higher for goats per unit of body weight than for dairy cows which causes cows and dairy goats to be equal in efficiency of milk production.

The breeds of dairy goats used in the United States are shown in Figures 30-2, 30-3, 30-4, 30-5, and 30-6.

The *Toggenburg* is medium in size, is vigorous, and produces much milk (Figure 30-2). The highest record of milk production for one lactation by a Toggenburg female is 5,750 lbs.

The *Saanen* is a large breed, capable of producing much milk (Figure 30-3). The record milk production for one lactation by a Saanen female is 5,496 lbs.

The *French Alpine* is a large, somewhat rangy goat (Figure 30-4). French Alpines also produce much milk; the record for one lactation of a female is 4,826 lbs.

The *Nubian* is a large proud-looking goat that differs from other goats in having long, wide, pendulous ears (Figure 30-5). The record milk production of a Nubian female is 4,420 lbs which is lower than for the three breeds previously described. However, Nubian milk is distinctive for its high milk fat content.

The *American LaMancha* goat differs from the other breeds in either having no external ears ("gopher" ears) or extremely short ears ("cookie" ears) (Figure 30-6). Milk production is relatively low; the record for one lactation for a LaMancha female is 3,408 lbs. It should be pointed out that relatively few females of this breed have been officially tested; therefore this low record may not indicate the breed's true capacity.

One should keep in mind that the high records for each of the breeds do not indicate that the breed averages are that high. The average for a breed may be estimated at 60% of the outstanding record for a female of that breed.

CARE AND MANAGEMENT

Simple housing for goats is adequate in areas where the weather is mild because goats do best when they are outside on pasture or in an area where they can exercise freely. If the

Figure 30-2. An ideal Toggenburg doe, Lotta Daisies Dorli M. (Permanent grand champion). Note the fore attachment of the udder and the well-placed teats.

Figure 30-3. A Saanen doe, Foolish Question.

Figure 30-4. A French Alpine doe, Lotta Daisies Danica M., a top 10 milker in the 1970s.

Figure 30-5. A Nubian doe. Note the ear size and shape.

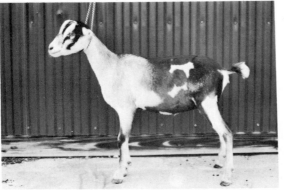

Figure 30-6. A LaMancha doe. Note the earless condition. This individual is homozygous for earlessness.

weather is wet at times, an open shed with a good roof is ideal. Lactating goats can be fed in an open shed in rainy winters and taken to the milking parlor for milking. If the weather is severely cold, an enclosed barn with ample space is needed. At least 20 square feet per goat is needed if goats are to be housed in a barn, but 16 square feet per goat may suffice in open sheds.

The use of several small pastures permits goats to be moved about, thus increasing grazing efficiency and reducing the risk of infestation by parasites. Fences must be properly constructed with woven wire and six-inch stays. Fencing for goats is different from fencing for cows; a woven-wire fence is best for containing goats, which can climb a rail fence. In addition, a special type of bracing at corners of the fence is necessary because a goat can walk up a brace pole and jump over the fence. Electric fences are sometimes used. It may be necessary to use two electric wires—one located low enough to keep goats from going underneath and one high enough to keep them from jumping out. However, an electric fence will not contain bucks of any breed, because they go through the fence in spite of the shock.

Several pieces of equipment are needed for normal care; a rotary tattoo set, hoof trimmers, a knife, a grooming brush, an emasculator, and a balling gun for administering boluses (large pills for dosing animals).

All of the breeds of dairy goats are available in polled strains. The polled condition has much to commend it because a polled goat is less likely to catch its head in a woven-wire fence than is a horned goat. Horned animals should be disbudded at, or shortly following, birth by use of caustic potash or an electric dehorning iron. For many years, goat breeders were plagued by a genetic linkage between hermaphroditism and polledness. This problem has apparently been solved, because polled goats of high fertility are now the general rule.

A goat's feet should be kept properly trimmed by qualified personnel to prevent deformities and footrot. A good pruning shear is ideal for leveling and shaping the hoof, but final trimming can be done with a rasp or a pocket knife. Footrot should be treated with formaldehyde, a copper solution, or iodine. Affected goats should be isolated from the others and placed on clean ground following treatment for footrot.

All kids should be identified by an ear tag and an ear tattoo, or by a tail-web tattoo. The ear tag is easy to read, and the tattoo serves as permanent identification in case an ear tag is lost. All registered goats must be tattooed because neither of the three goat registry associations accept ear-tag identification.

Male kids that are not to be considered for breeding should be castrated. This can be done by constricting the blood circulation to the testicles by use of an elastrator, by surgically removing the testicles, or by destroying the testicles with an emasculatome. Some experts crush the cord to the testicle without surgery. After the blood supply to the testicles is discontinued, the testicle atrophies and is later absorbed. One problem in using the emasculator without surgery is that circulation may not be prevented and the testicles, as a result, do not atrophy. Although the male in this instance is sterile, he continues to produce testosterone and therefore has a normal sex drive. A problem with the elastrator method is that as the tissue dies below the elastrator band, tetanus organisms can enter. Surgical removal of the testicles creates a wound that attracts flies; therefore, a fly repellant should be applied around the wound. It is best to engage the services of a veterinarian unless the owner is properly trained to castrate goats. Whatever method is chosen, adequate skill and caution must be used to prevent infection, gangrene, and other complications.

It is important that the udder of dairy goats have strong fore and rear attachments and that the teats be well spaced. Goats are milked by hand or by a milking machine. The stanchion should be high enough to place the goat at a convenient level for milking (Figure 30-7). It is important that the milking stanchion, the milking machine, and the goat be clean. The udder and teats should be washed and dried prior to milking and the teats should be dipped in a weak iodine solution afterward.

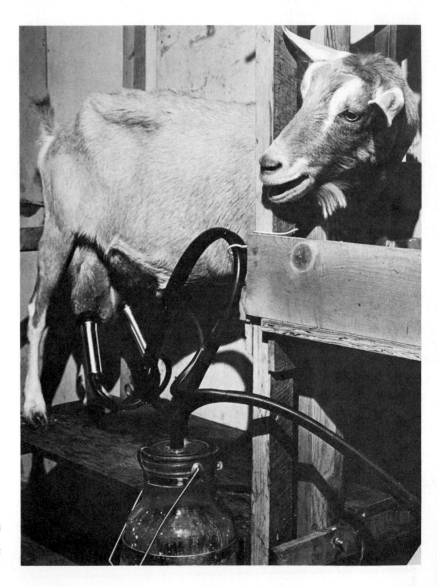

Figure 30-7. A goat being milked by machine in a stanchion at a convenient level for milking.

SPECIAL CARE AND MANAGEMENT OF DAIRY GOATS

Considerations of special importance in the management of dairy goats are care at time of breeding, care at time of kidding, and the feeding of the animals.

Time of breeding. The female goat comes into heat at intervals of three weeks (18 to 21 days) until she becomes pregnant. Young does can be bred first when they weigh 85 to 95 lbs at about 10 months of age, which means that does in good general condition may be bred to kid (have young) at one year of age. Goats are seasonal breeders, with the normal breeding season occurring in September, October, and November. If no effort is made to breed does at other times, most of the young will be born in February, March, or April. Because the lactation period is about 10 months in duration, a period of two months when no goats are lactating might occur, but if the breeding of the does is staggered, the kidding season can be somewhat extended. Housing goats in the dark for several hours each day in the spring and summer months causes some of them to come into heat earlier in the year than usual. Conversely, artificial light in the goat barn may serve to delay estrus in the autumn.

One service is all that is needed to obtain pregnancy, but it is generally wise to delay breeding for a day after the goat first shows signs of heat. The doe stays in heat from one to three days, but the optimum period of standing heat may last only a few hours. A female in heat is often noisy and restless and her milk production may decrease sharply. The female in heat may disturb the other goats in the herd, so it is advisable to keep her in a separate stall until she goes out of heat.

Male goats often have a rank odor about them, especially in the breeding season. It is best to house the males apart from the does in a direction away from the prevailing winds so that odors of the males will be less disturbing. Cleanliness and plenty of green feed help reduce "bucky" odors.

Careful records should be kept that indicate the dates of breeding and the bucks used. With the date of breeding on record, it is possible to calculate when the doe should deliver kids since the gestation period is about 150 days. Detailed breeding records are essential for knowing the lineage, which in turn helps in the selection of males to produce desired traits in offspring.

More males are conceived and born than females; the ratio is approximately 115:100. Delivery of twins is common and delivery of triplets occurs on occasion, particularly among mature does; many mature does average two young while younger does average 1.5 young. Kids at birth weigh 5 to 8 lbs, but singles are heavier than twins or triplets and males are usually heavier than females.

Time of kidding. A doe almost ready to deliver young should be placed in a clean pen that is well bedded with clean straw or shavings. The doe should have plenty of clean water and some laxative feed (such as wheat bran) as well as fresh legume hay. She should be carefully observed but not disturbed unless assistance is necessary, as indicated by excessive straining for three hours or more. If the kid presents the front feet and head, delivery should be easy. If only the front feet, but not the head, are presented, the kid should be pushed back enough to bring the head forward in line with the front feet. Breech presentations are common, but it is important that these deliveries be rapid so as to prevent the kid from suffocating. If assistance is needed, a qualified attendant should pull when contractions occur. Gentleness is essential and harsh or ill-timed pulling can cause severe internal damage.

As soon as the young arrives, its mouth and nostrils should be wiped clean of membranes or mucus. In cold weather, it may be necessary to take the newborn to a heated room for drying. A chilled kid can be helped to regain its body temperature by immersing its body up to the chin in water that is as hot as can be tolerated when the attendant's elbow is immersed in it for two minutes. The kid should be encouraged to nurse as soon as it is dry. Nursing helps the newborn kid to keep warm.

Difficulty in kidding may be caused by diseases such as brucellosis, whose microorganisms may coincidentally also affect the human attendant. It is advisable to use rubber or plastic gloves when assisting with delivery and to wash and disinfect the gloves before removing them.

Feeding. Because goats are ruminants, they can digest roughage effectively. However, types and proportions of feed should be related to the functions of the goats. For example, dry does and bucks that are not actively breeding do well on ample browse, good pasture, or good-quality legume hay. If the grass is short or if the hay is of poor quality, the feeding of supplemental concentrates may be necessary. However, overfeeding of supplemental concentrates can cause obesity, which interferes with reproduction and subsequent lactation.

Heavily lactating does and young does that must continue to grow while they are lactating should be given all the good-quality legume hay and concentrates they can consume. Good-quality pasture can be used to supplement the hay ration, but even with lush pasture, lactating goats require some amount of roughage in the form of hay to prevent scouring.

Pregnant does should be fed to gain some weight to assure the adequate nutrition of the young. A doe should be in good flesh but not fat when she kids, because she draws from her body reserves for the production of milk.

Grains such as corn, oats, barley, and milo may be fed whole because goats crack grains by chewing. If the protein is mixed with the grain or if the feed mix is being pelleted, the grain should be rolled, cracked, or coarsely ground. Some people prefer to mix the hay and grain and prepare the ration in pellet form. There is added expense in pelleting, but less feed is wasted.

For some areas, deficiencies in minerals such as phosphorus and iodine may exist. The use of iodized or trace-mineralized salt along with dicalcium phosphate or steamed bone meal usually provides enough minerals if legume hay or good pastures are available, because calcium is usually present in sufficient quantity in legume hay and phosphorus is adequately provided in grain feeds. To be safe, it is best to provide a mixture of trace-mineralized salt and bone meal free choice.

Young, growing goats and lactating goats need more protein than do bucks or dry does. A ration containing 12% to 15% protein is desirable for bucks and dry does, but 15% to 18% protein may be better for the young and for does that are producing much milk.

Generally, kids are allowed to nurse their dams to obtain colostrum. After three days, kids may be removed from their mothers and given milk replacer by means of a lamb feeder or hand-held bottle until they are large enough to eat hay and concentrates. It is advisable to encourage young kids to eat solid feed at an early age by having leafy legume hay and a palatable concentrate such as rolled grain available at all times. Solid feeds are less expensive than milk replacers, and when the kids can do well on solid feeds, milk replacers should not be fed. Small kids need concentrates until their rumens are sufficiently developed to digest enough roughage to meet all their nutritional needs. At five to six months of age, young goats can do well on good pasture or good quality legume hay.

CONTROLLING DISEASES AND PARASITES

The major diseases affecting goats are mastitis, brucellosis, footrot, **ketosis**, and milk fever.

Mastitis is an inflammation of the udder predisposed by bruising, lack of proper sanitation, and improper milking. The milk becomes curdled and stringy. It is advisable to engage the services of a veterinarian for treating mastitic udders. An udder may be treated with an antibiotic by sliding a special dull needle up the teat canal into the udder cistern and depositing the antibiotic. This procedure may cause the doe to cease giving milk. Some people apply hot packs and give care in the hope that the udder will heal spontaneously. In case this is done, antibiotic treatment may be applied later when the doe is not milking.

Brucellosis is common in goats, and it is harmful both to goats and to humans. All goats should be tested for brucellosis and reactor animals or any others suspected of having it should be slaughtered. Milk from goats should be pasteurized to ensure that it is safe from brucellosis.

Footrot occurs when goats are kept on wet land. It is highly contagious, but the bacterial organism causing it does not live long in the soil. If footrot is absent from a herd, extreme care should be taken to avoid introducing it. Any animal that is to be introduced into a herd should be isolated for 30 to 40 days to see if it develops footrot before it joins the herd. Severe trimming of infected feet, followed by treating with formaldehyde (as was described in Chapter 25) cures this disease.

Ketosis, or pregnancy disease, also occurs in goats. Affected goats are in agony and cannot walk. Exercising goats that are pregnant, feeding them well in the latter part of the pregnancy, and being certain that they have water at all times helps prevent ketosis.

Milk fever can occur in goats that are lactating heavily because they deplete their calcium stores. They have a high fever and cannot stand. An intravenous injection of calcium gluconate results in rapid recovery. In fact, the goat may be up and normal in less than an hour after the calcium gluconate injection.

Goats are subject to external and internal parasites. The external parasites include **lice** and **mites**. The animals may be sprayed, dipped, or dusted with an appropriate insecticide. The producer should consult an expert for a recommended treatment.

Internal parasites of goats include stomach worms and **coccidia**. The same treatment for stomach worms that is effective in sheep can be used (see Chapter 25). **Coccidiosis**, which is caused by a protozoan organism in the intestinal tract, is found mostly in young goats. Drenching the animal with sulfa drugs or certain antibiotic drugs helps to control coccidiosis. Routine treatment with thibenzole or phenothiazine, or an alternation of these each six months, helps to control most internal parasites. Because goats may inhale a drench, it is advisable to secure professional help in administering these drugs.

THE ANGORA GOAT

Angora goats (Figure 30-8) are raised chiefly to produce mohair, which is made into clothing. These goats are produced largely in the Southwest (Texas has the most Angora goats of any state), the Ozarks of Missouri and Arkansas, and the Pacific Northwest. They do well on browse, so they are most commonly produced on rough, brushy areas. They are poorly adapted for grazing on grassy pastures that are used for intensive grazing because they prefer broadleaf plants over grasses, and they are most likely to become parasitized when grazing near the soil level.

Good Angora goats are shorn once or twice a year. Some of them have a tendency to shed their mohair in the spring but selection against this trait has brought about improvement. In fact, some Angora goats are not shorn for two to three years, which allows the mohair to grow to lengths of one to two feet. This special mohair brings a high price. It is used for making such items as wigs and doll hair. Shearing is usually done in spring. The clip from does weighs 4 to 6 lbs, whereas bucks and wethers may shear 5 to 8 lbs. Highly improved goats often shear 10 to 15 lbs of mohair per year. Three classes of mohair are produced: the tight lock, which has ringlets the full length of the fibers; the flat lock, which is wavy; and the fluffy, or open, fleece. The tight-lock fleeces are fine in texture but

Figure 30-8. An Angora goat showing a full fleece of mohair. Angora goats are useful for meat production, brush clearing, and mohair production.

low in yield. The flat lock lacks the fineness of the tight lock, but it is satisfactory for making cloth and its yield is high. The fluffy fleece is of low quality, often quite coarse, and is easily caught in brush and lost.

The fleeces should be sorted at shearing time and coarse fleeces and fleeces with burs should be kept separate from the good fleeces. The clip from each fleece should be tied separately so that each fleece can be properly graded.

Fertility is usually low in Angora goats and losses of kids are often high. Because these goats are run in brushy areas, losses from predatory animals also are high. Generally, the kid crop that is raised to weaning is about 80% of the number conceived as estimated on the basis of the number of does bred. Losses of 15 to 20% or more may occur when proper care is lacking at the time kids are born.

Angora goats have horns. When they are in good condition on the range, mature bucks and wethers weigh 150 to 200 lbs. Mature does weigh 90 to 110 lbs but well fed show animals may weigh more. Kids are usually weaned at five months of age, which allows the doe to gain in weight at breeding time. Breeding over a period of days is usually done in late September and October. Since the gestation is 147 to 152 days, this time of breeding results in the kids arriving in the spring after the weather has moderated. Kidding may take place in open sheds when the weather is severe or outside when the weather is favorable. Young does are usually first bred to kid at the age of two years, but, if they are properly developed, they may first be bred to kid when they are yearlings.

In some areas, Angora or cull dairy goats are used to kill out brushy plants. Two main methods that employ goats are used to destroy brush. One method is to stock the area heavily with goats and give them some feed so that their health is not impaired. Keeping the goats hungry for green feed causes all brush to be destroyed in two years because the goats nip off the buds of new growth, thus starving the plant. It usually requires two years to starve hardy plants, but some of the less hardy ones will succumb the first year of starvation. The other method is to run goats along with cattle for 6 to 10 years. The goats tend to eat leaves of brushy plants while the cattle consume the grasses.

Angora goats need a shelter in rainy periods or wet snows, but an open shed is sufficient. The long mohair coats of Angoras keep them warm when the weather is cold, but they may nevertheless crowd together and smother in severe, cold snowstorms if no shelter is provided.

Angora goats are subject to the same parasites and diseases as dairy goats and the same treatments are effective. In most respects, the care given Angora goats resembles the care given sheep rather than that given dairy goats.

THE PYGMY GOAT

Pygmy goats are found in children's petting zoos. They are extremely active and stylish (see Figure 30-9). Some Pygmy goats (crossbred Pygmy dairy goats) are raised for the production of milk and the meat of the Pygmy goat is also excellent. Because Pygmy are small in size, the cost of keeping them is minimal; for these reasons they are used as laboratory animals.

Pygmy goats vary in color from black to white with a dosal dark stripe down the back and dark on the legs. On all Pygmy goats except black, the muzzle, forehead, eyes, and ears are accented in tones lighter than the dark portion of the body.

Figure 30-9. Eight-year-old Tina Van Geest and people of all ages like goats because they are such affectionate animals. The goat is an ideal animal for 4-H and FFA projects. Young people can learn responsibility by doing a project in which they work with goats while gaining some knowledge about reproduction, nutrition, and lactation. Miss Van Geest is Ralph Bogart's foster grand-daughter.

STUDY QUESTIONS

1. Name and describe three breeds of dairy goats.
2. How can you identify each goat in your herd?
3. Name three diseases of goats and tell how each can be controlled.
4. For what purposes are Angora goats used?
5. Tell how you would supply the nutrition needed by a female dairy goat in her lactation period.
6. How would you provide a means for containing goats in an enclosure?
7. How do dairy goats and dairy cows compare in total milk production and milk production per unit of body weight?
8. If goats are used to clear land of brush, how does one manage the goats to keep them in good health but also force them to eat brushy plants?
9. Why are dairy goats particularly useful as project animals for young people?
10. How many upper front teeth does a goat have?

SELECTED REFERENCES

American Dairy Goat Association. (No date.) *Own a Dairy Goat.* Spindale, North Carolina, P.O. Box 865: American Dairy Goat Association.

Brannon, W. F. (No date.) *The Dairy Goat.* Ithaca, New York: New York Agricultural Exp. Station Information Bulletin 78.

Brody, S. 1938. *Growth and Development.* XLIX. *Growth, Milk Production, Energy Metabolism, and Energetic Efficiency of Milk Production in Goats.* Columbia: Missouri Agricultural Experiment Station Research Bulletin.

Colby, B. E., Evans, D. A., Lyford, Jr., S. A., Nutting, W. B., and Stern, D. N. 1972. *Dairy Goats — Breeding, Feeding and Management.* Spindale, North Carolina: American Dairy Goat Association (distrib.)

Extension Service (1978). *Dairy Goats for Family Milk Supply.* Corvallis, Oregon: Oregon Agricultural Extension Service Circular No. 866.

Gray, J. A. (No date.) *Selecting Angora Goats for Increased Mohair, Kid Production.* College Station, Texas: Texas A & M Extension Publication MP-385.

Gray, J. A. (No date.) *Texas Angora Goat Production*. College Station, Texas: Texas A & M Agri. Extension Publication B-926.

Groff, J. L. and Kensing, R. H. (No date.) *Factors Affecting Profitable Angora Goat Production in Texas*. College Station, Texas: Texas A & M Agri. Extension Fact Sheet.

Magee, A. C. 1957. *Goats Pay for Clearing Grand Prairie Rangelands*. College Station, Texas: Texas A & M Agri. Exp. Station Publication MP-206.

Mohair Council of America. 1976. *Guidelines for Improving the Kid Crop from Angora Goats*. San Angelo, Texas: Mohair Council of America.

31

Behavior of Animals

Knowledge of animal behavior is important in managing animals and in the design of shelters and equipment. From a study of behavior of sheep, for example, the Australians were able to design yards, chutes, and gates through which sheep are efficiently worked. In addition, deviation from normal behavior for an animal indicates that something is wrong and that attention is needed. Thus, illness can be detected in an early stage and the afflicted animal can be treated in time or, if appropriate, isolated to prevent the contagion from spreading.

Behavior is complex because it results from the interaction of inherited activities (heredity) with experiences to which the individual has been subjected (environment). A behavior pattern is an organized segment of behavior having a special function. The primary function of behavior is to enable the animal to adjust to some internal or external change in conditions and improve its chance of survival.

Instinct (reflexes and responses) is what the animal has at birth. All mammals know at birth how to nurse even though they must first learn where the teat is located. Shortly after leaving the egg, young birds begin pecking to obtain food. Newborn pigs soon start to nuzzle whatever object is before them in an attempt to locate the teat of the mother. As soon as the teat is located, the baby pig starts to nurse. The calf shortly following birth also seeks the teat and starts to nurse. Baby lambs try to nurse any protruding object of proper size; therefore, it is essential that dung tags be removed from ewes prior to lambing.

Habituation is learning to respond without thinking. Response to a certain stimulus is established as a result of habituation. **Conditioning** is learning to respond in a particular way to a stimulus as a result of **reinforcement** when the proper response is made. Reinforcement is a reward for making the proper response. Trial and error is the performance of different responses to a stimulus until the correct response is performed, at which time the animal receives a reward. For example, newborn mammals soon become hungry and want to nurse. They search for some place to nurse on any part of the mother's body until they find the teat. This is trial and error until the teat is located; then, when the young nurses, it receives milk as its reward. Soon the young learns where the teat is located and finds it without having to go through trial and error. Thus, the young has become conditioned for nursing through reinforcement.

Reasoning is the ability to respond correctly to a stimulus the first time that a new situation is presented. **Intelligence** is the ability to learn to adjust successfully to certain situations. Both short-term and long-term memory are part of intelligence. Horses, dairy goats, and swine are more intelligent than other farm animals.

SYSTEMS OF BEHAVIOR

Farm animals show ten major systems of behavior, including ingestive, eliminative, sexual, care-giving, care-soliciting, **agonistic** (combat), **allelomimetic** (doing the same thing), shelter-seeking, investigative, and maladaptive.

Ingestive behavior. Ingestive behavior is exhibited by farm animals when they eat and drink. Cattle, sheep, and goats have no upper front teeth. Their lower front teeth can be pressed against the dental pad, which exists in the place of the upper front teeth. In grazing, these animals catch the forage between the lower front teeth and the dental pad. They then move the mouth forward so that the teeth cut off the grass. Cattle wrap their tongues about the forage so that, once it is cut, they can chew and swallow it directly. Because it is difficult for cattle to wrap their tongues around low grazing forage, they graze best on forage that is at least 6-in. high. Sheep, by contrast, can graze on low-growing forage because they cut it without wrapping their tongues around it and then gather it with their lips.

Cattle tend to graze well on grasses, whereas sheep and goats enjoy leafy plants. Sheep and goats tend to be more selective in their grazing than cattle. Sheep and goats eat large amounts of forbs (leafy weeds and other plants) and browse (brushy plants). Sheep are used effectively to control weeds on pastures and in corn fields. They will eat many plants such as poison oak and tansy ragwort, which are not eaten by or are poisonous to other animals.

When sheep are "set stocked" (stocked according to a given number per acre), they tend to stake out a territory and stay in that area most of the time. The effect is that they overgraze certain areas and undergraze other areas. Undergrazing and overgrazing can be prevented by use of temporary fencing that can be moved to force the use of undergrazed areas.

Cattle will usually not go more than four to five miles from water to graze. When they are on a large range, they tend to overgraze near the water area to avoid areas far removed from water. Development of watering places and proper fencing can increase the use of forage on large range areas and prevent severe overgrazing of certain areas.

Rather than initially chewing their feed thoroughly, ruminants swallow it as soon as it is well lubricated with saliva. After they have consumed a certain amount, they ruminate, that is, rest and regurgitate the feed for chewing. Cattle graze for four to nine hours a day and sheep and goats graze for nine to 11 hours a day. Grazing is usually done in periods followed by rest and **rumination**. Sheep rest and ruminate more frequently when grazing than do cattle—cattle ruminate from four to nine hours per day; sheep from seven to 10 hours per day. A cow may regurgitate and chew between 300 and 400 **boluses** of feed per day; sheep between 400 and 600 boluses per day. Cattle consume 13 to 27 lbs of dry matter per day when grazing; sheep, 2 to 6 lbs per day. Including the water present in the forage, ruminants consume more than 10% of their body weight per day in feed. Cattle normally cover less than three miles a day when grazing, but sheep travel as much as eight miles a day.

Horses have both upper and lower front teeth, so they can bite off grass. They chew what they eat, both when grazing and when fed hay or grain. They do not ruminate. Because they bite off the forage, horses can graze very close to the ground. Horses prefer pasture plants to forbs and browse, but, like donkeys, they eat bushy plants when there is a shortage of grass.

Geese graze grasses readily. In some areas where strawberries are grown, geese are used to control grasses that might otherwise compete with strawberries. The geese must be removed when the berries start to bloom, because they will pull off the blossoms even though they do not eat them.

Most farm animals avoid grazing where feces and urine contaminate the forage. In fact, large areas of green forage in a well-grazed pasture may be ungrazed because they are contaminated. Unwanted plants may make seed in such places. All new growth will be grazed if the pasture is mowed and the ground is harrowed to spread feces and contaminated forage.

Eliminative behavior. Cattle, sheep, goats, and chickens void their feces and urine indiscriminately. Horses and hogs, by contrast, defecate in definite areas of the pasture. Cattle, sheep, goats, and swine usually defecate while standing or walking. All these animals urinate while standing, but not usually when walking. Cattle defecate 12 to 18 times a day; horses five to 12 times. Cattle and horses urinate seven to 11 times per day. Animals on lush pasture drink less water than when they consume dry feeds; therefore, the amount of urine voided may not differ greatly under these two types of feed conditions. All farm animals urinate and defecate more frequently and void more excreta than normal when they are stressed or excited. They often lose 3% of their weight when transported by truck to market. Much of the shrinkage in transit occurs in the first hour, so considerable weight loss occurs even when animals are transported for short distances.

Sexual behavior. Observations made on the sexual behavior of female farm animals are useful in carrying out breeding programs. Cows that are in heat, for example, allow themselves to be mounted by others. Operators note this condition of "standing heat" and select the cows they desire to breed. Ewes in heat are not mounted by other ewes, but they can be detected by having vasectomized rams locate them.

Manifestations of sexual activity in male farm mammals occur in the following order: courtship, erection and protrusion, mounting, intromission (inserting penis into vagina), ejaculation, and dismounting.

The bull follows a cow that is coming into heat, smells and licks her external genitalia, and puts his chin on her rump. When the cow is in standing heat, she stands still when the bull chins her rump. When she reaches full heat, she allows the bull to mount from the rear. He thrusts his penis into the vagina, **ejaculates** almost immediately, and dismounts.

Cows are receptive for varying periods; dairy cows are usually in heat for 18 hours or more, whereas beef cows may be in heat for only eight to 12 hours.

Vigorous bulls breed several times a day. If more than one cow is in heat at the same time, bulls tend to mate with one cow once or a few times and then go to others.

The ram will chase a ewe that is coming into heat. The ram champs and licks, puts his head on the side of the ewe, and strikes with his foot. When the ewe reaches standing heat, she stands when approached by the ram.

Rams have a strong sex drive and mate several times a day. One ram has been observed to ejaculate 17 times in an hour. As do bulls, rams leave ewes that they have served to go to other ewes in heat.

The boar does not seem to detect a sow that is in heat by smelling or seeing. If introduced into a group of sows, a boar will chase any sow in the group. The sow that is in heat will seek out the boar for mating and when the boar is located she stands still and flicks her ears. Ejaculation requires several minutes for boars in contrast to an instantaneous ejaculation by rams and bulls. Boars do not usually serve as often as bulls and rams, but on occasion serve a sow a second time without dismounting.

The penis of bulls, rams, and boars is S-shaped when retracted. A retractor muscle is used to pull the penis into its sheath, and holds it there. At breeding, the penis is elongated by straightening of the S-curve as the retractor muscle relaxes rather than by gorging it with blood.

The stallion approaches a mare from the front and a mare not in heat runs and kicks at the stallion. When the mare is in standing heat she stands, squats somewhat, and urinates as he approaches. Her vulva "winks" (opens and closes) when she is in heat.

Erection in the stallion is caused by gorging the penis with blood, and the stallion usually has a complete erection when he mounts.

The buck goat does considerable snorting and blowing when he detects a doe in heat. The doe shows unrest and may be fought by other does. Mating in goats is similar to that in sheep.

In chickens and turkeys, a courtship sequence between the male and female usually takes place. If either individual does not respond to the other's previous signal, the courtship does not proceed further. After the courtship has developed properly, some females run from the rooster, who chases them until they stop and squat for mating. The male chicken or turkey stands on a squatting female and ejaculates semen as his rear descends toward the female's cloaca. Semen is ejaculated at the cloaca and the female draws it into her reproductive tract.

Male chickens and turkeys show a preference for certain females and may even refuse to mate with other females. Likewise, female chickens and turkeys may refuse to mate with certain males. This is a serious problem when pen matings of one male and 10 to 15 females are practiced. The eggs of some females may be infertile. Even when several turkey males are run with a group of hens, artificial insemination once or twice in the breeding season may increase hatchability by 10%.

Female rabbits do not show estrous cycles as other farm mammals, but have, instead, extended periods when they can be bred. The buck mounts the doe, and the doe lifts her rear to assist in copulation. When the penis enters the vagina, the male throws his rear forward with his hind legs going up and ejaculates suddenly. The doe does not ovulate unless she is mated. When mated, she ovulates within 10 hours.

Little relationship appears to exist between sex drive and fertility in male farm animals. In fact, some males that show extreme sex drive have reduced fertility because of frequent ejaculations that result in semen with a low number of sperm.

Care-giving behavior. When the young of cattle, sheep, goats, and horses are born, the mothers clean the young by licking them and encourage them to stand and to nurse. Sows do not clean their newborn, but encourage them to nurse by lying down and moving their feet as the young approach the udder region. They thus help the young to the teats. Sows (and to a degree, cows) tend to fight intruders, especially if the young squeal or bawl.

Strong attachments exist between mother and newborn young, particularly between ewe and lamb and cow and calf. Beef cows diminish their output of milk after about 100 to 120 days following birth of young, and ewes do the same after 60 to 75 days. This reduction in milk forces the young to search for forage, the consumption of which stimulates rumen development. It is at this time that care-giving by the mother declines.

If young pigs have a high-energy feed available at all times, they nurse less frequently. Without a strong stimulus of nursing, sows reduce their output of milk. Some sows may wean their pigs and show little concern for them after 40 to 50 days. Pigs are normally weaned at 56 days of age, but some producers wean them as early as 35 to 40 days.

Care-soliciting behavior. Young animals cry for help when disturbed or distressed. Lambs bleat, calves bawl, pigs squeal, and chicks chirp.

Even adult animals call for help when under stress. Because many animals associate closely with people, animals may call for help from people. An example is that of a mule that caught its foot in a barbed-wire fence. This mule brayed longingly for help until its owner came and released it.

Agonistic behavior. Includes behavior activities of fight and flight and those of aggressive and passive behavior when an animal is in contact (physically) with another animal.

Unless castrated when young, the males of all farm animals fight when they meet other unfamiliar males of the same species. This behavior has great practical implications for management of farm animals. Male farm animals are often run singly with a group of females in breeding season, but it is often necessary to keep males together in a group at times other than the breeding season. The typical farmer simply cannot afford to provide a separate lot for each male.

When two bulls that are strangers are put together, they approach each other, bellowing as they come. They usually stop when they are a few yards apart, arch their necks, and blow loudly through their nostrils for several minutes. The apparent objective is to make the opponent give in without fighting. When the fight starts, the bulls meet head on and push; they push and try to throw the other off balance until one bull admits defeat and runs.

The best place to put bulls together at the close of the breeding season is an open pasture. Two people on horses are needed for this operation. One bull is taken to the center area of the pasture and is kept there by one of the persons on horseback. The other person brings another bull to the open area and lets them fight. This process is repeated if several bulls are to be put together, each bull is brought in separately because the bulls will start to fight immediately when put together and will continue fighting with another until a definite dominance order is established.

Bulls exercise vigorously when fighting and thus generate much heat. Therefore, bulls should be put together either early in the morning or late in the evening when the environmental temperature is lower than at midday. If a mature bull is put with young bulls, little fighting usually occurs between the mature bull and the young bulls because the younger ones concede to the mature one. After the fighting is over, bulls may start mounting one another (one form of homosexuality).

Rams that are strangers will fight when they come together. Each one backs away from the other and then both run toward each other. When they are about 10 feet apart, both jump so that they meet head-on with a terrific thud. Occasionally, a ram breaks his neck in a fight. Usually one ram will give in to the other within 15 to 20 minutes.

Buck goats fight differently from rams. They stand on their hind feet and swing their heads sideways as a means of striking one another with their heads or horns, until one retreats in 10 to 15 minutes.

Fighting boars swing their heads from a lower to a higher position in an attempt to cut their opponent with their tusks. Because boars sometimes attack people in this same way, tusks should be removed once or twice each year.

Stallions that are strangers fight when they come together. They may strike with the front feet and bite one another or they may turn rear-to-rear and kick one another severely until one of them concedes. Donkeys do more biting and may maintain a fight for an hour or more.

Cows, sows, and mares usually develop a peck order, but fight less intensely than males. Sows that are strangers to each other sometimes fight. Ewes seldom, if ever, fight, so ewes that are strangers can be grouped together without harm.

Allelomimetic behavior. Animals of a species tend to do the same thing at the same time. Cattle and sheep tend to graze at the same time, and rest and ruminate at the same time. On a cattle ranch, cows from over the range gather at the watering place at about the same time each day because one follows another. This behavior is of practical importance because the attendant can then observe the cow herd with little difficulty, notice anything that is wrong with an animal, and have the animal brought in for treatment. If one is artificially inseminating beef cattle, the best time to locate cows in heat is when they gather at the watering place.

Shelter-seeking behavior. Animal species vary greatly in the degree to which they seek shelter. Cattle and sheep seek a shady area for rest and rumination if the weather is hot, and pigs try to find a wet area. When the weather is cold, pigs crowd against one another when they are lying down to keep each other warm. In snow and cold winds, sheep and goats often crowd together. In extreme situations they pile up to the extent that some of them smother. Unless the weather is cold and windy, cattle and horses often seek the shelter of trees when it is raining. This may be hazardous where strong electrical storms occur because animals under a tree are more likely to be killed by lightning than those in the open.

Investigative behavior. Pigs, horses, and dairy goats are highly curious and will investigate any strange object. They usually approach carefully and slowly, sniffing and looking as they approach. Cattle also do a certain amount of investigating (Figure 31-1). Sheep are less curious than some of the other farm animals and are more timid. They may notice a strange object, become excited, and run away from it.

Maladaptive behavior. When a motivation occurs to which an animal cannot adapt, the animal will exhibit inappropriate behavior. The motivation can be the result of training or internal or external stimuli. The best examples of such inappropriate behavior in farm animals are **masturbation** and homosexuality in males. Male animals that cannot adapt to

Figure 31-1. Cattle expressing investigative behavior as a photographer is taking the picture.

being separated from females either masturbate or attempt sexual intercourse with other males. Some submissive males may be highly tormented by sexually aggressive males, making it necessary to separate them from the others.

SOCIALIZATION

Primary social relationships are important in farm animals; they are usually formed in early life as parent-offspring relationships, but strong relationships may exist between animals and people. Birds, for example, when exposed only to people for the first three days of life, will show more attachment to people than to other birds of the same species. Learning associated with maturational readiness is called imprinting. If a lamb is kept by itself and cared for by a person, it will become highly attached to people and will subsequently avoid other sheep. Such an individual is unlikely to become a breeding animal, thus lessening its value to its owner. Most farm animals respond to people in the same manner as do sheep.

It is best that young animals be kept with other young of the same species, but they should also associate with people early in life. In this way, they associate first with other animals but also with people. Horses, especially, need association with people so that a relationship can be established in which the person is the master and the horse is the subordinate. Horses that have associated with people from early life are usually trained early and will fully respond to commands.

CONTINUOUS HOUSING AND OVERCROWDING

One of the problems in the production of animals that are often kept in continuous housing, such as poultry and swine, is that continuous housing and overcrowding exist simultaneously to cut costs of housing per animal. Quite frequently, both chickens and swine will resort to cannibalism, which leads to the loss of many animals if precautions are not taken to prevent it. Some swine producers remove the tails of baby pigs to prevent tail chewing because tail chewing can cause bleeding and whenever bleeding occurs, the pigs are likely to become cannibals.

Some anomalous behavioral forms. Abnormal behavior symptoms in a farm animal may arise from stimuli the animal has received but they may also indicate an abnormal health condition. For example, cows may chew on bones that are on the range. The chewing on bones likely is caused by a deficiency of phosphorous and proper mineral supplementation may be indicated.

STUDY QUESTIONS

1. Compare the horse, cow, goat, and sheep in terms of their preferences for grass, browse, and forbs.
2. If cattle, sheep, and horses are running on a pasture, which of them need fairly tall grass for grazing? Which can graze short grass? Which are most likely to overgraze a pasture?
3. Compare horses, pigs, cattle, and sheep in terms of their eliminative behavior.
4. Some farm animals require only an extremely short time to copulate. Others require much more time. Compare cattle, sheep, horses, and swine in terms of time required for copulation.
5. Why can one increase hatchability of eggs from turkeys by artificial insemination even though a fertile male is running with the hens?
6. How should one group several bulls together after breeding season has been completed?
7. Which of the farm animals are most curious (investigative)? Which types of farm animals tend to do the same thing that others of the same species are doing?
8. Describe some examples of anomalous behavior and their causes.

SELECTED REFERENCES

Fraser, A. F. (Ed.) *Applied Animal Ethology.* 1980. New York: Elsevier North Holland Publishing Co.
_____. 1974. *Farm Animal Behavior.* Baltimore: The William and Wilkins Co.
Hafez, E. S. E. 1975. *The Behavior of Domestic Animals.* 3rd edition. Baltimore: The Williams and Wilkins Co.

32

Making Effective Management Decisions

Effective management of livestock operations implies that available resources are used to maximize net profit while the same resources are conserved or improved. Available resources include fixed resources (land, labor, capital, and management) and renewable biological resources (animals and plants). Effective management requires a manager who knows how to make timely decisions based on a careful assessment of management alternatives. Modern technology is providing useful tools to make more rapid and accurate management decisions.

In previous chapters, we have discussed how biological principles determine the efficiency of animal production. It is important to identify some of the other resources which, combined with the efficiency of animal production, will determine the profitability of an operation.

THE MANAGER

The manager is the individual responsible for planning and decision-making. Simplified, the management process is to plan, act, and evaluate. This process is described in more detail in Figure 32-1.

It is imperative that producers of farm animals conduct their operation as a business, and not just as a way of life. Current economic pressures associated with keener competition are forcing more producers to manage their operations as business enterprises.

The manager may be an owner-operator with minimum additional labor, or the manager of a more complex organizational structure involving several other individuals (Figure 32-2). An effective manager, whether involved in a one-person operation or a complex organizational structure, needs to (1) be profit-oriented, (2) keep abreast of the current knowledge related to the operation, (3) know how to use time effectively, (4) attend to the physical, emotional, and financial needs of those employed in the operation, (5) incorporate incentive programs to motivate employees to perform at their full capacity each day,

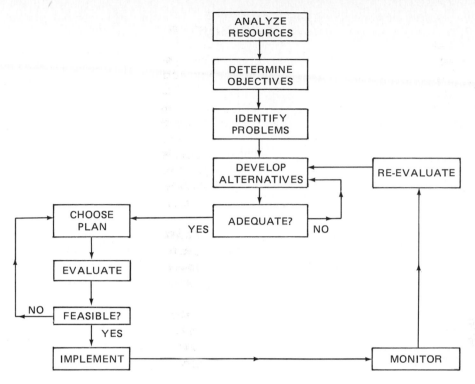

Figure 32-1. Major component parts of the planning process.

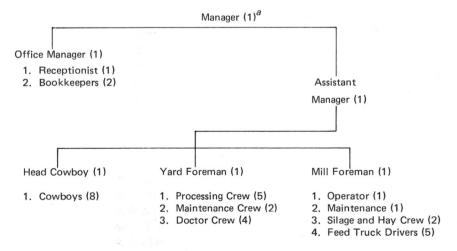

[a]Number in parenthesis indicates number of employees in each position

Figure 32-2. Organizational structure of a large commercial cattle feeding operation.

(6) have honest business dealings, (7) effectively communicate responsibilities to all employees and make employees feel they are part of the operation, (8) know what needs to be done and at what time, (9) be a self-starter, (10) set priorities and allocate resources accordingly, (11) remove or alleviate high risks, (12) manage oneself so that others can see and follow a good example, and (13) identify objectives and form both short-term and long-range goals to achieve those objectives.

FINANCIAL MANAGEMENT

Costs, returns, and profitability of a livestock operation can only be assessed critically with a meaningful set of records. A financial record system should include a financial statement, a yearly budget of anticipated income and expenses, adequate income and expense records to complete the IRS tax forms (primary form shown in Figure 32-3), and a cash-flow statement that shows the anticipated income and expenses by months. The latter would show how adequate the business' operating money is, and knowing this could save the business interest because the manager could borrow and repay at certain critical times of the year. Cash-flow statements can also identify growth-limiting factors in the operation and determine whether or not a particular investment decision should be accepted or rejected.

Profits should be determined by evaluating all cash costs of the operation, all inventory increases or decreases, and the value of opportunity costs. Opportunity costs represent the returns that would be forfeited if the debt-free resources (such as owned land, livestock, and equipment) were used in their next best level of employment; for example, the value of the pastureland if it were leased, the returns if all capital represented by equipment and livestock were invested in a certificate of deposit (or similar investment), and the returns if all family labor and management were utilized in other employment. By calculating this, a price can be established for living on a farm or ranch, and a goal established for increasing profits.

Credit and money management become crucial during periods of inflation, high interest rates, and relatively low livestock prices. The prudent use of credit can enable a livestock operation to grow more rapidly than it could through the use of reinvested earnings and savings, so long as borrowed funds return more over time than they cost. Thus, farmers and ranchers have to look to credit as a financial tool and learn to use it effectively.

The net cost of credit may be less than the interest rate depending on your taxable income bracket (Table 32-1).

INCOME TAX CONSIDERATIONS

In addition to keeping accurate records for income tax purposes, there are some other management considerations regarding taxes that are important. A few of these are briefly mentioned for your awareness. Their details and utilization should be more critically evaluated, preferably with a tax accountant.

1. Capital gains can be taken on livestock held for breeding purposes. For example, of the $500 slaughter value of a breeding cow, only 40% or $200 is taxable. Management decisions are needed to determine number of replacement females saved, age of

SCHEDULE F
(Form 1040)

Department of the Treasury
Internal Revenue Service

Farm Income and Expenses

▶ Attach to Form 1040, Form 1041, or Form 1065.
▶ See Instructions for Schedule F (Form 1040).

OMB No. 1545-0074

1981
16

Name of proprietor(s)   *JAMES A. BROWN*

Farm name and address ▶ *JAMES A. BROWN, RR #1, BOX 25*
HOMETOWN   YOUR STATE   02115

Social security number
579 28 6685

Employer identification number
71 9367974

Part I Farm Income—Cash Method		
Do not include sales of livestock held for draft, breeding, sport, or dairy purposes; report these sales on Form 4797.		

Sales of Livestock and Other Items You Bought for Resale

a. Description	b. Amount	c. Cost or other basis
1 Livestock ▶ *HEIFERS BOUGHT FOR RESALE*	*6,900*	*4,000*
2 Other items ▶		
3 Totals	*6,900*	*4,000*
4 Profit or (loss), subtract line 3, column c, from line 3, column b . . . . ▶	*2,900*	

Sales of Livestock and Produce You Raised and Other Farm Income

Kind	Amount
5 Cattle and calves	*3,316*
6 Sheep	
7 Swine	
8 Poultry	
9 Dairy products	*91,454*
10 Eggs	
11 Wool	
12 Cotton	
13 Tobacco	
14 Vegetables	
15 Soybeans	
16 Corn	
17 Other grains	
18 Hay and straw	*2,266*
19 Fruits and nuts	
20 Machine work	*1,258*
21 a Patronage dividends . *272*	
b Less: Nonincome items *22*	
c Net patronage dividends	*250*
22 Per-unit retains	
23 Nonpatronage distributions from exempt cooperatives . .	
24 Agricultural program payments: a Cash . . .	*80*
b Materials and services . .	*324*
25 Commodity credit loans under election (or forfeited) . .	*550*
26 Federal gasoline tax credit	*100*
27 State gasoline tax refund	
28 Crop insurance proceeds	*624*
29 Other (specify) ▶ *FAIR PRIZES*	*700*
30 Add amounts in column for lines 5 through 29 .	*100,922*
31 **Gross profits*** (add lines 4 and 30) . . . . ▶	*103,822*

Part II Farm Deductions—Cash and Accrual Method		**F**
Do not include personal or living expenses (such as taxes, insurance, repairs, etc., on your home), which do not produce farm income. Reduce the amount of your farm deductions by any reimbursement before entering the deduction below.		

Items	Amount
32 a Labor hired	*4,438*
b Jobs credit	
c WIN credit	
d Total credits	
e Balance (subtract line 32d from line 32a) . .	*4,438*
33 Repairs, maintenance . .	*3,236*
34 Interest	*4,124*
35 Rent of farm, pasture . .	*1,200*
36 Feed purchased	*24,182*
37 Seeds, plants purchased .	*2,286*
38 Fertilizers, lime, chemicals .	*4,498*
39 Machine hire	*1,200*
40 Supplies purchased . . .	*3,204*
41 Breeding fees	*1,078*
42 Veterinary fees, medicine .	*2,020*
43 Gasoline, fuel, oil	*3,070*
44 Storage, warehousing . .	
45 Taxes	*3,050*
46 Insurance	*2,106*
47 Utilities	*2,050*
48 Freight, trucking	*2,056*
49 Conservation expenses . .	*1,580*
50 Land clearing expenses . .	
51 Pension and profit-sharing plans	
52 Employee benefit programs other than line 51 . . .	
53 Other (specify) ▶ *ADVERTISING*	*468*
FINANCIAL RECORDS	*166*
FARM ORG. DUES	*350*
DEATH LOSS - HEIFER BOUGHT FOR RESALE	*450*
54 Add lines 32e through 53 .	*66,812*
55 Depreciation (from Form 4562)	*8,483*
56 **Total deductions** (add lines 54 and 55) . . . . . ▶	*75,295*

57 Net farm profit or (loss) (subtract line 56 from line 31). If a profit, enter on Form 1040, line 18, and on Schedule SE, Part I, line 1a. If a loss, go on to line 58. (Fiduciaries and partnerships, see the Instructions.) . . . . . . | 57 | *28,527*

58 If you have a loss, do you have amounts for which you are not "at risk" in this farm (see instructions)? . . . ☐ Yes ☐ No
If you checked "No," enter the loss on Form 1040, line 18, and on Schedule SE, Part I, line 1a.

*Use amount on line 31 for optional method of computing net earnings from self-employment. (See Schedule SE, Part I, line 3.)

For Paperwork Reduction Act Notice, see Form 1040 Instructions.

Figure 32-3. Internal Revenue Service form showing income and expense items for which documented records must be kept.

Table 32-1. Net Cost of Borrowed Dollars

Income Tax Bracket (%)	Annual Rate of Interest Loan (%)				
	6	8	10	12	14
	Net rate of interest after tax savings (%)				
19	4.86	6.48	8.10	9.72	11.34
25	4.50	6.00	7.50	9.00	10.50
32	4.08	5.44	6.80	8.16	9.52
39	3.66	4.88	6.10	7.32	8.54
48	3.12	4.16	5.20	6.24	7.28

breeding animals in the herd, and culling decisions based on the level of productivity of each individual breeding animal.

2. A spouse can be paid a salary. This may save social security taxes, enhance retirement and estate planning, and provide eligibility for child-care tax credit. It may have the disadvantage of reducing social security benefits.

3. An investment credit can be taken. A 10% deduction can be used on the amount paid for new or remodeled livestock confinement buildings, milking parlors, other single purpose livestock structures, fences, paved roads, water wells, and drainage tiles.

4. Depreciation can be taken on breeding livestock, barns, sheds, grain storage bins, corrals, fences, irrigation wells and pumps, silos, farm lighting systems, and several others. For examples, a dairy cow purchased for $1000 could be depreciated $200 per year over the expected productive lifetime of five years. Her slaughter value would have to be claimed as income when sold.

5. Children working on a family farm can be paid up to $3,300 per year without having to pay any federal income tax.

6. Mileage costs or actual expenses of operating and maintaining the farm share of the automobile or pickup can be deducted.

ESTATE AND GIFT TAX PLANNING

Many livestock operations have a large amount of debt-free capital invested in land, livestock, buildings, and equipment. Adequate knowledge and proper planning is necessary for farmers and ranchers so they may pass on viable economic units to their heirs.

In 1981, the Economic Recovery Tax Act made major revisions to the previous tax code. When the new estate tax amendments are fully phased in and implemented, a husband and wife who own farm or ranch property other than in joint tenancy will be able to transfer $2.7 million of this property (fair market value) to their children free of federal estate taxes. Also under the new law, each parent will be permitted to transfer $10,000 to each child and his or her spouse (up to $40,000 for each child and spouse, to a maximum of $160,000 per parent couple of property annually) without incurring any federal gift tax.

Livestock producers should review their tax plans—both estate, gift, and income—in light of the amendments contained in the 1981 tax bill, and consult with the professionals who work with them.

COMPUTERS AND PROGRAMMABLE CALCULATORS

Livestock operators need to keep a voluminous amount of information for financial and production records, inventories, and to more critically assess management alternatives. Excellent managers, with conventional records systems, often have an inate ability to "use their heads" to make good management decisions with a high degree of accuracy. Evidently they have a unique mind (the most complex computer) that functions exceptionally well in assimilating, storing, and recalling useful information.

Small computers (microcomputers) and programmable calculators are becoming more frequently used as a management tool by farmers and ranchers. Producers must recognize what computers and calculators are, and what they can and cannot do, at the present time and in the future. The technology of computers is changing rapidly; therefore, producers must be ready to accept the frustration of change if they become involved with computers.

Every computer, regardless of size and shape, is nothing more than a fast adding machine and an electronic filing cabinet. In comparison, the programmable calculator (Figure 32-4) is a fast adding machine with a relatively small filing cabinet. Programs for these calculators have been developed at Iowa State University to assist cattle feeders in analyzing their rations. The feeder enters the type and weight of cattle being fed and the amount and price of each feed included in the ration. The calculator then calculates: (1) what the cattle should gain, (2) the feed per pound of gain, and (3) total cost per pound of gain. These same calculations could be done with a pencil, long-hand; however it would take 20 minutes instead of 2 minutes.

The programs for the programmable calculator are stored on small strips of magnetic tape. Each tape serves as a storage mechanism for a program, much like a tape recording.

Figure 32-4. A programmable calculator with attached printer.

A microcomputer system consists primarily of a system of hardware and software. The hardware for microcomputers (Figure 32-5) consists of a terminal (a typewriter keyboard and an optional tv screen), a printer, data storage devices (which are the filing cabinets), main memory, and a central processing unit (CPU), which is the adding machine. The data storage devices can be either magnetic tape (similar to cassette tapes), floppy discs (look like 45 rpm records) or hard discs (large storage capacity in one unit). The devices that read and write the information on these discs are called "disc drives" or just "drives."

Software are the programs that make the hardware function. Each program typically consists of thousands of minutely detailed instructions that tell the computer exactly what data to manipulate mathematically, print, store, or display.

Good software (programs) are the most limiting factor in microcomputer utilization by livestock producers. Software developed for one microcomputer currently will not function on a different brand machine. Computer utilization is expected to increase rapidly as more compatible programs are developed. The four ways producers can obtain software are: (1) purchase a complete commercial package, (2) hire someone to do the programming, (3) learn to program the computer themselves, or (4) obtain it from their land grant university.

The three major areas where microcomputers have the largest potential for livestock producers are: (1) business accounting, (2) herd performance, and (3) financial management. A computer to handle these three functions would need a relatively large memory system and would cost approximately 4000 to 5000 dollars. Some producers have access to computer terminals with a telephone hook-up to a large main-frame computer. An example of such a system is AGNET (Agricultural Computer Network) which functions through the joint efforts of the Agricultural Extension Services of Wyoming, Nebraska, South Dakota, North Dakota, and Washington. An example of one of approximately 250 AGNET programs is COWCOST, which evaluates the cost and returns for beef cow-calf enterprises.

Figure 32-5. Microcomputers are becoming more commonly used by livestock producers.

Computers do not simplify a livestock business or cut overhead costs. They can make a business more complex and more costly. If wisely acquired and properly utilized, a computer can increase profits, but usually not by reducing expenses. The computer will not solve the problem of a poorly organized operation. Well-planned and efficient office management practices must exist and function smoothly before the addition of a computer will do anything but aggravate the organizational problem of the business. The computer cannot assist the producer in managing the business effectively unless the computer receives all the required data in precisely the prescribed manner. Unless the computer can provide the information needed by producers at the appropriate time and in an easily understood form, it should not be part of a management program, regardless of price.

A producer should follow six steps before deciding to purchase a computer. They are: (1) determine what are the most important management decisions that need to be made and what information is needed to effectively make these decisions, (2) see what programs are available to meet the management information needs, (3) decide what are the hardware needs to utilize the needed software, (4) contact local computer hardware dealers and evaluate the various alternatives and service, (5) estimate the cost/benefit ratio of the proposed management information system, and (6) make the final decision after evaluating the experiences of other producers who are using computers in their operations. Some producers are cautiously initiating computer-supported management by starting with a programmable calculator, with the intent of later expanding to a microcomputer.

MANAGEMENT SYSTEMS

Management systems analysis provides a method of systematically organizing the information needed to make valid management decisions. It permits variables to be more critically assessed and analyzed as regards their contribution to the total or the desired end point. Individuals who have been educated to think broadly in the framework of management systems can often make valid management decisions without the use of data processing equipment. A pencil, hand calculator, and a well-trained mind are the primary components required for making competent management decisions. Without question, however, the use of the computer in synthesizing voluminous amounts of information enhances the management system.

Resources are different for each operation; no fixed recipe exists for successful livestock production. The uniqueness of each operation results from variables such as different levels of forage production, varying marketing alternatives, varying energy costs, different types of animals, environmental differences, feed nutrients at various costs, varying levels of competence in labor and management, and others too numerous to mention. All of these variables and their interactions pose challenges to the producer who needs to combine them into sound management decisions for a specific operation.

It is a common practice to increase or maximize production of animals by using known biological relationships. Animal production typically has been maximized without careful consideration of cost benefit ratios and how the increased productivity relates to land, feed, and management resources. However, recognition is now being given to the need for optimization rather than maximization of animal productivity. Thus, there is an increased interest in management systems, which attempt to optimize production with net profit being the primary (and possibly the only) goal involved. As a result, some basic biological truths will not be useful or applicable because an economic analysis will prevent their inclusion in sound management decisions. For example, it is a well-known biological fact that calves born earlier in the calving season will have heavier weaning weights. Because of this relationship, some ranchers move the calving season to a period earlier in the year. However, management evaluation for a specific ranch demonstrated that by changing the calving season to match forage availability, profits were increased significantly. Changing calving season alone was estimated to increase the ranch's carrying capacity by 30 to 40%, in terms of animal units. In addition, the annual cow cost was reduced 18% per cow because of reduction in winter feed requirements.

During the past several years, research workers have increasingly applied management systems to livestock management. After a system is conceptualized, it is usually described by a set of mathematical equations and called a model. Models are constructed to simulate real-life situations, and attempts are made to validate them on that basis. Obviously, the models are no better than the data used to construct the models, and output data depends on the validity of the input data.

The farmer and rancher are limited in utilizing the computer in management systems because of the lack of software. Few programs being used combine production data with an economic assessment. However there is increased discussion about and movement toward making more management system software available to producers. Several projections indicate that programs will be used commonly in the next few years. Producers have to make management decisions now. To be competitive, livestock operations will have to be operated more like a business. This means keeping useful and accurate financial and production records to be used in making management decisions. The computer will become more important as a tool to improve the effectiveness of the management-decision process. However, until more usable computer programs become available, producers must use the traditional tools (pencil, calculator, and a keen mind) more effectively than in the past.

STUDY QUESTIONS AND SUGGESTIONS

1. What are the personal attributes of an effective manager?
2. Why should management plans be put on paper rather than kept in one's head?
3. What are the detailed steps in the planning process?
4. What are the important parts of financial planning for livestock producers?
5. What is the meaning of computer hardware and software?
6. Why does the same management plan not work for different operations?
7. How should producers decide whether to incorporate a computer into their operations?
8. What is the meaning of management systems?
9. How do producers decide which management alternatives are most useful to their specific operations?
10. What is the difference between a programmable calculator and a microcomputer?

SELECTED REFERENCES

Bartlett, E. T., and Clawson, W. J. 1978. Profit, meat production, or efficient use of energy in ranching. *Journal of Animal Science* 46:812.

Cook, C. W., Bartlett, E. T., and Evans, G. R. 1974. *A Systems Approach to Range Beef Production.* Colorado State University Range Science Department Science Series No. 15.

Evans, G. R. 1978. Systems approach for land resource analysis and planning of limited renewable natural resources. *Journal of Animal Science* 46:819.

Fitzhugh, H. A. 1978. Bioeconomic analysis of ruminant production systems. *Journal of Animal Science* 46:797.

Harl, N. 1980. *Farm Estate and Business Planning.* 6th Edition. Skokie, IL: Century Communications, Inc.

Hughes, H. 1981. *The Computer Explosion.* Proceedings, The Range Beef Cow, A Symposium on Production, VII. Rapid City, SD, December.

Hughes, H. 1981. Six steps to take in making a decision to buy a computer. *BEEF,* December.

Joandet, G. E., and Cartwright, T. C. 1975. Modeling beef production systems. *Journal of Animal Science* 41:1238.

Killcreas, W. E., and Hickel. R. 1982. *A Set of Microcomputer Programs for Swine Record Keeping and Production Management.* Mississippi Agricultural and Forestry Experiment Station AEC Tech. Public. No. 35.

Klinefelter, D. A., and Hottel, B. 1981. *Farmers' and Ranchers' Guide to Borrowing Money.* Texas Agric. Expt. Station MP-1494.

Luft, L. D. Sources of credit and the cost of credit. *Great Plains Beef Cattle Handbook,* GPE-4351 and 4352.

Maddux, J. 1981. *The Man in Management.* Proceedings, The Range Beef Cow, A Symposium on Production, VII. Rapid City, SD, December.

McGilliard, M. L. 1979. Relationships among genetic goals and herd-breeding expenses. *Journal of Dairy Science* 62:85.

Miller, W. C., Brinks, J. S., and Sutherland, T. M. 1978. Computer-assisted management decisions for beef production systems. *International Journal of Agricultural Systems* 3:147.

Natter, D. R., Sanders, J. O., Dickerson, Smith, G. R., and Cartwright, T. C. 1979. Simulated efficiency of beef production for a midwestern cow-calf-feedlot management system. III. Crossbreeding systems. *Journal of Animal Science* 49:92.

Nelson, T. 1981. *Software — Is It Available?* Proceedings, The Range Beef Cow, A Symposium on Production, VII. Rapid City, SD, December.

Ott, G. Planning for profit with partial budgeting. *Great Plains Beef Cattle Handbook,* GPE-4551.

Smith, G. M., and Harrison, V. L. 1978. The future of livestock systems analysis. *Journal of Animal Science* 46:807.

Smith, N. E. and Larue, E. L. 1974. *Interface of Animal Production and Economic Systems.* Federation Proceedings 33:198.

Stonaker, H. H. 1975. Beef production systems in the tropics. I. Extensive production systems on infertile soils. *Journal of Animal Science* 41:1218.

Trede, L. D., Boehlje, M. D., and Geasler, M. R. 1977. Systems approach to management for the cattle feeder. *Journal of Animal Science* 45:1213.

Whitson, R. E., and Kay, R. D. 1978. Beef cattle forage systems analysis under variable prices and forage conditions. *Journal of Animal Science* 46:823.

Wilt, R. W. Jr., and Bell, S. D. 1977. *Use of Financial Cash Flow Statements as a Financial Management Tool.* Alabama Agric. Expt. Station Bulletin 487.

_____. 1981. Examples of benefits to farmers and ranchers resulting from new tax legislation. *National Cattlemen's Association Beef Business Bulletin,* September 4.

_____. 1981. What new tax legislation means to farmers and ranchers. *National Cattlemen's Association Beef Business Bulletin,* August 14.

33

Careers and Career Preparation in the Animal Sciences

The millions of domestic animals that provide food, fiber, and recreation for mankind create many and varied types of career opportunities. A placement survey of approximately 25,000 agriculture graduates of state colleges and universities is shown in Tables 33-1 and 33-2. Animal sciences placement data follow a similar pattern because they compose a high percent of the graduates shown in these tables.

Beef cattle, dairy cattle, horses, poultry, sheep, and swine are the animals of primary importance in animal science curricula, whereas reproduction, nutrition, breeding (genetics), meats, and live animal appraisal are the specialized topics typically covered in animal science courses. A few animal science programs include studies of pet and companion animals. Most college and university curricula in the animal sciences are designed to assist students in the broad career areas of production, science, agribusiness, and the food industry. The major careers in each of these areas are shown in Table 33-3.

Table 33-1. Post-graduation activities of bachelor degree recipients in agriculture

Post-graduation activity	Percent of graduates
Agribusiness (industry)	33
Graduate and professional study	22
Farming and ranching	13
Education, including extension	8
Government (national, state, and local)	11
Miscellaneous (not placed or not seeking employment)	13
	100

Source: National Association of State Universities and Land Grant Colleges, 1980.

372

Table 33-2. Average starting salaries of agriculture graduates from state universities

Graduate	Average annual starting salary
Bachelor of Science (B.S.)	$12,900
Master of Science (M.S.)	15,600
Doctor of Philosophy (Ph.D.)	20,700

Source: National Association of State Universities and Land Grant Colleges, 1980.

Table 33-3. Animal sciences careers in production, science, agribusiness, and the food industry

Animal Sciences			
Production	**Science**	**Agribusiness**	**Food Industry**
feedlot positions	To graduate school for MS and Ph.D degrees	sales and management positions with feed companies, packing companies, drug and pharmacy companies, equipment companies, etc.	food processing plants
livestock production operations (beef, dairy, swine, sheep and horses)	research (university or industry in nutrition, reproduction, breeding and genetics, products, and production management		food ingredient plants
ranch positions			food manufacturing plants
breed associations		livestock publications	government — protection and regulatory agencies
AI studs and breeding	university or college teaching	advertising and promotion	government — Dept. of Defense (food supply and food service)
livestock buyers for feeders and packers	university extension and area extension	finance (PCAs, banks, etc.)	government — Dept. of Agriculture (Research and Information)
county extension agents	management positions in industry	public relations	university research, teaching, and extension
meat grading and handling distribution	government work	meat grading (federal government)	positions in food companies
marketing (auctions, Cattle Fax, livestock sales management, etc.)	international opportunities	international opportunities	research and development with food companies
international opportunities	consulting	graduate school in business administration	
livestock and meat market reporting (government)	to veterinary school for DVM	foreign agriculture	
riding instructors	private practice	technical sales and service	
feed manufacturing	consulting	positions in poultry production units	
	university teaching and research		
	meat inspection		

PRODUCTION

Many individuals are intrigued with animal production because of the possibility of being their own boss and working directly with animals. These ambitions may be somewhat idealistic because the financial investment required in land is typically in the hun-

Figure 33-1. There are numerous career opportunities in the production of the various species of farm animals. Some students interested in the production area will not be directly involved in animal production. These students will work for companies, associations, or groups closely associated with the animal producers.

dreds of thousands of dollars. Beef cattle, dairy, sheep, and swine production operations continue to become fewer in number, more specialized, and larger in terms of the amount of capital invested. Many of those who find a career in the production area have a family operation to which they can return. A few others have the needed capital to invest (Figure 33-1).

Some students who seek employment in animal production are disillusioned with the base salary, benefits, and working hours when production work is compared with jobs in the agribusinesses. However, there are those who are anxious and willing to obtain experience, prove they can work, and eventually locate permanent employment in production operations. Some producers of cattle, sheep, swine, and horse operations allow an employee to buy into an operation on a limited basis or own some of the animals after a year or two in the business.

Certain careers require education and experience in the production area even though the individual will not be producing animals directly. Career opportunities such as breed association and publication, fieldmen, extension agents, and livestock marketing require an understanding of production as one works directly with livestock producers.

SCIENCE

An animal science student who concentrates heavily in science courses is usually preparing for further academic work with a goal of achieving advanced degrees. A minimum grade average of B is usually required for admission into graduate school or a professional veterinary medicine program. Advanced degrees are usually obtained after entrance into graduate school or after being admitted to a professional veterinary medicine program where a Doctor of Veterinary Medicine (D.V.M.) degree is awarded. Students accepted into graduate school progress toward a Master of Science (M.S.) and possibly further progress towards a Doctor of Philosophy (Ph.D.) degree. While the two-year college certificate or Bachelor of Science (B.S.) degree allows breadth of education, advanced degrees are generally directed to a specialization. These specialized animal science areas are typically in nutrition, reproduction, breeding (genetics and statistics), animal products, and less frequently in production (management) (Figures 33-2, 33-3, 33-4, and 33-5).

AGRIBUSINESS

Although the number of individuals who work to produce livestock and poultry production has decreased, the number of individuals and businesses serving producers has greatly increased. Positions in sales, management, finance, advertising, public relations, and publications are prevalent in feed, drug, equipment, packing, and livestock organizations. Agribusiness careers which relate to the livestock industry require a student to have a foundation of knowledge of livestock along with an excellent comprehension of business, economics, and effective communication. He or she should understand people, know how to communicate with people, and enjoy working with people. Extracurricular activities that give students experience in leadership and working with people are invaluable for meaningful career preparation in the agribusiness area.

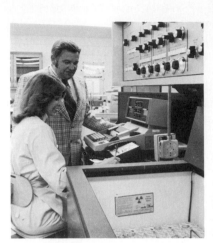
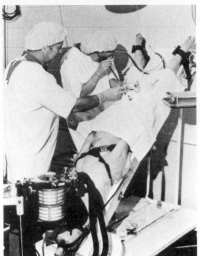

Figure 33-2. Many students concentrating in the sciences are preparing themselves for entrance into graduate school. There they will complete M.S. and Ph.D. degrees, which are primarily research-oriented degrees. A few examples are depicted here.

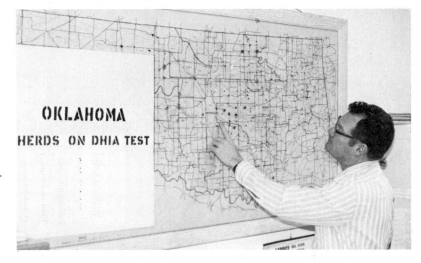

Figure 33-3. A Ph.D. is usually required for teaching in a college or university.

Figure 33-4. University extension personnel need a Ph.D. degree so they can interpret the current research and apply it in off-campus educational programs.

Figure 33-5. Students will usually have a B.S. degree before applying for admission to the College of Veterinary Medicine. Students with high academic performance in the science aspect of animal science are typically well prepared for completing the D.V.M. degree.

Figure 33-6. The demand for animal products creates many industries that support the animals and process the products for human consumption.

FOOD INDUSTRY

Red meats, poultry, milk, and eggs, the primary end products of animal production, provide basic nutrition and eating enjoyment for millions of consumers. Career opportunities are numerous in providing one of mankind's basic needs—food. Processing, packaging, and distribution of food are important components of the food production chain. New food products are continually being produced, and new methods of manufacturing and fabricating food are being developed. Electrical stimulation of carcasses for increased tenderization, vacuum packaging of primal cuts for a longer shelf life, and blade tenderization are examples of recent innovations in meat processing. A vital and continuing challenge over the next several decades will be to provide a food supply that is nutritious, safe, convenient, attractive, and economical while providing the desired eating satisfaction to the consumer (Figure 33-6).

INTERNATIONAL OPPORTUNITIES

Much has been written and said during the past decade of the challenge to provide adequate nutrition to an ever-expanding world population. Many countries have tremendous natural resources for expanded food production but lack technical knowledge and adequate capital to develop these resources. Federal government programs, designed to help foreign countries help themselves, offer several career opportunities in organizations such as the Peace Corps, Vista, and USAID. Many of these opportunities of assisting people in agricultural production in developing countries are open to individuals educated and experienced in the animal sciences.

The Foreign Agricultural Service of the USDA employs individuals in animal economics, marketing, administration, as attachés, and as international secretaries. Animal science students should take several courses in economics, marketing, foreign languages, and business administration if they wish to qualify for these positions.

Certain private industries offer opportunities in animal production and related businesses in foreign countries. Multi-national firms that develop livestock feed companies, drug and pharmaceutical companies, companies that export and import animals and animal products, and consulting companies are some examples.

MANAGEMENT POSITIONS

Management and administrative positions exist in production, science, agribusiness, and food industry areas of animal science. Because management generally implies less work and more pay than other positions that require more physical labor, young men and women typically desire management positions. Nevertheless, although management and administrative salaries are usually higher, the work load and pressures are usually much greater.

Many college graduates have the impression that once the degree is in hand, it automatically qualifies them for a management position. In actuality, management positions are usually earned based on an individual's proven ability to solve problems and make effective decisions. After being employed for a few years, individuals having previously learned the academic principles must also experience the component parts of the operation or business as well. For example, presidents or vice-presidents of feed companies have typically started their initial careers in feed sales. Most successful managers have had a continual series of learning experiences since the end of their formal education. Managers need to understand all aspects of the business they are managing, particularly the products that are produced and sold.

WOMEN IN ANIMAL SCIENCE CAREERS

Women are pursuing and finding job opportunities in all areas of the animal sciences. The number of women majoring in animal science has been growing steadily in most colleges and universities. In the early 1970s, the number of female graduates in the animal sciences was 20 to 25% of the total graduates in most schools. At the present time, nearly one-half of the graduates are women (Figure 33-7).

This trend is not so much a part of the Women's Liberation Movement as it is an equal opportunity crusade. City-raised women as well as those raised on a farm are looking for career opportunities in which they can work with animals or in some business associated with the livestock and poultry industries. Many agricultural employers, especially in the agribusinesses, are now actively recruiting women with animal science degrees.

During the 1970s, some women found career opportunities that many felt were reserved for men only. Although some men expressed some fear and had serious reservations, many of these women proved they could perform these jobs quite competently. It does, however, appear that some employers in private production agriculture will continue to be reluctant to hire women on the same basis as men. Nevertheless, women interested in careers in the animal sciences should prepare themselves adequately in terms of both education and experience. Challenging career opportunities await both women and men who are adequately prepared and are willing to work hard.

Recent observations in a California study showed that women graduates in agriculture were at a disadvantage in salary and job status when compared to men graduating with similar degrees. However, a Colorado State University survey of its women graduates in animal sciences since 1979, revealed that many of them were in high-paying careers. The survey clearly demonstrated that women desiring employment in the production area had struggled in identifying meaningful employment. The agribusiness area (e.g., Farmer's Home Administration, feed companies, food processing companies, and drug and pharmaceutical companies) had provided attractive careers for women graduates, as several reported annual salaries above $20,000.

CAREER PREPARATION

An education should allow an individual to obtain the tools necessary to earn a living, continue learning, and live a full, productive life even beyond the typical retirement years. A broadbased education is important because an individual may be preparing for a job that does not even exist at the present time. Change has occurred and will no doubt continue to occur, so a person needs to be flexible and adaptable and must be willing to continue learning throughout life.

Occasionally a person chooses a career for which he or she has little real talent or for which she or he is not qualified or properly motivated. This person may spend years in frustrating preparation or achieving partial success when he or she could have been

A

B

C

Figure 33-7. Women are seeking and finding career opportunities in the animal sciences. Some examples shown are: **(A)** Teaching equitation; **(B)** Veterinarian's assistant; **(C)** Feedlot manager; **(D)** Researcher; **(E)** Livestock photographer.

D

E

outstanding in another area. In animal sciences, students preparing for acceptance into veterinary medicine, graduate school, or another professional school seem to experience this frustration the most. Most people are capable of success in several different careers if they receive the appropriate education and experience. A career choice should be consistent with a person's interest, desire, and motivation. A career goal should be established with flexibility for change if it proves to be unrealistic.

After tentatively choosing a career area, it should be pursued vigorously with sustained personal motivation. Students should write and visit with individuals currently working in the student's chosen career area. Every personal visit or written letter should request another lead or source of further information. Find out as early as possible if your chosen career area is in reality what you originally perceived it to be. Think realistically about the facts you find out and do not get caught up in the glamor, glory, or imaginations you might have about your career choice. Career interest tests can be helpful in identifying areas to pursue in more detail. However, these tests are not infallible. They are only helpful tools to give direction but not to give the final answer. You should always temper the results with additional information from several sources before you make a decision.

A nationwide survey conducted by an American Society of Animal Science committee identified the primary concerns that employers had concerning animal science graduates as the graduates started on their first job. In descending order of importance, these concerns were: lack of practical knowledge and experience, lack of communicative skills, lack of drive and initiative, overly high expectations, failure to assume responsibility, and lack of business knowledge. Students preparing for an animal science career should obtain knowledge and experiences that will allow them to overcome these common deficiencies.

Employers hire people, not a college major or area of study. Your educational background is only part of you. Therefore, in addition to academic preparation, students should develop their marketable personal assets such as leadership, problem solving, ability to communicate and work with people, and a desire to work hard consistently. One develops many of these personal characteristics through personal experience, part-time and summer jobs, and extracurricular activities. Internships are especially valuable in obtaining exposure to a career area and developing personal abilities and skills through practical experience. A combination of useful courses, practical experience, effective communication, and excellent work skills is the best assurance of finding and keeping a meaningful career in the animal sciences.

SELECTED REFERENCES

Coulter, K. J. *Graduates of Higher Education in the Food and Agricultural Sciences.* USDA Miscellaneous Publication Number 1385. July 1980.

Wood, J. B., Dupre, D. H., and Thompson, O. E. *Women in the Agricultural Labor Market.* Sept.-Oct. 1981, p. 16.

_____. *Employer Reactions to Animal Science Graduates and Their Qualifications.* Report II. National Survey by the ASAS Committee on Future Career Opportunities for Animal Science Graduates. July 29, 1979.

_____. 1979 *Survey of Agricultural Degrees Granted and Post Graduation Activities of Graduates.* National Association of State Universities and Land Grant Colleges. July 1980.

Glossary

abomasum: One of the stomach components of ruminant animals that corresponds to the true stomach of monogastric animals.

abortion: To give birth before the fetus is viable.

absorption: The passage of liquid and digested (soluble) food across the gut wall.

accessory organs: The seminal vesicles, prostate, and Cowper's glands in the male. These glands add their secretions to the sperm to form semen.

ad libitum: Allowing animals to eat all they want at all times.

agnostic behavior: Combat, or fighting, behavior, which includes defense (submission), offense (aggression), escape, and passiveness.

AI: Abbreviation for artificial insemination.

alleles: Genes occupying corresponding loci in homologous chromosomes that affect the same hereditary trait but in different ways.

allelomimetic behavior: Doing the same thing.

amino acid: An organic acid in which one or more of the hydrogen atoms has been replaced by the amino group ($-NH_2$). Amino acids are the building blocks in the formation of proteins.

ampulla: The dilated or enlarged upper portion of the vas deferens in bulls, bucks, and rams, where sperm are stored for sudden release at ejaculation.

anabolic: A constructive, or "building up," process.

anatomy: The science of the structure of the animal body and the relation of its parts.

androgen: A male sex hormone, such as testosterone.

anemia: A condition in which the blood is deficient in quantity of red blood cells or in its hemoglobin content.

anterior: Situated in front of, or toward the front part of, a point of reference. Toward the head end of an animal.

antibiotic: A product produced by living organisms, such as yeast, which destroys or inhibits the growth of other organisms, especially bacteria.

antibody: A specific protein substance formed by and in an animal as a reaction to the presence of an antigen. Antibodies may cause flocculation, lysis, or inactivation of antigens.

antigen: A substance which, when introduced into the blood or tissues, causes the formation of antibodies. Antigens may be toxins or native proteins.

antihormone: An antibodylike substance developed when protein hormones are injected into an animal over a period of time. Antihormones inactivate hormones.

arteritis: An inflammation of the arteries.

artificial insemination: The introduction of semen into the reproductive tract (usually the cervix or uterus) of the female by a technician.

artificial vagina: A device used to collect semen from a male while he mounts in a normal manner to copulate. The male ejaculates into this device, which simulates the vagina of the female in pressure, temperature, and sensation to the penis.

ascaris: Any of the genus (Ascaris) of parasitic roundworms.

assimilation: The process of transforming food into living tissue.

barrow: A male swine that was castrated prior to reaching puberty.

basal metabolism: The chemical changes that occur in an animal's body when the animal is in a thermoneutral environment, resting, and in a postabsorptive state. It is usually determined by measuring oxygen consumption and carbon dioxide production.

beef: The meat from cattle (bovine species) other than calves (the meat from calves is called veal).

behavior: The reaction of an animal to internal and external stimuli, which is an attempt by the animal to adjust or adapt to the situation.

beri-beri: A disease caused by a deficiency of Vitamin B_1.

blemish: Any defect or injury that mars the appearance of, but does not impair the usefulness of, an animal.

blowflies: Any of the family of two-winged flies that deposit eggs or living larvae in open wounds.

bots: Any of a number of related flies whose larvae are parasitic in horses and sheep.

B.L.U.P.:   Best Linear Unbiased Prediction method for estimating breeding value of dairy bulls.

boar: A male swine of breeding age. This term sometimes denotes a male pig, which is called a boar pig.

bog spavin: A soft enlargement of the anterior, inner aspect of the hock.

bolus: (1) Regurgitated food. (2) A large pill for dosing animals.

bone spavin: A bony (hard) enlargement of the inner aspect of the hock.

break joint: Denotes a lamb carcass where the foot and pastern are removed at the cartilaginous junction of the front leg.

breech: The buttocks. A breech presentation at birth is one in which the rear portion of the fetus is presented first.

breeding value: A measure of genetic potential of an animal by combining into one number several performance values, on one trait, that have been accumulated on an animal and the animal's relatives.

brisket disease: A non-infectious disease of cattle characterized by congestive right heart failure. It affects animals residing at high altitudes (usually above 7,000 ft).

British thermal unit: The quantity of heat required to raise the temperature of 1 lb of water 1 °F at or near 39.2 °F.

broiler: A young chicken of either sex (usually 9 to 12 weeks of age) that is tender meated and has smooth-textured skin and flexible breast-bone cartilage.

broodiness: The desire of a female bird to sit on eggs (incubate).

browse: Woody or brushy plants. Livestock feed on tender shoots or twigs.

buck: A male sheep or goat. This term usually denotes animals of breeding age.

bulbourethral (Cowper's) gland: An accessory gland of the male that secretes a fluid which constitutes a portion of the semen.

bull: A bovine male. The term usually denotes animals of breeding age.

bulldog: Large bulging head and bulging through the shoulders.

bullock: A young bull, typically under 20 months of age.

buttermilk: The fluid remaining after butter has been made from cream and removed. By use of bacteria, cultured buttermilk is also produced from milk.

calcification: Deposits of calcium in the tissues.

calf: A young male or female bovine animal under a year of age.

calorie: The amount of heat required to raise the temperature of 1 g of water from 15 °C to 16 °C.

candling: The shining of a bright light through an egg to see if it contains a live embryo.

cannibalism: The eating by animals of other animals of the same kind.

canter: A slow, easy gallop.

capped hocks: Hocks that have hard growths that cover, or "cap," their points.

carbohydrates: Any foods, including starches, sugars, celluloses, and gums, that are broken down to simple sugars through digestion.

carcass merit: The value of a carcass for consumption.

carnivorous: Subsisting or feeding on animal tissues.

carotene: The orange pigment found in carrots, leafy plants, yellow corn, and other feeds, which can be broken down to form two molecules of vitamin A.

casein: The major protein of milk.

cash flow statement: A form showing the monthly itemized income and expenses (cash in and cash out) for an operation or business.

castrate: **(1)** To remove the testicles. **(2)** An animal that has had its testicles removed.

cataract: An opacity of the crystalline eye lens or its capsule.

cecum: A blind pouch at the anterior end of the colon.

cervix: The portion of the female reproductive tract between the vagina and the uterus. It is usually sealed by thick mucus except when the female is in estrus or delivering young.

chromosome: A rodlike or stringlike body found in the nucleus of the cell that is darkly stained by chrome dyes.

chyme: The thick, liquid mixture of food that passes from the stomach to the small intestine.

chymotrypsin: A milk-digesting enzyme contained in the pancreas.

clitoris: An organ in females inside the ventral part of the vulva which is homologous to the penis in the male. It is highly sensory.

clutch: Eggs layed by a hen on consecutive days.

coccidia: A protozoan organism that causes an intestinal disease called coccidiosis.

coccidiosis: A morbid state caused by the presence of organisms called coccidia, which belong to a class of sporozoans.

cod: Scrotal area of steer remaining after castration.

coefficient of determination: A percent of variation in one trait that is accounted for by variation in another trait.

colon: The large intestine from the end of the ileum and beginning with the cecum to the anus.

colostrum: The first milk given by a female following delivery of her young. It is high in antibodies that give the young protection from invading microorganisms.

colt: A young male of the horse or donkey species.

compensatory growth: Increased growth rate in response to previous undernourishment that an animal has recently experienced.

computer: An electronic machine which by means of stored instructions and information performs rapid, often complex, calculations, or compiles, correlates, and selects data.

concentrate: A feed that is high in energy, low in fiber content, and highly digestible.

conditioning: The treatment of animals by vaccination and other means prior to putting them in the feedlot.

conformation: The physical form of an animal; its shape and arrangement of parts.

contracted heels: A condition in which the heels of a horse are pulled in so that expansion of the heel when the foot strikes the ground cannot occur.

convection: The cooling or warming of an animal by air currents.

correlation coefficient: A measure of the association of one trait with another.

corpus luteum: A yellowish body in the mammalian ovary. The cells that were follicular cells develop into the corpus luteum, which secretes progesterone. It becomes yellow in color from the yellow lipids that are in the cells.

cow: A sexually mature female bovine animal.

cow hocks: A condition in which the hocks are close together but the feet stand apart.

cotyledon: An area where the placenta and the uterine lining are in close association such that nutrients can pass to and wastes can pass from the circulation of the developing young.

creep: An enclosure in which young can enter to obtain food but larger animals cannot enter.

crimp: The waves, or kinks, in a wool fiber.

crossbreeding: A term generally used to mean the crossing of two breeds of farm animals. It sometimes denotes the crossing of lines, breeds, or species of animals.

cryptorchidism: The retention of one or both testicles in the abdominal cavity in animals that typically have the testicles hanging in a scrotal sac.

cull: To eliminate from breeding or to prevent from leaving genes in the population.

curb: A hard swelling that occurs just below the point of the hock.

curd: Coagulated milk.

cutability: Fat, lean, and bone composition of meat animals. Used interchangeably with yield grade.

cutting chute: A narrow chute that allows animals to go through in single file with gates such that animals can be directed into pens along the side of the chute.

debeaking: The removal of a portion of the beak of chickens.

delayed implantation: The delayed attachment of a fertilized egg to the wall of the uterus after fertilization takes place.

deoxyribonucleic acid (DNA):   A complex molecule consisting of deoxyribose, phosphoric acid, and four bases (two purines, adenine and guanine and two pyrimidines, thymine and cytosine). It is the coding mechanism of inheritance.

digestibility: The quality of being digestible. If a high percentage of a given food taken into the digestive tract is absorbed into the body, that food is said to have high digestibility.

digestion: The reduction in particle size of food so that the food becomes soluble and can pass across the gut wall into the vascular or lymph system.

diploid: Having the normal, paired chromosomes of somatic tissue as produced by the doubling of the primary chromosomes of the germ cells at fertilization.

disease: Any deviation from a normal state of health.

disinfectant: A chemical that destroys disease-producing microorganisms or parasites.

dock: **(1)** To cut off the tail. **(2)** The remaining portion of the tail of a sheep that has been docked.

doe: A female goat.

dominance: A situation in which one gene of an allelic pair prevents the phenotypic expression of the other member of the allelic pair.

dominant gene: A gene that overpowers and prevents the expression of its recessive allele when the two alleles are present in a heterozygous individual.

dressing percentage: The percentage of the live animal that becomes the carcass at slaughter. It is determined by subtracting the losses due to removal of blood, hide, and intestines from live-animal weight, dividing that quantity by live-animal weight, and multiplying by 100.

dry cow: A cow that is not presently producing milk.

dubbing: The removal of a part or all of the soft tissues (comb and wattles) of chickens.

dung: The feces (manure) of farm animals.

dwarfism: The state of being abnormally undersized. Two kinds of dwarfs are recognized; one is proportionate and the other is disproportionate.

dystocia: Difficult birth where the female will require some assistance at parturition.

ectoderm: The outer of the three layers of the primitive embryo.

edema: The presence of abnormally large amounts of fluid in the intercellular tissue spaces of the body.

ejaculation: A discharge of semen from the male.

endocrine gland: A ductless gland that secretes a hormone into the bloodstream.

endoderm: The inner of the three layers of the primitive embryo.

energy: The force, or power, that is used to drive a wide variety of systems. It can be used as motive power in animals, but most of it is used as chemical energy to drive reactions necessary to convert feed into animal products and to keep the animal warm.

enterotoxemia: A toxin-caused disease of the intestinal tract caused by bacteria. Its symptoms are characteristic of food poisoning.

entropion: Turned-in eyelids.

environment: The sum total of all external conditions that affect the well-being and performance of an animal.

enzyme: A complex protein produced by living cells that causes changes in other substances in the body without being changed itself and without becoming a part of the product.

epididymis: The long, coiled tubule leading from the testis to the vas deferens.

epiphysis: A piece of bone separated from a long bone in early life by cartilage, which later becomes part of the larger bone.

epistatic: A term designating a gene that interacts with other genes with which it is not allelic.

equine encephalomyelitis: An inflammation of the brain of horses.

eruction: The elimination of gas by belching.

essential nutrient: A nutrient that cannot be synthesized by the body, but must be supplied in the ration.

estrogen: Any hormone (including estradiol, estriol, and estrone) that causes the female to come physiologically into heat and to be receptive to the male. Estrogens are produced by the follicle of the ovary and by the placenta.

estrous: An adjective meaning "heat," which modifies such words as "cycle." The estrous cycle is the heat cycle, or time from one heat to the next. The Americanized term is estrual (see estrus).

estrous synchronization: Controlling the estrous cycle so that a high percentage of the females in the herd express estrus at approximately the same time.

estrus: The period of mating activity in the female mammal. The time when the female has a strong sexual urge. It is also called heat.

eviserate: The removal of the internal organs during the slaughtering process.

ewe: A sexually mature female sheep. A ewe lamb is a female sheep prior to sexual maturity.

family selection: Selection based on performance of a family.

farrow: To deliver, or give birth to, pigs.

fat: Adipose tissue.

feather picking: The picking of feathers from one bird by another.

feed bunk: A manger arrangement for feeding (usually hay) to farm animals.

feed efficiency: **(1)** The amount of feed required to produce a unit of gain in weight; for poultry, this term can also denote the amount of feed required to produce a given quantity of eggs. **(2)** The amount of gain made per unit of feed.

felting: The intermingling of wool fibers to produce woolen cloth.

fertility: The capacity to initiate, sustain, and support reproduction. With reference to poultry, the term typically refers to the percentage of eggs which, when incubated, show some degree of embryonic development.

fertilization: The process in which a sperm unites with an egg to produce a zygote.

fetus: A young organism in the uterus from the time that the organ systems have been formed until it is born.

fill: The contests of the digestive tract.

filly: A young female horse.

finish: The degree of fatness of an animal.

fistula: A running sore at the top of the withers of a horse, resulting from a bruise followed by invasion of microorganisms.

fleece: The wool from all parts of a sheep.

flushing: Having females fed to gain in condition for stimulating greater rates of ovulation and conception.

fly strike: An infestation with large numbers of maggots hatched from eggs laid by blowflies.

foal: A young male or female horse.

follicle: A blisterlike, fluid-filled sac in the ovary that contains the egg.

follicle-stimulating hormone: A hormone produced and released by the anterior pituitary that stimulates the development of the follicle in the ovary.

footrot: A disease of the foot in sheep and cattle. In sheep it causes rotting of tissue between the horny part of the foot and the soft tissue underneath.

forb: Weedy or broad-leaf plants, as contrasted to grasses, that serve as pasture for animals.

forging: The striking of the heel of the front foot with the toe of the hind foot by a horse in action.

freshen: To give birth to young and initiate milk production. This term is usually used with reference to dairy cattle.

gallop: A three-beat gait in which each of the two front feet and both of the hind feet strike the ground at different times.

gametes: Male and female reproductive cells. The sperm and the egg.

gametogenesis: The process by which sperm and eggs are produced.

gelding: A male horse that has been castrated.

gene: An active area in the chromosome that codes for a trait and determines how a trait will develop.

general combining ability: The ability of individuals of one line or population to combine favorably or unfavorably with individuals of several other lines or populations.

genotype: The genetic constitution, or makeup, of an individual. For any pair of alleles, three genotypes (e.g. *AA, Aa,* and *aa*) are possible.

gestation: The time from breeding or conception of a female until she gives birth to her young.

gilt: A young female swine prior to the time that she has produced her first litter.

globulin: A protein characterized by being insoluble in water and alcohol but soluble in a 0.5% to 1.0% solution of a neutral salt, from which solution it can be precipitated by heat.

goiter: Enlargement of the thyroid gland, usually caused by iodine deficient diets.

gonad: The testis of the male; the ovary of the female.

grading up: The continued use of purebred sires of the same breed in a grade herd or flock.

gross energy: The amount of heat, measured in calories, produced when a substance is completely oxidized. It does not reveal the amount of energy that an animal could derive from eating the substance.

growth: The increase in protein over its loss in the animal body. Growth occurs by increases in cell numbers, cell size, or both.

habituation: The gradual adaptation to a stimulus or to the environment.

hand mating: Bringing a female to a male for service (breeding), after which she is removed from the area where the male is located.

hank: A measurement of the fineness of wool. A hank is 560 yds of yarn. Fine wools spin more than coarse wools; thus, the spinning count for fine wools is high.

haploid: One-half of the diploid number of chromosomes for a given species, as found in the germ cells.

hatchability: A term that indicates the percentage of a given number of eggs set from which viable young hatch, sometimes calculated specifically from the number of fertile eggs set.

heat increment: The increase in heat production following consumption of feed when an animal is in a thermoneutral environment. It includes additional heat generated in fermentation, digestion, and nutrient metabolism.

heaves: A respiratory defect in horses in which the animal has difficulty completing the exhalation of inhaled air.

heifer: A young female bovine cow prior to the time that she has produced her first calf.

hemoglobin: The iron-containing pigment of the red blood cells. It carries oxygen from the lungs to the tissues.

hen: An adult female domestic fowl, such as a chicken or a turkey.

herbivorous: Subsisting or feeding on plants.

heritability: The portion of phenotypic variation that is accounted for by additive gene action.

hernia: The protrusion of some of the intestine through an opening in the body wall (also commonly called rupture). Two types, umbilical and scrotal, occur in farm animals.

heterosis: The amount by which the F_1 generation exceeds the P_1 generation for a certain trait (also called hybrid vigor). The animal breeder usually speaks of the amount of superiority the crossbred has over the straightbreds.

heterozygous: A term designating an individual that possesses unlike genes for a particular trait.

hinny: The offspring that results from crossing a stallion with a female donkey (jenny).

hobble: To tie the front legs of an animal together, or to tie the hind legs to a rope run between the front legs and over the shoulder. An animal is hobbled to prevent it from kicking.

homeotherm: A "warm-blooded" animal. An animal that maintains its characteristic body temperature even though environmental temperature varies.

homogenized: Milk that has had the fat droplets broken into very small particles so that the milk fat stays in suspension in the milk fluids.

homologous: Corresponding in type of structure and derived from a common primitive origin.

homologous chromosomes: Chromosomes having the same size and shape that contain genes affecting the same characters. Homologous chromosomes occur in pairs in typical diploid cells.

homozygous: A term designating an individual whose genes for a particular trait are alike.

hormone: A chemical substance secreted by a ductless gland. This substance is usually carried by the bloodstream to other places in the body where it has its specific effect.

hydrocephalus: A condition characterized by an abnormal increase in the amount of cerebral fluid, accompanied by dilation of the cerebral ventricles.

hypertension: High blood pressure.

hypothalamus: A portion of the brain found in the floor of the third ventricle. It regulates body temperature and has other functions.

hypoxia: A condition resulting from deficient oxygenation of the blood.

immunity: The ability of an animal to resist or overcome an infection to which most members of its species are susceptible.

imprinting: Learning associated with maturational readiness.

inbreeding: The mating of individuals who are more closely related than the average individuals in a population. Inbreeding increases homozygosity in the population but it does not change gene frequency.

incubation period: The time that elapses from the time an egg is placed into an incubator until the young is hatched.

index: **(1)** An overall merit rating of an animal. **(2)** A method of predicting the milk-producing ability that a bull will transmit to his daughters.

influenza: A virus disease characterized by inflammation of the respiratory tract, high fever, and muscular pain.

inheritance: The transmission of genes from parents to offspring.

instinct: Inborn behavior.

integration: The bringing together of all segments of a livestock production program under one centrally organized unit.

intelligence: The ability to learn to adjust successfully to situations.

interference: The striking of the supporting leg by the foot of the striding leg by a horse in action.

interstitial cells: The cells between the seminiferous tubules of the testicle that produce testosterone.

intravenous: Within the vein. An intravenous injection is an injection into a vein.

jenny: The female donkey.

ked: An external parasite that affects sheep. Although commonly called "sheep tick," it is actually a wingless fly.

kemp: Coarse, hairlike fibers in wool.

ketosis: A condition (also called acetonemia) which is characterized by high concentration of ketone bodies in the body tissues and fluids.

kilocalorie (kcal, Kcal):   An amount of heat equal to 1,000 calories (*see calorie*). Also called Calorie.

lactalbumin: A nutritive protein of milk.

lactation: The secretion and production of milk.

lactose: Milk sugar. When digested, it is broken down into one molecule of glucose and one of galactose.

lamb: **(1)** A young male or female sheep, usually an individual less than 10 months of age. **(2)** To deliver, or give birth to, a lamb.

lambing jug: A small pen in which a ewe is put for lambing. It is also used for containing the ewe and her lamb until the lamb is strong enough to run with other ewes and lambs.

legume: Any plant of the family *Leguminosae,* such as pea, bean, alfalfa, and clover.

leucocytes: White blood cells.

libido: Sex drive or the desire to mate on the part of the male.

lice: Small, flat, wingless insect with sucking mouth parts that is parasitic on the skin of animals.

line crossing: The crossing of inbred lines.

lipid: An organic substance that is soluble in alcohol or ether but insoluble in water, used interchangeably with the term fat.

litter: The young produced by multiparous females such as swine. The young in a litter are called litter mates.

liver flukes: A parasitic flatworm found in the liver.

locus: The place on a chromosome where a gene is located.

luteinizing hormone: A protein hormone, produced and released by the anterior pituitary which stimulates the formation and retention of the corpus luteum.

macroclimate: The large, general climate in which an animal exists.

macromineral: A mineral that is needed in the diet in relatively large amounts.

maintenance: A condition in which the body is maintained without an increase or decrease in body weight and with no production or work being done.

mammal: Warm-blooded animals that suckle their young.

management: The act, art, or manner of managing, handling, controlling, or directing a resource or integrating several resources.

management systems: Methods of systematically organizing information to assist in making more effective management decisions.

marbling: The distribution of fat in muscular tissue; intramuscular fat.

mare: A sexually developed female horse.

market class: Animals grouped according to the use to which they will be put, such as slaughter, feeder, or stocker.

market grade: Animals grouped within a market class according to their value.

masticate: To chew food.

mastitis: An inflammation of the mammary gland.

masturbation: Ejaculation by a male without involving a female.

maternal breeding value (MBV):   A breeding value which measures primarily milk production in beef cattle.

meat: The tissues of the animal body that are used for food.

meiosis: A special type of cell nuclear division that is undergone in the production of gametes (sperm in the male, ova in the female). As a result of meiosis, each gamete carries half the number of chromosomes of a typical body cell in that species.

mesoderm: The middle of the three layers of the primitive embryo.

messenger RNA: The ribonucleic acid that is the carrier of genetic information from nuclear DNA that is important in protein synthesis.

metabolism: **(1)** The sum total of chemical changes in the body, including the "building up" and "breaking down" processes. **(2)** The transformation by which energy is made available for body uses.

metabolizable energy: Gross energy minus the sum of energy in feces, gaseous products of digestion, and energy in urine.

microclimate: A small, special climate within a macroclimate created by the use of such devices as shelters, heat lamps, and bedding.

microcomputer: A small computer which has a smaller memory capacity than a larger or main-frame computer.

micromineral: A mineral that is needed in the diet in relatively small amounts. The quantity needed is so small that such a mineral is often called a trace mineral.

milk fat: The fat in milk.

milk letdown: The squeezing of milk out of the udder tissue into the gland and teat cisterns.

mineral: An inorganic substance required in the diet. Minerals are classified as either macrominerals or microminerals.

minimum culling level: A selection method in which an animal must meet minimum standards for each trait desired in order to qualify for being retained for breeding purposes.

mites: Very small arachnids that are often parasitic upon animals.

mitosis: A process in which a cell divides to produce two daughter cells, each of which contains the same chromosome complement as the mother cell from which they came.

modifying genes: Genes that modify the expression of other genes.

mohair: The fibers produced from the skins of Angora goats.

monogastric: Having only one stomach or only one compartment in the stomach. Examples are swine, mink, and rabbits.

monoparous: A term designating animals that usually produce only one offspring at each pregnancy. Horses and cattle are monoparous.

moon blindness: Periodic blindness that occurs in horses.

mule: The hybrid that is produced by mating a male donkey with a female horse.

mulefoot: Having one instead of two toes, on one or more of the feet.

multiparous: A term that designates animals that usually produce several young at each pregnancy. Swine are multiparous.

mutation: A change in a gene.

mutton: The meat from mature sheep.

navel: The area where the umbilical cord was formerly attached to the body of the offspring.

net energy: Metabolizable energy minus heat increment. The energy available to the animal for maintenance and production.

nicking: The way in which certain lines, strains, or breeds perform when mated together. When outstanding offspring result, the parents are said to have nicked well.

nodular worm: An internal parasitic worm that causes the formation of nodules in the intestines.

nucleotide: Compound composed of phosphoric acid, sugar, and a base (purine or pyrimadine) all of which constitute a structural unit of nucleic acid.

nutrient: **(1)** A substance that nourishes the metabolic processes of the body. **(2)** The end product of digestion.

obesity: An excessive accumulation of fat in the body.

omasum: One of the stomach components of ruminant animals that has many folds.

omnivorous: Feeding on both animal and vegetable substances.

on full feed: A term that refers to animals that are receiving all the feed they will consume.

oogenesis: The process by which eggs, or ova, are produced.

opportunity costs: Returns given up if debt-free resources (for example, land, livestock, equipment) were used in their next best level of employment.

outbreeding: The process of continuously mating females of the herd to unrelated males of the same breed.

outcrossing: The mating of an individual to another in the same breed which is not related to it.

ova: Plural of ovum, meaning eggs.

ovary: The female reproductive gland in which the eggs are formed and progesterone and estrogenic hormones are produced.

overeating disease: A toxic condition caused by the presence of undigested carbohydrates in the intestine, which stimulates harmful bacteria to multiply. When the bacteria die, they release toxins.

oviduct: A duct leading from the ovary to the horn of the uterus.

ovulation: The shedding, or release, of the egg from the follicle of the ovary.

ovum: The egg produced by the female.

pace: A lateral two-beat gait in which the right rear and front feet hit the ground at one time and the left rear and front feet strike the ground at another time.

paddling: The outward swinging of the front feet of a horse that toes in.

parasite: An organism that lives a part of its life cycle in or on, and at the expense of, another organism. Parasites of farm animals live at the expense of the farm animals.

pasteurization: The process of heating milk to 161 °F and holding it at that temperature for 15 seconds to destroy pathogenic microorganisms.

pasture rotation: The rotation of animals from one pasture to another so that some pasture areas have no livestock on them in certain periods.

pedigree: The record of the ancestry of an animal.

pellagra: A disease caused by a deficiency of nicotinic acid.

pelt: The natural, whole skin covering, including the wool, hair, or fur.

pendulous: Hanging loosely.

penis: The male organ of copulation. It serves both as a channel for passage of urine from the bladder as an extension of the urethra, and as a copulatory organ through which sperm are deposited into the female reproductive tract.

performance test: The evaluation of an animal by its own performance.

pernicious anemia: A chronic mycrocitic anemia caused by a deficiency of vitamin B_{12} or a failure of intestinal absorption of vitamin B_{12}.

phenotype: The characteristics of an animal that can be seen and/or measured. For example, the presence or absence of horns, the color, or the weight of an animal.

physiology: The science that pertains to the functions of organs, organ systems, or the entire animal.

picking: The removal of feathers in dressing poultry.

pink tooth: Congenital porphyria, teeth are pink gray and the animals tend to sunburn easily.

pinworms: A small nematode worm with unsegmented body found as a parasite in the rectum and large intestine of animals.

pneumonia: Inflammation or infection of alveoli of the lungs caused by either bacteria or viruses.

poikilotherm: A "cold-blooded" animal. An animal whose body temperature varies with that of the environment.

polled: Naturally or genetically hornless.

poll evil: An abscess behind the ears of a horse.

pork: The meat from swine.

postgastric fermentation: The fermentation of feed that occurs in the cecum, behind the area where digestion has occurred.

poultry: This term includes chickens, turkeys, geese, pigeons, peafowls, guineas, and game birds.

pregastric fermentation: Fermentation that occurs in the rumen of ruminant animals. It occurs before feed passes into the portion of the digestive tract in which digestion actually occurs.

prenatal: Prior to being born. Before birth.

probe: A device used to measure backfat thickness in pigs and cattle.

production testing: An evaluation of an animal based on its production record.

progeny testing: An evaluation of an animal on the basis of the performance of its offspring.

progesterone: A hormone produced by the corpus luteum that stimulates progestational proliferation in the uterus of the female.

program: A set of coded instructions directing the computer to perform a particular function.

prolapse: Abnormal protrusion of part of an organ.

prostaglandin: A product injected into cycling females to synchronize estrus.

prostate: A gland of the male reproductive tract that is located just back of the bladder. It secretes a fluid that becomes a part of semen at ejaculation.

protein: A substance made up of amino acids that contains approximately 16% nitrogen (based on molecular weight).

puberty: The age at which the reproductive organs become functionally operative.

purebred: An animal that meets the standard of a recognized breed and whose ancestors are registered in the herd book of that breed.

quality grades: Animals grouped according to value as prime, choice, etc., based on conformation and fatness of the animals.

ram: A male sheep that is sexually mature.

ram epididymitis: An inflammation of the epididymis that occurs in rams.

realized heritability: The portion obtained of what is reached for in selection.

reasoning: The ability of an animal to respond correctly to a stimulus the first time that the animal encounters a new situation.

recessive gene: A gene that has its phenotype masked by its dominant allele when the two genes are present together in an individual.

reciprocal recurrent selection: The selection of breeding animals in two populations based on the performance of their offspring after animals from two populations are crossed.

recurrent selection: Selection for general combining ability by selecting males that sire outstanding offspring when mated to females from varying genetic backgrounds.

regurgitate: To cast up undigested food to the mouth as is done by ruminants.

reinforcement: A reward for making the proper response to a stimulus or condition.

replicate: To duplicate, or make another exactly alike, the original.

reproduction: The production of live, normal offspring.

reticulum: One of the stomach components of ruminant animals that is lined with small compartments giving a honeycomb appearance.

ribonucleic acid: An essential component of living cells, composed of long chains of phosphate, ribose sugar, and several bases.

rhinopneumionitis: Equine herpes virus-1. It produces acute catarrh upon primary infection.

rickets: A disease of disturbed ossification of the bones caused by a lack of vitamin D or unbalanced calcium/phosphorus ratio.

ringbone: An ossification of the lateral cartilage of the foot of a horse all around the foot.

roughage: A feed that is high in fiber, low in digestible nutrients, and low in energy. Such feeds as hay, straw, silage, and pasture are examples.

rumen: The large fermentation pouch of the ruminant animal where bacteria and protozoa break down fibrous plant material that is swallowed by the animal, sometimes referred to as the paunch.

ruminant: A mammal whose stomach has four parts (rumen, reticulum, omasum, and abomasum).

rumination: The regurgitation of undigested food and chewing of it for a second time, after which it is again swallowed.

salmonella: Gram-positive, rod-shaped bacteria that cause various diseases such as food poisoning in animals.

screwworms: Larvae of several American flies that infest wounds of animals.

scrotum: A pouch which contains the testes. It is also a thermoregulatory organ that contracts when cold and relaxes when warm, thus tending to keep the testes at a lower temperature than that of the body.

scurvy: A deficiency disease in humans which causes spongy gums and loose teeth. It is caused by a lack of vitamin C (ascorbic acid).

selection: Differentially reproducing what one wants in a herd or flock.

selection differential: The difference between records of animals selected and records of the population from which the selected animals were chosen.

semen: The fluid containing the sperm that is ejaculated by the male. Secretions from the seminal vesicles, the prostate gland, the bulbourethral glands, and the urethral glands provide most of the fluid.

seminal vesicles: Accessory sex glands of the male that provide a portion of the fluid of semen.

seminiferous tubules: Minute tubules in the testicles in which sperm are produced.

sheep bot: Any of a number of related flies whose larvae are parasitic in sheep. They usually are found in the sinuses.

sickle hocks: Hocks which have too much set, causing the hind feet to be too far forward and too far under the animal.

side bones: Ossification of the lateral cartilages of the foot of a horse.

sigmoid flexure: The S-curve in the penis of boars, rams, bucks, and bulls.

silage: Forage, corn fodder, or sorghum preserved by fermentation that produces acids similar to the acids that are used to make pickled foods for people.

sleeping sickness: An infectious disease common in tropical Africa and transmitted by the bite of a tsetse fly.

software: Program instructions to make computer hardware function.

soilage: Green forage that is cut and brought to animals as food.

solids-non-fat: Total solids minus fat. It includes protein, lactose, and minerals.

somatotropin: The growth hormone from the anterior pituitary that stimulates nitrogen retention and growth.

sonoray: A machine that is used to measure fat thickness and ribeye area in swine and cattle. The machine sends sound waves into the back of the animal and records these waves as they bounce off the tissues. Different wave lengths are recorded for fat than for lean.

sow: A female swine that is sexually mature.

specific combining ability: The ability of a line or population to exhibit superiority or inferiority when combined with other lines or populations.

spermatid: The haploid germ cell prior to spermiogenesis.

spermatogenesis: The process by which spermatids that carralf as many chromosomes as a typical body cell of the individual are produced in the seminiferous tubules of the testis.

spermiogenesis: The process by which the spermatid loses most of its cytoplasm and develops a tail to become a mature sperm.

spinning count: The number of hanks of yarn that can be spun from a pound of clean wool. One method of evaluating fineness of wool.

spool joint: Denotes a mutton carcass where the foot and pastern are removed from the front leg.

stags: Castrated male sheep, cattle, goats, or swine that have reached sexual maturity prior to castration.

stallion: A sexually mature male horse.

steer: A castrated bovine male that was castrated early in life before puberty.

stomach worms: *Haemonchus contortus,* or worms of the stomach of sheep and goats. The adults produce many eggs. These worms have a debilitating effect on the animals.

strangles: An infectious disease of horses, characterized by inflammation of the mucous membranes of the respiratory tract.

streptococcus: Sperical, gram-positive bacteria that divide in only one plane and occur in chains. Some species cause serious disease.

stringhalt: A sudden and extreme flexion of the back of a horse, producing a jerking motion of the hind leg in walking.

strongyles: Any of various soundworms living as parasites, especially in domestic animals.

suckling gain: The gain that a young animal makes from birth until it is weaned.

subcutaneous: Situated beneath, or occurring beneath, the skin. A subcutaneous injection is an injection made under the skin.

superovulation: The ovulation of a greater than normal number of eggs, stimulated by hormone injections.

T.D.N.:   Total Digestible Nutrients; it includes the total amounts of digestible protein, nitrogen-free extract, fiber, and fat (multiplied by 2.25) all summed together.

T.P.I.:   Total Prediction Index used in dairy cattle breeding. It includes the predicted differences for milk production, fat percentage, and type into one figure in a ratio of milk production × 3: fat percentage × 1: type × 1.

tandem selection: Selection for one trait for a given period of time followed by selection for a second trait and continuing in this way until all important traits are selected.

teaser ram: A ram made incapable of impregnating a ewe by vasectomy or by use of an apron to prevent copulation, which is used to find ewes in heat.

testicle: The male sex gland that produces sperm and testosterone.

testosterone: The male sex hormone that stimulates the accessory sex glands, causes the male sex drive, and causes the development of masculine characteristics.

tetanus: An acute infectious disease caused by toxin elaborated by the bacterium *Clostridium tetani,* in which tonic spasms of some of the voluntary muscles occur.

tetrad: A group of four similar chromotids formed by the splitting longitudinally of a pair of homologous chromosomes during meiotic prophase.

thiaminase: An enzyme that breaks down thiamine into simpler compounds having no vitamin activity. It is present in certain fishes and in bracken fern.

thoroughpin: A hard swelling that is located between the Achilles tendon and the bone of the hock joint.

transcription: The synthesis of an RNA from DNA in the nucleus by matching the sequences of the bases.

trot: A diagonal two-beat gait in which the right front and left rear feet strike the ground in unison, and the left front and right rear feet strike the ground in unison.

twist: Vertical measurement from top of the rump to point where hind legs separate.

twitch: To tightly squeeze the upper lip of a horse by means of a small rope that is twisted.

type: **(1)** The physical conformation of an animal. **(2)** All those physical attributes that contribute to the value of an animal for a specific purpose.

udder: The encased group of mammary glands of mammals. Each mammary gland is provided with a nipple or teat.

umbilical cord: A cord through which arteries and veins travel from the fetus to and from the placenta, respectively. This cord is broken when the young are born.

unsoundness: Any defect or injury that interferes with the usefulness of an animal.

uterus: That portion of the female reproductive tract where the young develop during pregnancy.

vagina: The copulatory portion of the female's reproductive tract. The vestibule portion of the vagina also serves for passage of urine during urination. The vagina also serves as a canal through which young pass when born.

vas deferens: Ducts that carry sperm from the epididymis to the urethra.

vasectomy: The removal of a portion of the vas deferens. As a result, sperm are prevented from traveling from the testicles to become part of the semen.

veal: The meat from very young cattle, under three months of age.

vermifuge: A chemical substance given to the animals to kill internal parasitic worms.

vitamin: An organic catalyst, or component thereof, which facilitates specific and necessary functions.

vulva: The external genitalia of a female mammal.

walk: A four-beat gait of a horse in which each foot strikes the ground at a time different from each of the other three feet.

warble: The larval stage of the heel fly that burrows out through the hide of cattle in springtime.

weaning: Taking a young animal from its dam so that it can no longer suckle from the dam.

weaning breeding value (WBV): A breeding value which measures primarily pre-weaning growth in beef cattle.

white muscle disease: A disease caused by a deficiency of selenium in which there is calcification of the muscle tissues.

wool: The fibers that grow from the skin of sheep.

woolens: Cloth made from short wool fibers that are intermingled in the making of the cloth by carding.

worsteds: Cloth made from wool that is long enough to comb and spin into yarn. The finish of worsteds is harder than woolens and worsted clothes also hold a press better.

yearling breeding value (YBV): A breeding value which measures primarily post-weaning growth in beef cattle.

yield grades: The grouping of animals according to the estimated trimmed lean meat that their carcass would provide.

zone of thermoneutrality: The environmental temperature (about 65 °F) at which heat production and heat elimination are approximately equal for most farm animals.

zygote: **(1)** A cell formed by the union of two gametes. **(2)** An individual from the time of fertilization until death.

CREDITS (continued)

State University. *Figure 1-8:* (A) Courtesy of Winrock International; (B) Courtesy of Dr. Budi S. Nara; (C) FAO photo, courtesy of C. N. Coombes. *Figure 1-9:* (A) Courtesy of Guide Dogs for the Blind, San Rafael, California; (B) Courtesy of *The Western Horseman*; (C) Courtesy of the Kentucky Derby; (D) and (E) Courtesy of Larry Thomas; (F) Courtesy of José Rafael Cortes, editor of "La Nación" newspaper, San Cristobal, Venezuela.

Chapter 2. *Figures 2-1 and 2-2:* Courtesy of W. D. Frischknecht, Oregon Agricultural Extension Service. *Figure 2-3:* Courtesy of Dr. R. W. Henderson, Oregon Agricultural Experiment Station. *Figures 2-4, 2-5, and 2-6:* Drawing adapted from National Live Stock and Meat Board, copyright © 1973. Photograph by Dr. R. W. Henderson, Oregon Agricultural Experiment Station. *Figures 2-7 and 2-8:* Courtesy of Mr. W. D. Frischknecht, Oregon Agricultural Extension Service.

Chapter 4. *Figure 4-1:* Courtesy of Dr. R. W. Henderson, Oregon Agricultural Experiment Station. *Figure 4-2:* Courtesy of Dr. J. H. Landers, Extension Animal Science, Oregon State University.

Chapter 5. *Figures 5-1 and 5-2:* Courtesy of the USDA. *Figures 5-4 and 5-7:* Courtesy of the National Live Stock and Meat Board. *Figures 5-5 and 5-6:* Drawn by Dennis Giddings. *Figures 5-8 and 5-9:* Courtesy of USDA. *Figure 5-10:* Courtesy of Iowa State University. *Figures 5-12, 5-13, 5-14, and 5-15:* Courtesy of USDA. *Figures 5-16 and 5-17:* Courtesy of the National Live Stock and Meat Board. *Figure 5-18:* Courtesy of A. R. Gulich, USDA.

Chapter 6. *Figure 6-1:* Courtesy of the Chester White Swine Record Association. *Figure 6-2:* Courtesy of the American Angus Association. *Figure 6-3:* A *Sheep Breeder and Sheepman Magazine* photo. *Figures 6-4, 6-5, 6-6, and 6-7:* Drawn by Dennis Giddings.

Chapter 7. *Figures 7-1 and 7-2:* Drawn by Dennis Giddings. *Figure 7-3 and 7-4:* Courtesy of the Animal Reproduction Laboratory, Colorado State University. *Figure 7-5:* From Bone, J. F., *Animal Anatomy and Physiology*, 4th ed., Corvallis: Oregon State University Book Stores, copyright © 1975. *Figures 7-6, 7-7, 7-8, and 7-9 :* Drawn by Dennis Giddings. *Figure 7-10:* Courtesy of Dr. Arthur S. H. Wu, Oregon State University. *Figure 7-11:* From Wu, A. S. H., and McKenzie, F. F. "Microstructure of Spermatozoa After Denudation as Revealed by Electron Microscope," *Journal of Animal Science* 14(4): 1151–66, 1955. *Figure 7-12:* Courtesy of Dr. Arthur S. H. Wu, Oregon State University. *Figure 7-13:* Courtesy of the Animal Reproduction Laboratory, Colorado State University. *Figure 7-14:* Drawn by Dennis Giddings. *Figure 7-15:* Courtesy of the Animal Reproduction Laboratory, Colorado State University. *Figure 7-16:* From *Animal Agriculture*, 2d ed., edited by H. H. Cole and W. N. Garrett, W. H. Freeman and Company, copyright © 1980. *Figure 7-17:* From Battaglia and Mayrose, *Handbook of Livestock Management Techniques*, Burgess Publishing Company, Minneapolis, 1981, pp. 131, 134, 135. *Figure 7-18:* (A) Courtesy of Colorado State University; (B) Courtesy of D. C. England, Oregon State University; (C) Courtesy of *The Western Horseman*; (D) Courtesy of Colorado State University; (E) Courtesy of *Poultry Digest*; (F) Courtesy of the American Polled Hereford Association.

Chapter 8. *Figure 8-1:* Drawn by Dennis Giddings. *Figures 8-2, 8-3, and 8-4:* Courtesy of the Animal Reproduction Laboratory, Colorado State University. *Figure 8-5:* Courtesy of Lake, P. E., and Stewart, J. M., *Artificial Insemination in Poultry*, 1978, Ministry of Agriculture, Fisheries, and Food, No. 213, London: Her Majesty's Stationery Office, British Crown Copyright. *Figure 8-6:* (A) and (B) Courtesy of Animal Reproduction Laboratory, Colorado State University; (C) Courtesy of Lake, P. E., and Stewart, J. M., *Artificial Insemination in Poultry*, 1978, Ministry of Agriculture, Fisheries, and Food. No. 213, London: Her Majesty's Stationery Office, British Crown Copyright. *Figure 8-8:* Drawn by Dennis Giddings. *Figure 8-9:* Courtesy of Lake, P. E., and Stewart, J. M., *Artificial Insemination in Poultry*, 1978, Ministry of Agriculture, Fisheries, and Food, No. 213, London: Her Majesty's Stationery Office, British Crown Copyright. *Figure 8-10:* Courtesy of Colorado State University.

Chapter 9. *Figure 9-2:* Courtesy of Dr. David M. Young, Montana State University.

Chapter 10. *Figure 10-2:* Courtesy of *Hoard's Dairyman*.

Chapter 12. *Figures 12-1 and 12-2:* Reprinted with permission from Church and Pond, *Basic Animal Nutrition and Feeding*, Corvallis, Oregon: published by D. C. Church, copyright © 1974. *Figure 12-3:* Courtesy of J. E. Parker, Oregon State University.

Chapter 14. *Table 14-1:* Courtesy of William M. Burns, National Academy Press, National Academy of Sciences.

Chapter 18. *Figure 18-3:*   Data summarized by G. E. Shook, University of Wisconsin, from DHIA records. Courtesy of H. H. Cole and Magnar Ronning, *Animal Agriculture,* W. H. Freeman and Company, copyright © 1974, p. 271. *Figure 18-4:* Courtesy of Beefmaster Breeders Universal (Beefmaster), USDA (Devon), and Curtiss Breeding Industries, Inc. (all others). *Figure 18-5:* Courtesy of North American Limousin Foundation, American Pinzgauer Association, American Murray Grey Association, American Galloway Breeders Association, Texas Longhorn Breeders Association, and Curtiss Breeding Industries, Inc. (Gelbvieh, Hereford, and Maine-Anjou). *Figure 18-6:* Courtesy of American Shorthorn Association, American Tarentaise Association, USDA (Red Poll, Scotch Highland), and Curtiss Breeding Industries, Inc. (all others). *Figure 18-7:* Adapted from The USDA Farmer's Bulletin No. 2228, *Beef Cattle Breeds.*

Chapter 19. *Figures 19-2 and 19-3:*   Courtesy of Western Livestock Roundup. *Figure 19-4:* Courtesy of Norden Laboratories. *Figure 19-5:* Courtesy of *The Charolais Banner. Figure 19-6:* Courtesy of Dr. A. T. Ralston, Oregon State University. *Figure 19-7:* Photo by Duane Dailey, University of Missouri. *Figure 19-8:* Courtesy of BEEF. *Figure 19-9:* Courtesy of USDA. *Figures 19-10 and 19-11:* Courtesy of Western Livestock Roundup. *Figures 19-12 and 19-13:* Courtesy of BEEF. *Figures 19-14 and 19-15:* Courtesy of International Minerals and Chemical Corporation.

Chapter 20. *Figure 20-1:*   Adapted from National Milk Producers Federation, Dairy Producers Highlights and Milk Production, Distribution and Income, USDA Economics and Statistics (DA1-Z). *Figure 20-2:* Courtesy of *Hoard's Dairyman* (Jersey); and Danny Weaver and Jim Miller, photographers, Agri-Graphic Services, Cary, IL (others). Composite photograph by Dr. R. W. Henderson, Oregon Agricultural Experiment Station. *Figure 20-3:* Courtesy of *Hoard's Dairyman.*

Chapter 21. *Figures 21-1 and 21-2:*   Courtesy of Dr. Lloyd Swanson, Animal Science Department, Oregon State University. *Figure 21-3:* Courtesy of De Laval Agricultural Division, Alfa-Laval, Inc., Poughkeepsie, NY. *Figure 21-4:* Courtesy of Dr. Lloyd Swanson, Oregon State University. *Figure 21-5:* Courtesy of Dr. H. P. Adams, Oregon State University.

Chapter 22. *Figure 22-1:*   The major breeds of swine in the United States. Courtesy of Poland China Record Association, Knoxville, Illinois (Poland China); National Spotted Swine Record, Bainbridge, Indiana (Spotted); American Landrace, Lebanon, Indiana (Landrace); USDA (Hampshire and Chester White); American Yorkshire Club, West Lafayette, Indiana (Yorkshire); American Berkshire Association, Springfield, Illinois (Berkshire); United Duroc Swine Registry, Peoria, Illinois (Duroc); and Boar Power (boar/gilt breeding system by Monsanto), Des Moines, Iowa (typical hybrid boar). *Figure 22-2:* Courtesy of D. C. England, Oregon State University. *Figure 22-3:* Courtesy of Iowa State University. *Figure 22-4:* From Battaglia and Mayrose, *Handbook of Livestock Management Techniques,* Burgess Publishing Company, 1981, p. 264. *Figure 22-5:* Courtesy of The Hampshire Swine Registry. *Figure 22-6:* Courtesy of The National Spotted Swine Record, Inc. *Figures 22-7 and 22-8:* Courtesy of Pork Industry Handbook.

Chapter 23. *Figure 23-1:*   Courtesy of Western Livestock Roundup. *Figures 23-2 and 23-3:* Courtesy of the University of Illinois. *Figure 23-4:* Courtesy of R. W. Henderson, Oregon State University. *Figure 23-5:* Drawn by Dennis Giddings. *Figure 23-6:* Courtesy of E. R. Miller, Michigan State University. *Figures 23-7 and 23-8:* Courtesy of the University of Illinois.

Chapter 24. *Figure 24-1:*   Courtesy of Continental Dorset Club, Hudson, IA (Dorset and Polled Dorset); Montadale Sheep Breeders' Association, Indianapolis, IN (Montadale), American Cheviot Sheep Society, Lebanon, VA (Cheviot), American Hampshire Sheep Association, Columbia, MO (Hampshire), and National Suffolk Sheep Association, Logan, UT (Suffolk). *Figure 24-2:* Courtesy of American Rambouillet Sheep Breeders Association, San Angelo, TX (Rambouillet); American Cotswold Record Association, Rochester, NH (Cotswold); National Lincoln Sheep Breeders' Association, West Milton, OH (Lincoln); American Romney Breeders Association, Corvallis, OR (Romney); USDA, ARS, Western Region, U. S. Sheep Experiment Station, Dubois, ID (Targhee); and Animal Science Department, University of Minnesota (Finnsheep). *Figure 24-3:* Courtesy of Dwight and Mae Holaway.

Chapter 25. *Figure 25-1:*   Adapted from USDA Crop Reporting Board E55, Sheep and Goats. *Figures 25-4 and 25-6:* Courtesy of Mr. John H. Pedersen, Midwest Plan Service, Agricultural Engineering, Iowa State University. *Figure 25-7:* From Battaglia and Mayrose, *Handbook of Livestock Management Techniques,* Burgess Publishing Company, 1981, p. 381. *Figure 25-8:* Courtesy of Mr. John H. Pedersen, Midwest Plan Service, Agricultural Engineering, Iowa State University.

Chapter 26. *Figure 26-1:*   Courtesy of Halbach Poultry Farm, Waterford, WI. *Figures 26-3 and 26-4:* Courtesy of Nicholas Turkey Breeding Farms, Sonoma, CA.

Chapter 27. *Figure 27-1:*   Courtesy of Chick Master Incubator Co., Medina, OH. *Figures 27-2 and 27-3:* Courtesy of A. R. Wood Manufacturing Co., Northco Systems, Luverne, MN. *Figures 27-4 and 27-5:* Courtesy of Big Dutchman, a division of U. S. Industries, Atlanta, GA.

Chapter 28. *Figure 28-1:*   Courtesy of Appaloosa Horse Club, Moscow, ID. *Figure 28-2:* Courtesy of New York Racing Association and Mr. Louis Weintraub, Photo Communications Co., and the Jockey Club, New York, NY. *Figure 28-3:* Courtesy of American Quarter Horse Association, Amarillo, TX (portrait by Orren Mixer); United States Trotting Association, Columbus, OH; Appaloosa Horse Club, Moscow, ID; Pinto Horse Association, San Diego, CA; Palomino Horse Breeders of America, Mineral Wells, TX (portrait by Orren Mixer); American White Horse Registry, Crabtree, OR. *Figures 28-5, 28-6, and 28-7:* From Bone, J. F., *Animal Anatomy and Physiology*, 4th ed., Corvallis: Oregon State University Book Stores, copyright © 1975.

Chapter 29. *Figure 29-1:*   Courtesy of USDA. *Figures 29-2 and 29-3:* From Battaglia and Mayrose, *Handbook of Livestock Management Techniques*, Burgess Publishing Company, 1981. *Figure 29-4:* Courtesy of Dr. Robert Henderson, Oregon Agricultural Experiment Station. *Figures 29-5, 29-6, and 29-7:* From Battaglia and Mayrose, *Handbook of Livestock Management Techniques*, Burgess Publishing Company, 1981.

Chapter 30. *Figure 30-1:*   Courtesy of Mr. R. F. Crawford and Mr. Ted Edwards, Emerald Dairy Goat Association Newsletter. Photograph by Mr. Ole Hoskinson. *Figure 30-2:* Courtesy of Mrs. Eva Rappaport. *Figure 30-3:* Courtesy of Nancy Lee Owen. *Figure 30-4:* Courtesy of Mrs. Eva Rappaport. *Figures 30-5 and 30-6:* Courtesy of Cindy Schneider. *Figure 30-7:* Courtesy of Mrs. Eva Rappaport. *Figure 30-8:* Courtesy of Mr. and Mrs. Don F. Kessi.

Chapter 31. *Figure 31-1:*   Courtesy of R. W. Henderson.

Chapter 32. *Figure 32-4:*   Courtesy of Iowa State University. *Figure 32-5:* Courtesy of BEEF.

Chapter 33. *Figure 33-1:*   Institute of Agriculture and Natural Resources, University of Nebraska; lower left photo courtesy of University of Illinois. *Figure 33-2:* Institute of Agriculture and Natural Resources, University of Nebraska photo; lower left photo courtesy of West Virginia University. *Figures 33-3 and 33-4:* Courtesy of Oklahoma State University. *Figure 33-5:* Courtesy of Colorado State University. *Figure 33-6:* Courtesy of the University of Nebraska; Colorado State University; and Oklahoma State University. *Figure 33-7:* Courtesy of Colorado State University; University of Illinois, Office of Agricultural Communications; BEEF; University of Wyoming; *The American Hereford Journal* and Oklahoma State University.

Index

Student Survey

Scientific Farm Animal Production, 2d ed.
Ralph Bogart and Robert Taylor

Students, send us your ideas!

The authors and the publisher want to know how well this book served you and what can be done to improve it for those who will use it in the future. By completing and returning this questionnaire, you can help us develop better textbooks. We value your opinion and want to hear your comments. Thank you.

Your name (optional) _____ School _____

Your mailing address _____

City _____ State _____ ZIP _____

Instructor's name (optional) _____ Course title _____

1. How does this book compare with other texts you have used? (Check one)
   ☐ Better than any other   ☐ Better than most
   ☐ About the same as the rest   ☐ Not as good as most

2. Circle those chapters you especially liked:
   Chapters   1   2   3   4   5   6   7   8   9   10   11   12   13   14   15   16   17
           18   19   20   21   22   23   24   25   26   27   28   29   30   31   32   33
   Comments:

3. Circle those chapters you think could be improved:
   Chapters   1   2   3   4   5   6   7   8   9   10   11   12   13   14   15   16   17
           18   19   20   21   22   23   24   25   26   27   28   29   30   31   32   33
   Comments:

4. Please give us your impressions of the text. (Check your rating below)

	Excellent	Good	Average	Poor
Local organization	( )	( )	( )	( )
Readability of text material	( )	( )	( )	( )
General layout and design	( )	( )	( )	( )
Match with instructor's course organization	( )	( )	( )	( )
Illustrations that clarify the text	( )	( )	( )	( )
Up-to-date treatment of subject	( )	( )	( )	( )
Explanation of difficult concepts	( )	( )	( )	( )
Selection of topics in the text	( )	( )	( )	( )

OVER PLEASE

5. Please list any chapters that your instructor did not assign. _____

_____

6. What additional topics did your instructor discuss that were not covered in the text? _____

_____

7. Did you buy this book new or used?   □ New   □ Used

Do you plan to keep the book or sell it?   □ Keep it   □ Sell it

Do you think your instructor should continue to assign this book?   □ Yes   □ No

8. Before taking this course, what was your level of knowledge about farm animals? (Please mark the line below.)

No Previous
   Knowledge   Moderate   Extensive

└_____|_____|_____|_____┘

How suitable was the book for your level of expertise?

Overly
Technical   Appropriate   Not
Technical Enough

└_____|_____|_____|_____┘

9. Are you interested in taking more courses in this field?   □ Yes   □ No

Are you a major in animal science?   □ Yes   □ No

10. Is a lab offered in conjunction with the course?   □ Yes   □ No

If so, author and title of the lab book used: _____

Kindly rate the coordination of the lab book with the text.

        □ Excellent   □ Good   □ Average   □ Poor   □ Very poor

11. **GENERAL COMMENTS:**

May we quote you in our advertising?   □ Yes   □ No

To mail, remove this page and mail to: Mary L. Paulson
                                      Burgess Publishing Company
                                      7108 Ohms Lane
                                      Minneapolis, MN 55435

THANK YOU!